AUTONOMY AND CHOICE IN CONTEXT: AN INTERNATIONAL PERSPECTIVE

AUTONOMY AND CHOICE IN CONTEXT: AN INTERNATIONAL PERSPECTIVE

Edited by

Professor Rina Shapira
Department of Sociology and School of Education,
Tel Aviv University, Israel

and

Professor Peter W. Cookson, Jr
Center for Educational Outreach and Innovation Teachers College,
Columbia University, New York

PERGAMON

U.K.	Elsevier Science Ltd, The Boulevard, Langford Lane, Kidlington, Oxford OX5 1GB, U.K.
U.S.A.	Elsevier Science Inc. New York, 656 Avenue of the Americas, New York, NY 10010, USA.
JAPAN	Elsevier Science Japan, Tsunashima Building Annex, 3-30-12 Yushima, Bunkyo-ku, Tokyo 113, Japan

First edition 1997

Library of Congresss Cataloging in Publication Data

A catalogue record for this book is available from the Library of Congress

British Library Cataloguing in Publication Data

A catalogue record for this book is available from the British Library

ISBN 0 08 042777 4

Printed and bound in Great Britain by Alden Press Ltd, Oxford

Dedication

To Uri and Susan with love.

TABLE OF CONTENTS

Foreword ix

Introduction 1

Major Summary Points 9

ISRAEL

1 Ayalon, H. 13
School Autonomy in a Centralized System:
The Case of Israeli Secondary Education

2 Benyamini, K. 31
Sponsored Autonomy and Misguided Choice:
The Rise and Fall of an Educational Endeavor

3 Chen, M. 41
School Choice as a Bargain in a Sectarian Educational System

4 Haymann, F., Golan, H. and Shapira, R. 77
School Autonomy and Parental Choice:
Steps in Local Educational Planning

5 Horowitz, T. 109
Parental Choice as Leverage Toward Social Change in Education: Two Schools in Jerusalem

6 Ichilov, O. and Mazawi, A. E. 135
Educational Opportunities, Parental Choice, Community Structure, and Mobility: The Case of the Arab Community of Jaffa

7 Nevo, D. 161
The Function of Evaluation in School Autonomy

8 Yogev, A. 177
Autonomy and Choice as School Strategies for Peripheral Communities in Israel

UNITED STATES

9 Chubb, J. E. and Moe, T. M. 203
Politics, Markets, and Equality in Schools

10 Coleman, J. S. 249
The Design of Schools as Output-Driven Organizations

11 Cookson, P. W. Jr 271
School Choice and the Creation of Community

12 Fliegel, S. 297
Creative Non-Compliance

13 Glenn, C. L. 323
Schools of Choice and the Revival of Urban Community

14 Goldring, E., Hawley, W., Saffold, R. and Smrekar, C. 353
Parental Choice: Consequences for Students, Families and Schools

15 Nathan, J. 389
Possibilities, Problems, and Progress: Early Lessons from the Charter Movement

GREAT BRITAIN

16 Williams, V. 407
School Autonomy and Parental Choice: Circumscribed Realities of Reform

THE NETHERLANDS

17 Lodewijks, H. G. L. C. 439
Raising the Effectiveness of Schooling and Learning by Enlarging School Autonomy: The Case of The Netherlands

CANADA

18 King, A. J. C. 457
Constraints on Parents' Involvement in School

Author Index **477**

Subject Index **481**

FOREWORD

RINA SHAPIRA AND PETER W. COOKSON, JR.

There are few educational reform movements that are international in scope. Even though we are living in an increasingly globalized and interdependent world, school reform tends to be local, although educational politics are often national. Aside from international comparisons of student achievement, systematic attempts to describe and analyze educational reforms across national boundaries are rare. In effect, the dialogue between practitioners and policy makers stops short of national boundaries, thereby limiting our knowledge of productive innovations and possible pitfalls.

In the 1980s one reform movement — parental choice, or school choice as it is called in some countries — became increasingly popular at the grassroot and policy levels. This reform is conceptually straightforward — empowering families to make choices about their children's education — but practically, financially, politically, and pedagogically, the implications of this educational "paradigm shift" are enormous. For this reason the parental choice movement has been surrounded by controversy. The need for international data and dialogue about the causes and consequences of allowing families to choose their own schools is compelling. Will parental choice lead to greater educational equity, or will it increase stratification? Is therc a proven relationship between parental choice and higher student achievement? What role should the state take in educating the young? How expensive is parental choice compared to traditional forms of school governance? What are the implications of strengthening the private school sector at the expense of public or state education?

These and other questions guided the workshop participants, who gathered in Israel in June 1993 to discuss, analyze, and debate the issues generated by the school choice movement in Israel, the United States, Great Britain, the Netherlands, and Canada. To our knowledge the workshop at the University of Tel Aviv was the first of its type. Recognizing the need for international comparisons and led by Rina Shapira and James Coleman, the committee that organized the parental choice workshop sought a diversity of views and research strategies. The issues raised by the participants went to the very heart of the personal and social purposes of education. The workshop addressed these issues in the spirit of truth-seeking and the struggle for core understandings; the chapters in this volume capture much of the workshop in terms of its comprehensive approach and openness to differing perspectives. While the results are not definitive, of course, they are suggestive of the problems and possibilities of this important reform. While the school choice movement has grown in the three years since the workshop, we feel confident that the issues raised in this volume are central to the debate and remain contemporary.

We would like to thank the sponsors of the workshop for their generosity. We were fortunate to have support from the following organizations: Tel Aviv University, School of Education and the Unit of Sociology of Education and Community; The CRB Foundation; Tel Aviv Municipality; the Israeli Ministry of Education and Culture; and the Israeli Center for Social and Economic Progress. Several individuals contributed a great deal to the workshop. Particular thanks go to Mrs Daphna Kariv, workshop coordinator, Mrs Flor Haymann, Mrs Rona Shavit, Mrs Margalit Arnon,

Mrs Hagar Shohan and Professors David Nevo and Ellen Goldring who provided critical and intellectual support.

One of the major intended participants was Professor James Coleman of the University of Chicago. Professor Coleman could not attend the workshop, but we have included his paper and listed him as an editor. His recent death is a great blow to the scholarly world; his contributions to our understanding of school organization and school choice are unparalleled.

INTRODUCTION

Rina Shapira* and Peter W. Cookson, Jr**
Tel Aviv University
Columbia University

In early June 1993 a group of scholars from Israel, Great Britain, the United States, and Canada convened at the University of Tel Aviv in Israel to consider the most current research about the academic effects and policy implications of the school reform known as parental choice. This workshop was ground-breaking because it was the first time that empirical studies of school choice were examined in an international context. The guiding questions of the workshop were: What are the characteristics of parental choice programs in different societies? How do these choice programs affect school organization and student achievement? And what can we learn from parental choice in order to make schools more innovative, productive, and just? The workshop was highlighted by a variety of research findings and a diversity of viewpoints.

If we were to examine opinions about parental choice as ranging from very positive to very negative, we would discover that at the workshop there were a full range of opinions. Some participants view choice as a panacea for revolutionizing school systems, while others view choice as a prelude to the destruction of a democratic school system. In the main, however, there was a general feeling that parental choice is a worthwhile tactic of reform, but in all likelihood choice is inadequate as a main strategy of reform. Parental choice covers many educational reform policies that are quite different organizationally and, as a consequence, lead to different educational structures and school processes. The very definition of choice is a matter of some debate. For some scholars, choice implies the application to market principles to school organization, which in turn suggests that state education ought to be partially or completely deregulated. For other scholars, choice should be a reform alternative that strengthens public sector schools by avoiding privatizing solutions. In effect, there was little disagreement among the participants about the goals of educational reform, but there were significant differences of opinion about how to achieve these goals.

At the heart of the choice debate is the continuous, yet fundamentally healthy, tension between the rights of individuals and the rights of communities. To what degree should the state have the authority to compel students to attend particular schools? How much educational freedom can a society tolerate before the entire system becomes Balkanized and anarchic? Every educational system confronts these fundamental dilemmas and there is little reason to expect that there will ever be a complete resolution to the tensions between the competing rights of the individual and the community. Yet these issues must be addressed because in a democracy they raise fundamental questions about the nature of community and civic commitment. Thus, parental choice is a school reform that leads us to ask fundamental questions about the relationship of school and society and the individual and the collective.

Within this larger philosophical framework there are a variety of other questions related to school choice — questions of feasibility, accountability, and in some countries, the relationship between church and state. Is choice really an inexpensive magic bullet of reform or are there hidden costs to parental choice that place an even greater burden on public resources? How reasonable is it for children to travel great distances to attend school and what are the costs related to transportation? Who is to ensure that schools will "deliver" the educational processes and products they promise? Can any group for a school — no matter what their ideological persuasion? Who is to ensure that public monies are not misspent? Is choice a mechanism by which state funds can be channelled to religious organizations? Should public monies be used to support the views of those who are highly critical of state-mandated educational policies? In short, creating choice policies is not a simple task. In fact, as one moves farther away from the concepts of the neighborhood school and the separation of the public and private sectors, conventional ideas about school governance begin to fade and the full complexity of creating new kinds of schools becomes more apparent. A new paradigm for school organization is emerging, but there is little consensus about the nature of this paradigm.

Complexity should not be a deterrent to policy creation. Good public policies are those that acknowledge that the relationship between school and society is highly intricate and socially constructed: individual and organizational variables are interactive, interlinked, and virtually inseparable. Genuine educational reform does not lend itself to simplistic solutions; the debate about how to improve schools is intense and sometimes partisan. Israel provided a dramatic backdrop to the workshop because in that country the issue of school choice is galvanizing public opinion and leading educators and parents to ask fundamental questions about the nature of Israeli schools and Israeli society.

This volume includes the papers of the workshop participants. Some papers, among them those presented by John Chubb, Charles Glenn, Ellen Goldring, Peter Cookson and Joe Nathan offer overall approaches based on philosophical examinations of the assumptions underlying parental choice, as well as empirical studies on parental choice impact. Other papers describe and analyze the experience from several programs concerning school autonomy and parental choice taking place in different countries. These range from Seymour Fliegel's description of East Harlem choice schools and the Minnesota choice plan presented by Joe Nathan, through the Israeli experience as described by Kalman Benyamini, Michael Chen, Tamar Horowitz, Orit Ichilov and Andre Mazawi, Dan Inbar and Rina Shapira, and Flor Haymann, to an analysis of recent state reform policies in Great Britain and the Netherlands in one paper by Vivian Williams and Hans Lodewijks. Other presentations, like those of Hanna Ayalon, David Nevo and Alan King address issues of school autonomy, community involvement and their linkage to school accountability, educational improvement and social equity. The papers of all the participants mentioned in the above paragraph are not included in this book.

One of the premises underlying the different papers was that without equality of educational opportunity, parental choice could lead to an even greater stratification of schools and society. There was a general consensus that planned or managed choice is a policy option that has great potential for creating just and innovative schools. Some participants at the workshop, however, took issue with this perspective, arguing that state intervention undermines the basic principle of family empowerment. Most participants were of the opinion that unregulated market choice would be detrimental to genuine school improvement, and would be socially divisive. It was generally agreed that parental choice would not be particularly effective in creating innovative schools if it were not coupled with other reforms. Some market advocates have claimed that choice and choice alone is a panacea for reforming schools. While this is an intriguing concept, most workshop members felt that choice alone could not bring about the kind of educational innovation that is desired in virtually every country in the world. Research indicates, however, that under the right conditions, choice can lead to improved organizational climate, greater parental involvement, increased teacher empowerment, more social and ethnic integration, higher academic achievement, more school and student autonomy, and a better learning environment for "at-risk" students. The empirical evidence concerning the relationship between parental choice and student achievement, however, is ambiguous and it was generally acknowledged that the hard data required to test the "parental choice/higher student achievement" hypothesis were either unconvincing or lacking altogether.

It was generally agreed that the decision to implement managed choice should be taken collectively by the municipality, the community, the school, and other involved groups. Schools, however, must be autonomous when making decisions about curriculum, assessment, and school organization. It was emphasized that each school should develop its own ethos in stages. Each developmental stage must include students, teaching staff, parents, administrators, educational experts, and community members. One of the primary goals of empowering families is to create schools that are diverse and accountable to their communities. Choice creates diversity, but diversity creates meaningful choice. Diversity allows for distinctiveness and excellence which, in turn, creates strong school cultures. Children who attend schools with a clear ethos of achievement and community are more likely to achieve than children who attend schools with no particular identity.

The discussions also emphasized the importance of educational goals. Because schools play such a critical role in national and international economies, it is critical that school reform takes into account emerging issues of literacy, technical competency, and civil commitment. Within this context, schools should be free to organize their own specific curricula, extracurricular activities, teaching strategies, resource allocations, staff development, and evaluation techniques. Accountability is a crucial element in ensuring that school autonomy is meaningful. Before accountability can be realized, however, schools must have the necessary resources to implement their educational programs. If all children are to have equal opportunities to learn, the vast resource disparities between rich and poor schools must be minimized, if not eliminated. There was a consensus that there should not be tracking or ability grouping within schools of choice. Of critical importance is accurate and accessible information for parents; managed choice cannot be effective without information dissemination. Choice also changes the nature and type of information exchanged between schools and parents. One of the most dynamic and innovative aspects of parental choice is the necessity for accurate information which, in turn, creates a school dynamic of self-reflection and change. Parent information centers are intrinsically useful for school reform because they compel schools to examine curricula and student outcomes and parents are allowed to evaluate schools directly.

One of the most fascinating elements of the workshop was the sharing of ideas about parental choice as reflected through the lens of culture. Certain educational problems are transnational because they are structural to all school systems, but culture plays a critical role in how these structural problems are identified, interpreted, and resolved. Thus, in some countries, parental choice may lead to school differentiation, while in other countries this may not be the case. Cross-cultural studies of parental choice provide a rich resource for those who seek to understand policy development in an international context and in a world that is increasingly interdependent.

Parental choice is a worldwide movement. Under some political conditions this reform may rise to prominence, while under other political conditions its apparent importance may be diminished. Evidence from North America, Europe, Russia, and Israel, however, strongly points in the direction of increasing school autonomy and creating learning environments that are distinctive and tied to communities. Clearly, there is a great need for more research and more cross-cultural studies. This workshop was a first step in creating a community of scholars and a body of research that can be called upon by those interested in creating schools that are academically distinctive, responsive to families and organizationally flexible. Change requires schools to become increasingly adaptive; parental choice is one adaptive mechanism that holds great promise.

MAJOR SUMMARY POINTS

This summary presents the major points discussed at the workshop as they pertain to educational policy issues relevant today in many countries throughout the world.

(1) The topic of choice in education was discussed at the seminar in terms of why, where, when and how, if at all, it should be implemented. It was generally agreed that **when implemented it should be done in such a way that every individual and every social group is given equal access to choice and equal chances to attain the kind of education they think suits their youngsters** (various models of "controlled choices, planned, managed, guided choices" were mentioned in this connection). The possibility of free choice was not entertained, and was thought to be possibly detrimental to schools and school systems. If properly activated, socially controlled, and carefully designed, it could be a strong strategy; if not, it could be a faulty intervention that could lead to segregated schools. This would especially be the case if schools were allowed to choose students in a free choice system.

(2) The emphasis was on choice as **one of many strategies for school improvement**. There was agreement that choice could be effective if it was not the only strategy for school improvement, but used in

conjunction with other strategies. Controlled choice is one of the conditions necessary to realize improved educational outcomes, but certainly not enough.

(3) **Choice can be instrumental in achieving educational and social aims.** Research suggests these objectives include: improved organizational climate, fostering parental involvement, increasing teacher empowerment, promoting social and ethnic integration, improved academic achievement, fostering school and student autonomy, and advancing at-risk students. **Each community decides** to adopt a model of controlled choice depending on the objectives and goals it has set for itself.

(4) The **decision** to implement any model of controlled choice **should be a decision taken collectively by** the municipality, the community, the school and other involved groups. The characteristics of specific decisions must be school-based.

(5) The need for **each school to develop its own ethos in stages** was emphasized. These developmental stages must include the school staff with input from parents, subject experts and education experts. It is recommended that the unique school ethos should not be specifically geared to one subject (communication?), but include a broader orientation for **all** students.

(6) Controlled choice should be among **schools with differentiated characteristics**. The variance among schools leads to choice, and choice leads to variance.

(7) Numerous models of such types of school choice plans were presented and discussed, taking into account the unique setting and circumstances of each place. The crucial point of these models is the

school-based nature of developing the school ethos. Schools decide on the "how" and "what" of their distinctive program.

(8) The discussions emphasized the importance of educational goals and a clearly stated policy to encompass both **school autonomy and national policy**. This must be in place for **schools** to be able **to translate their credo into specific curricula**, extracurricular activities, teaching strategies, resource allocations, staff development, and evaluation techniques.

(9) Accountability is also a crucial part of school autonomy and choice plans. However, before accountability can be realized, **schools must have the necessary resources** to implement their educational programs.

(10) This process can be **successful** if initiatives are encouraged from a broad community of participants — parents as well as professionals — **in a collaborative partnership**.

(11) It is clear that **there should not be tracking or ability grouping** in these specialty schools.

(12) Emphasis was placed on the importance of **information for parents**. This type of controlled choice program cannot be effective without information dissemination. Choice also changes the nature and type of information exchanged between schools and parents. Parent centers, where parents learn about these schools and their choices, were discussed. A proper "fit" is encouraged between the school and the family.

Chapter One

SCHOOL AUTONOMY IN A CENTRALIZED SYSTEM: THE CASE OF ISRAELI SECONDARY EDUCATION

HANNA AYALON
Tel Aviv University

The study reported in this chapter evaluates potential social implications of school autonomy by analyzing the relationship between school characteristics and subjects offered to students preparing for the matriculation exams in Israel.

One of the main features of school autonomy is the freedom to construct the curriculum (Silberstein 1990). The curriculum of the autonomous school is expected to correspond to the interests, needs, and capacities of the community (Reshef 1990; Shapira *et al.* 1991).

The benefits of school autonomy notwithstanding (Reshef 1990), my purpose in the present study is to point at potential negative social consequences. By this I refer to the concern that the attempt to match the curriculum with the interests and assumed capacities of the students may result in the enhancement of educational inequality.

This concern is based on the assumption that the subjects taught in school are stratified (Apple 1990). The effort to match between students and school subjects may result in creating a correspondence between the stratification of subjects and the social stratification of the students who study them, thus enhancing educational inequality.

Studies on the knowledge offered to students of different social classes indicate that higher-status knowledge is considered inappropriate and irrelevant for underprivileged students (Anyon 1981a, b; Oakes 1985). This implies that in the process of shaping the curriculum of the autonomous school lower-status students may be denied access to higher-status knowledge.

The present study concentrates on secondary education in Israel. Israeli high schools are by no means autonomous (Pur 1989). However, academic high schools do have a certain degree of freedom in shaping the curriculum. The curriculum of the academic programs is composed of compulsory and optional subjects. The Ministry of Education and Culture offers a great variety of optional subjects, at different levels. The schools must offer the compulsory subjects, but are free to make their choice of the optional ones. Hence, Israeli high schools are partly autonomous in constructing the curriculum. The assumed capacities and interests of the population of students are probably among the factors which affect a school's choice of optional subjects. This supposition is based on practical reasons. Israeli high schools obtain significant prestige by enhancing the percentage of students who manage to acquire the matriculation diploma (a necessary prerequisite for higher education) upon graduation. The school principal, who is the dominant figure in constructing the curriculum, probably attempts to maximize the number of diploma owners. Decisions at the school level are based on perceptions about the capacities of the student body as a whole. It is likely, then, that the principal and other curriculum decision-makers try to match between subjects and students. Consequently, schools with higher percentages of underprivileged students prefer to include "easier" subjects in their curriculum.

The philosophy behind the matching between students and curriculum in the autonomous school is, of course, different. This philosophy stresses the legitimate variation in needs and capacities of students, and views all students as entitled to concentrate on different types of knowledge. However, the consequences of an ideologically based matching of students and curriculum may be similar to those based on practical considerations. Thus, an analysis of the social implications of the matching of students and school subjects in the Israeli academic high schools my prove most relevant for the evaluation of school autonomy.

The Stratification of School Subjects

School subjects are a form of organization of knowledge (Bernstein 1971). Every society defines some sorts of knowledge as superior, and specifies which social categories are entitled to access to it (Young 1971; Eggleston 1977; Labaree 1988). The hierarchy of school subjects is not restricted to a single dimension. School subjects may be ranked according to several criteria, each representing a different source of differentiation.

The most obvious source of stratification of school subjects is their academic versus their utilitarian character. Subjects viewed as utilitarian are less valued, receive fewer resources, and are offered to the less able students (Eggleston 1977; Goodson 1983, 1992). Utilitarian subjects are usually offered in vocational education. Since the present study concentrates on academic high schools, it does not refer to that aspect of the stratification of subjects.

The second source of stratification is between math and the scientific subjects, on the one hand, and the non-scientific subjects, on the other. The high status of the scientific subjects is a relatively newly established phenomenon. When first introduced into the school curriculum, science was considered a utilitarian subject, particularly relevant to lower-class students (Goodson 1992). Only after undergoing a notable change of content, did science establish itself as a highly prestigious academic subject most instrumental for students' future occupational achievements.

Apple (1990) compares the resources attached to the development of the curriculum in math and sciences to the lower investment in arts and the humanities, whose declining status in the modern curriculum is well documented (e.g., Kliebard 1992). Arts and the humanities are not considered instrumental for future occupational achievements, but they are highly valued according to the third source of subject differentiation — their use in the acquisition of cultural capital, in Bourdieu's (1977) terms.

The declining status of the humanities in the modern school curriculum is explained by the expansion of education and the incorporation of lower-status students into the education system. The humanities, defined as high culture appropriate to the elite only, and perceived as leisure activity, have been replaced by other, more practical subjects (Kliebard 1992; Labaree 1988). However, arts and humanities do serve as a means of social exclusion by differentiating between insiders and outsiders, in Weber's terms. Students may obtain the knowledge required for the acquisition of a high-status profession, but they need to prove familiarity with high culture to legitimize an elite position (DiMaggio & Useem 1982). This suggests that in spite of the general decline in the portion of knowledge defined as high culture in the school curriculum, it gets a significant share in the curriculum offered to students preparing for elite positions. Studies of elite boarding schools in the United States indeed indicate that, along with the instrumental subjects, these schools offer a rich curriculum in arts and humanities (Cookson & Persell 1985).

Inequality and Curriculum in Israeli Secondary Education

The Israeli setting provides a particularly appropriate arena for the study of the consequences of the matching between students and curriculum. This is mainly because of the high degree of curriculum differentiation within the academic programs prevailing in Israeli secondary education.

Curriculum differentiation in Israeli secondary educational follows a reform introduced in the late 1970s. Prior to the reform, academic-track students were divided in the 11th grade into strictly structured streams (**Megamot**), and took the matriculation exams accordingly. The reform eliminated the structured streams and enabled students to create their own combinations of subjects, enroll in the appropriate courses, and take examinations accordingly. High school curriculum is composed of compulsory and optional subjects. Each subject can be offered at different levels, ranging usually from one unit of study to five units. The number of units refers to the time devoted to each subject, which corresponds, of course to its level and degree of difficulty[1]. With the exception of English, which is compulsory at the four-unit level, subjects are compulsory at their lower level (one to three units) only. Subjects which are compulsory at the lower level may be offered as optional at higher levels.

The matriculation diploma is composed of "internal" grades, awarded by the school teachers, and "external" ones, based on the student's performance in standard national exams. To qualify for a matriculation diploma, a student has to pass national exams that total 20 units of study. A four- or five-unit optional subject is not a necessary component of the diploma. However, the universities demand at least one four-unit subject from their candidates, and offer bonuses for each subject taken at the four- or five-unit level.

The freedom of students in the process of choosing optional subjects is conditioned by the supply of subjects available in the school. As noted, the Ministry of Education and Culture offers a remarkable variety of subjects, at different levels, and the schools are free to make their choices. Israeli schools differ, then, in the content and the level of optional courses they offer.

The present study analyzes the correspondence between the supply of optional subjects and the ethnic composition of the school. The concentration on the ethnic factor stems from its centrality in the Israeli research on inequality in education. This research focuses on the two major Jewish ethnic groups — Jews of Asian or North African origin (hereafter **Mizrachim**, literally "easterners"), and Jews of European or American origin. The **Mizrachim**, the educationally disadvantaged group, are characterized by lower achievements, overrepresentation in vocational education, lower rates of possession of the matriculation diploma, and lower rates of enrollment in postsecondary education (Yogev 1981; Yogev & Ayalon 1986; Shavit 1990).

As in the United States and Europe, the scientific subjects are very highly valued in Israeli secondary education (Doron & Millin 1979; see also Mandler 1992). Scientific subjects, which are considered particularly difficult, are offered to and requested by the most able students. School principals report that availability of high-level scientific courses serves as a means of attracting able students (Mandler 1992). Due to the correlation between ethnic origin and achievements in Israeli education, **Mizrachim** students are perceived as less able, and hence unlikely clientele for the demanding scientific subjects.

"Cultural capital" subjects are not considered particularly difficult. However, based on the American and European reports on the assumed irrelevance of high culture knowledge to lower-status students, I hypothesize that "cultural capital" subjects are less often offered in **Mizrachim**-dominated schools, on the same grounds of irrelevance.

The third group of subjects included in the analysis are "regular" ones, perceived as less demanding, and hence appropriate for less capable students. The major hypothesis is that scientific and "cultural capital" subjects are offered less often in **Mizrachim**-dominated schools, whereas a school's ethnic composition has no effect on the availability of the "regular" subjects.

Data and Method

Data and sample

The data are based on a survey conducted by the Israel Ministry of Education and Culture in 1989. The survey included all 12th graders in the academic programs. A questionnaire was sent to the school principals, who were asked to provide information on the subjects studied by each student during all his or her years in high school. Eighty-five percent of the school principals responded to the questionnaire. The school-level data for the present study has been aggregated from the individual-level survey. The study refers to the data on Jewish students only[2]. The original data-set consists of 208 Jewish schools that offer matriculation-oriented academic programs. Seven schools, which did not provide information on the ethnic composition, were excluded from the analysis.

Variables

The dependent variable is the availability of scientific, "cultural capital”, and "regular" subjects in school. Each dependent variable is a construct composed of the appropriate subjects. The subjects which are included in the analysis were chosen according to two criteria: their popularity as higher-level optional subjects, and their assignability to one of the subject categories used in the study[3]. Each subject is dichotomized to: (1) offered at the higher level (at least four units of study for most subjects. For art, music, philosophy, and theater, which are relatively scarce and optional at all levels, higher level is defined as three or more units of study), and (0) otherwise.

The subjects which belong to each category are summed to the composite construct. The construct is coded 1 if at least one of its components equals one (namely, is offered at the higher level), and 0 otherwise. The subjects, their distributions, and the distributions of the constructs are presented in Table 1.1.

The definition of math, biology, physics, chemistry, and computer sciences as scientific subjects is straightforward[4], and so is the definition of art, music, theater and philosophy as high culture (e.g., Kliebard 1992). The inclusion of French in the "cultural capital" category is based on Israeli studies which report that French is perceived as a linguistic cultural capital, and as a highly valued resource (Ben-Rafael, Herzlich & Freund 1990).

Arabic, which is categorized as a "regular" subject, is usually considered as an alternative to French for students who elect to study a third language. However, the two languages are differently valued by students and parents

Table 1.1. Distribution of individual school subjects and of the composite constructs

Subject*	Mean	S.D.
Scientific	0.89	0.65
Math	0.88	0.33
Biology	0.77	0.42
Physics	0.61	0.49
Chemistry	0.42	0.49
Computer sciences	0.19	0.37
"Cultural Capital"	0.43	0.50
French	0.23	0.42
Art	0.15	0.36
Music	0.07	0.26
Theater	0.06	0.24
Philosophy	0.6	0.24
"Regular"	0.65	0.48
Arabic	0.33	0.47
Geography	0.29	0.46
Literature	0.56	0.50

*Dummy variables. Coded 1 if subject to group of subjects are offered at the highest level, or otherwise.

(Kraemer 1990). Unlike French, Arabic is regarded as non-prestigious, and of no importance as a means of communication (Ben-Rafael & Brosh 1991).

Geography is another lower prestige "regular" subject (Bar-Gal 1993), and so is literature. The inclusion of literature among the "regular" subjects may be surprising, since it is considered a part of high culture in Israel as elsewhere. However, there seems to be a split between literature as a component of high culture and literature offered to students in school. Israeli studies on the curriculum in literature indicate that it stresses the emotional side of literature and not its intellectual and cultural aspect (Yaoz & Iram 1987). This process is not unique to Israel. Kliebard (1992) claims that English literature has undergone an internal transformation in the modern American curriculum.

Explanatory variables

Five explanatory variables are included in the analysis. The major explanatory variable, **school ethnic composition** (hereafter %Mizrachim), is defined as the percentage of **Mizrachim** students among the 12th graders in the school. The additional explanatory variables — peripheral geographic location, sector, size, and sex composition, serve mainly as controls.

Periphery is a dummy variable coded 1 if a school is located in the peripheral areas of Israel, which are economically and educationally disadvantaged, and 0 otherwise.

Sector is a dummy variable which distinguishes between regular public education (coded 1) and religious public education (0). Eighty-seven of the 201 schools in the sample are religious[5].

Size, an indicator of school resources (e.g., Kilgore 1991), is defined as the number of 12th graders in school.

Sex composition (%male) is defined as the percentage of male students among the 12th graders in school. Sixty-three of the 87 religious schools in the sample are composed of

Table 1.2. Descriptive statistics and correlations among the independent variables

Variable	(2)	(3)	(4)	(5)	Mean	S.D.
%Mizrachim (1)	−0.37	−0.30	−0.03	−0.37	0.45	0.26
Size (2)		0.48	−0.02	−0.29	94.34	89.05
Sector (3)			−0.03	−0.03	0.57	0.50
%Male (4)				−0.02	0.39	0.30
Periphery (5)					0.29	0.47

single-sex students only (36 are all-female, and 27 all-male).

The descriptive statistics pertaining to these variables and their correlations are presented in Table 1.2.

Method

Since the dependent variables are dichotomous, the data are analyzed by logistic regression. Two equations are computed for each group of subjects. The first, reduced form equation, contains only the major explanatory variable, %Mizrachim. The second, extend form equation, adds periphery, sector, size, and %Male as controls. The extended equation takes the following form:

$$\log (p_i/1 - p_{ij}) = \beta_0 + \sum \beta_j X_{ij}, \qquad i = 1, 2, \ldots, 201.$$

where p_i is the probability that the ith school offers higher level subjects (scientific, "cultural capital" or "regular"); X_{ij} are the values of the explanatory variables for the ith school; and β_j are the parameters to be estimated by the data.

Results

The results of the logistic regressions are presented in Table 1.3. The parameter estimates in all three reduced form equations are negative and statistically significant, implying that students in **Mizrachim**-dominated schools are less often exposed to all forms of higher-level subjects.

The results are very clear at the descriptive level: higher-level subjects are less often offered by schools with higher rates of **Mizrachim** students. To what extent can we attribute this to the tendencies, revealed in Table 1.2, of **Mizrachim**-dominated schools to be religious, smaller, and located in the periphery? To address this question, I turn to the multivariate analyses.

For all three groups of subjects, %Mizrachim coefficients in the extended equations are lower than in the corresponding reduced ones, indicating an intervening effect of the control variables.

The effect of the control variables is marginal regarding the scientific subjects. In the equations pertaining to the sciences (columns 1 and 2), the modest change in the magnitude of the %Mizrachim coefficients is not followed by a meaningful change in the statistical significance.

The control variables are more significant in the "cultural capital" equation. The %Mizrachim coefficient still maintains its statistical significance, but at a different level ($p < 0.10$).

School's size, sector, and sex composition appear as dominant in the "regular" subjects equation. Here we find a particularly substantial reduction in the magnitude of the %Mizrachim coefficient which loses its statistical significance. The comparison between columns 5 and 6, and the correlations among the variables (presented in Table 1.2) indicate that student in **Mizrachim**-dominated schools are less often exposed to higher-level literature, Arabic, and geography because these schools tend to be smaller and religious.

To summarize, the findings of the multivariate analyses corroborate the hypotheses: After the inclusion of the control variables, schools characterized by higher rates of **Mizrachim** students less often offer scientific and, to a certain extent,

Table 1.3. Logistic regression of scientific, "cultural capital", and regular subjects (standard error)

	Scientific		"Cultural Capital"		"Regular"	
Independent variables	(1)	(2)	(3)	(4)	(5)	(6)
%Mizrachim	−3.66**	−2.66**	−2.54**	−1.22*	−2.02**	0.37
	(0.86)	(1.02)	(0.64)	(0.74)	(0.59)	(0.85)
%Male		0.86		−0.55		−3.17*
		(0.78)		(0.56)		(0.83)
Size		0.03**		0.01**		0.04**
		(0.01)		(0.00)		(0.01)
Sector		−0.63		0.81**		2.50**
		(0.57)		(0.36)		(0.51)
Periphery		0.77		−0.57		0.66
		(0.56)		(0.40)		(0.51)
Constant	4.09	1.70	0.84	−0.44	1.54	−2.13
	(0.62)	(0.86)	(0.30)	(0.51)	(0.59)	(0.73)

Note: All equations are significant at the $p < 0.05$ level.
* $p < 0.10$.
** $p < 0.05$.

"cultural capital" subjects at the highest level. This tendency cannot be attributed either to sector, size, sex composition, or peripheral location. The ethnic factor lacks a net effect on the probability of offering the "regular" subjects. Ethnic origin, then, retains an independent effect only regarding the subjects that represent a highly evaluated knowledge which is probably perceived as irrelevant and inappropriate for students of **Mizrachi** origin.

Concluding Remarks

The study demonstrates that students in **Mizrachim**-dominated schools are less often exposed to higher-level subjects. This stems partly from the concentration of these schools in the periphery, their smaller size, and their over-representation in religious education. However, the lower exposure to the scientific and "cultural capital" subjects is due to an additional factor not captured by the former three.

I believe that this factor represents curricular decisions based on the assumed capacities and interest of the students. American scholars report that principals and teachers consider some subjects too difficult and others irrelevant for underprivileged students (Anyon 1981a, b; Oakes 1985). Similar considerations probably exist in Israel. Thus the relative autonomy of the Israeli high schools results in reduced access of students originating from Asia and North-Africa to scientific and "cultural capital" subjects.

Do the current findings imply that school autonomy is all wrong and that the existing highly centralized system is preferable? Certainly not. The main implication of these findings is that in the process of constructing the curriculum of the autonomous school, principals, teachers and the community should consider the fact that knowledge is stratified. This stratification may be temporary and perhaps not accepted by the community. However, as long as certain knowledge constitutes a source of social power, its deprivation from weaker social groups reproduces existing inequalities.

Notes

1 One unit of study is defined as one weekly hour for three years, or three weekly hours in one year.

2 The Jewish and Arab education systems are completely segregated and differ in many aspects, including the curriculum. Hence there is no point in including them both in the same study.

3 History and social sciences, two popular optional subjects (higher-level history is offered in 37 percent of the schools included in the analysis; social sciences in 31 percent), were excluded from the analysis due to lack of information on their status in the Israeli curriculum, which caused difficulties in categorizing them. Bible and oral law are two additional popular subjects (offered in 49 percent and 31 percent of the schools, respectively). These subjects constitute a different category due to their particular value in religious education. Since the study does not concentrate on the differences between the two sectors, I preferred to exclude these subjects from the analysis.

4 The literature on school curriculum usually distinguishes between math and the scientific subjects (e.g., Apple 1990). Here, I follow the tradition of the Israeli education system defining math as a scientific subject (e.g., IBS 1989).

5 Religious education, which is an integral part of the Israeli education system, maintains an ideologically bound curriculum with particular emphasis on religious subjects (Schwarzwald 1990). The scientific subjects are less central in religious education (IBS 1989), probably due to conflicts between the scientific and the religious viewpoints.

References

Anyon, J. (1981a). Social class and school knowledge. *Curriculum Inquiry*, **11**(1), 3–42.

Anyon, J. (1981b). Elementary schooling and distinctions of social class. *Interchange*, **12,** 2–3.

Apple, M. W. (1990). *Ideology and curriculum* (2nd ed.). New York: Routledge.

Bar-Gal, Y. (1993). *Moledet and geography in hundred years of Zionist education.* Tel Aviv: Am Oved (in Hebrew).

Ben-Rafael, E. and Brosh, H. (1991). A sociological study of language diffusion: The obstacles to Arabic teaching in the Israeli schools. *Language Problems and Language Planning*, **15**(1), 1–23.

Ben-Rafael, E., Herzlich, R. and Freund, M. (1990). Symbole d'Identité ou capital symbolique: Le parcours social du Français en Israel. *Review Française de Sociologic*, **31,** 315–329.

Bernstein, B. (1971). On the classification and framing of knowledge. In M. F. D. Young (Ed.), *Knowledge and control*, pp. 47–69. London: Cassell and Collier McMillan.

Bourdieu, P. (1977). Cultural reproduction and social reproduction. In J. Karabel and A. H. Halsey (Eds), *Power and ideology in education*, pp. 487–511. New York: Oxford University Press.

Cookson, P. W. and Persell, C. H. (1985). *Preparing for power: American elite boarding schools.* New York: Basic Books.

DiMaggio, P. and Useem, M. (1982). The arts in class reproduction. In M. W. Apple (Ed.), *Cultural and economic reproduction in education*, pp. 181–201. London: Routledge and Kegan Paul.

Doron, R. and Millin, D. (1979). Prestige and status of "tracks" in the Israeli comprehensive school. *Megamot*, **24**(3), 387–399 (in Hebrew).

Eggleston, J. (1977). *The sociology of the school curriculum.* London: Routledge and Kegan Paul.

Goodson, I. F. (1983). *School subjects and curriculum change.* London: Falmer Press.

Goodson, I. F. (1992). On curriculum form: Notes toward a theory of curriculum. *Sociology of Education*, **65,** 66–75.

Israel Central Bureau of Statistics (IBS) (1989). *Subjects and courses of studies in general secondary education in Israel.* Jerusalem:

Central Bureau of Statistics, Special Series No. 844 (in Hebrew)

Kilgore, S. B. (1991). The organizational context of tracking in schools. *American Sociological Review*, **56**(2), 189–203.

Kliebard, H. M. (1992). The decline of humanistic studies in the American school curriculum. In H. M. Kliebard (Ed.), *Forging the American curriculum: Essays in curriculum history and theory*, pp. 3–26. New York: Routledge.

Kraemer, R. (1990). Social psychological factors related to the study of Arabic among Israeli Jewish high school students. Ph.D. thesis, School of Education, Tel Aviv University.

Labaree, D. F. (1988). *The making of an American high school*. New Haven: Yale University Press.

Mandler, N. (1992). A report on a conference of school principals in secondary education. *Ha'aretz, March 4, 1992*, p. A6 (in Hebrew).

Oakes, J. (1985). *Keeping track*. New Haven: Yale University Press.

Pur, D. (1989). *In favour of equality in education*. Jerusalem: Beit Berl Edition (in Hebrew).

Reshef, S. (1990). School autonomy — A new era in public education. In I. Friedman (Ed.), *Autonomy in education*, pp. 13–31. Jerusalem: The Henrietta Szold Institute (in Hebrew).

Schwarzwald, J. (1990). *A research perspective on religious public education in Israel*. Ramat-Gan: Bar-Ilan University (in Hebrew).

Shapira, R., Goldring, E., Heiman, F. and Shavit, R. (1991). *School autonomy in common registration areas: Controlled parent's choice in Israel*. Tel Aviv University, School of Education, The Unit of Sociology of Education and the Community, Position Paper No. 1-91 (in Hebrew).

Shavit, Y. (1990). Arab and Jewish minorities in Israeli education. *American Sociological Review*, **55**(1), 115–126.

Silberstein, M. (1990). The autonomous school. In I. Friedman (Ed.), *Autonomy in education*, pp. 100–129. Jerusalem: The Henrietta Szold Institute (in Hebrew).

Yaoz, H. and Iram, Y. (1987). Changes in the literature curriculum. *Studies in Education, 46/47*, 157–170 (in Hebrew).

Yogev, A. (1981). Determinants of early educational career in Israel: Further evidence for the sponsorship thesis. *Sociology of Education*, **54,** 181–195.

Yogev, A. and Ayalon, H. (1986). High school attendance in a sponsored multi-ethnic system: The case of Israel. In A. C. Kerckhoff (Ed.), *Research in Sociology of Education and Socialization* Vol. 6, pp. 45–78. Greenwich, Connecticut: JAI Press.

Young, M. F. D. (1971). An approach to the study of curricula as a socially organized knowledge. In M. F. D. Young (Ed.), *Knowledge and control*, pp. 19–46. London: Cassell and Collier McMillan.

Chapter Two

SPONSORED AUTONOMY AND MISGUIDED CHOICE: THE RISE AND FALL OF AN EDUCATIONAL ENDEAVOR

KALMAN BENYAMINI
Hebrew University of Jerusalem

In 1974 the Israeli Ministry of Education and culture proposed that the Department of Education of the City of Jerusalem establish special classes for gifted children in elementary schools. The proposal, which originated in the Ministry's Unit for Promotion of Gifted Children, sought to include Jerusalem in a program already in operation in other large cities of Israel.

From the outset, the City had reservations about grouping gifted children into a separate school or even separate classrooms. Public policy was an important reason: An effort was being made to improve the educational opportunities within the elementary educational system in the capital for the large number of children from underprivileged backgrounds, and to raise the level of special education. These efforts included planned integration of students from disparate social groups in elementary schools, and seemed socially and politically more significant than a specialized concern for the gifted.

An additional reason was that for about four years, the Department of Education had been involved in an extra-curricular enrichment program for gifted students. The operation of "Cheled" (Hebrew initials for "Group for Seekers of Knowledge") dealt at that time with 350 children between the ages of 5 and 12 who were selected and examined by the School Psychological Service and recommended to Cheled. The programs of Cheled included classes in sciences, arts and crafts, as well as group activities.

The proposal, however, was not rejected for two reasons: The persistence of the Ministry of Education, and our second thoughts about the proposal's potential. The fate of gifted students in the elementary school had for some time troubled some psychologists and educators. Previous ideas about giving adequate treatment to these children in their schools had been rejected by the educational authorities. In fact, Cheled had been created as an extracurricular program because of the system's lack of readiness to cope with the problems of the gifted within the framework of formal education. The new initiative of the Ministry of Education therefore represented an opportunity to make the treatment of these children part of the pedagogical responsibility of the regular school, as had been advocated in the literature of that time.

The experience gained in Cheled provided important guidelines towards the formation of a special project for the gifted in the schools. First, research and experience have shown that one could not make generalizations about gifted children, and that one should consider their great variability in many respects. Hence, concentrating gifted children into classes based only on their measured abilities would not at all ensure satisfactory development for each individual student. Indeed, in our experience, interpersonal and intra-personal tensions tended to disrupt the activity of formal groups of gifted children. In addition, it was learned that the separation of gifted children into groups of their own tended to generate some worrisome effects, namely, undesirable elitism, tensions among students, and pressures on the part of parents.

Beyond the welfare of the gifted, this undertaking had strategic aims. As psychologists pioneering in the educational system, we had a vision of promoting the "adjustment" of each elementary school to all of its students, by improving its capacity to offer significant educational approaches and settings to a great variety of students. The sociopolitical principle of school integration already prevailing in Jerusalem schools was thus extended by a psychoeducational principle of maximal accommodation of a wide array of specialized student needs and potentialities rather than referring children for all kinds of special education outside the school. Helping the original school cater to the gifted served as a welcome opportunity to implement this policy. This opportunity also gave rise to an expectation that specialized and individualized practices would gradually "spread" to the general student population, as well as to children with learning behavior and emotional problems. We assumed that such practices might have a better chance of "radiating" to the entire school when originating from a student group of high prestige.

These considerations helped determine the principles of the project proposed by the School Psychological Service to the City's Department of Education and the Ministry. Instead of creating separate classes in the elementary school, it was proposed to assign expert teachers for the gifted children in the school. The special teachers would work with gifted children directly, alone or in small groups, and indirectly through their homeroom teachers and parents. The work would be done largely during normal school hours, without removing the children from their regular classes. The enrichment programs for the children would be guided by the needs and abilities of each individual child, as learned from psychological evaluations, developmental and education assessments, and ongoing observations.

The proposal was aimed at those schools in which there were only a few gifted children and a distinct gap prevailed between their high level and the rest of the students. Because of their small numbers, we assumed that their special needs were not attended to and their proper development was thus in greater jeopardy. An additional consideration in choosing a school was the existence of an individualized approach to the children and a history of close cooperation with psychological and counselling services. It was, of course, necessary to consider the readiness of the institution to accommodate such a project. Moreover, in the general spirit of this endeavor, each school was expected to plan the contents and structure of the program on its own, which involved showing potential for autonomous decision making.

Towards implementation, eight eligible institutions were approached with the general ideas of the project and asked to submit lists of 4th–7th grade students who were "outstanding", "exceptional", "gifted", and "extraordinarily able". When six schools responded, group intelligence and creativity tests, as well as personal interest questionnaires, were administered to the students who had been named. On the basis of results of these preliminary examinations, four schools were recommended for the project. Individual psychological tests were then given each candidate. Eventually, the numbers of participants were 40 students in School A, 20 students in Schools B and C, and 30 in School D; 60 percent were boys and 40 percent girls. The Ministry of Education and Culture assigned special teaching positions: a full position to School A and a half position to each of the other schools. In two schools, the extra budget was allotted to a single teacher, while in others it was divided among a number of teachers both from within the school and outside staff. In three schools, volunteers, including parents, extended the activities beyond the budgetary limits. In addition, a project coordinator was appointed and the psychologists of the participating schools were allowed special paid time for consultation.

In the beginning of the school year of 1976/77, teachers interviewed the participating children in order to get acquainted and learn their individual interests. On the basis of this information, a number of programs and activity patterns were designed and introduced. One pattern which emerged in all schools was supervised individual work of children in their field of interest, either with new material or by enlarging on material covered in the regular class. Another pattern was the creation of study groups. In the various schools the subjects included science, humanities and social science, arts and crafts. Children from two schools cooperated in a study project on Jerusalem and attended a conference on the excavation site of the Second Temple. A select group from all four schools organized a literary club in cooperation with Writers' House, where students met with authors.

In addition, gifted children participated in activities which benefited their schools. These service activities included tutoring younger students with learning difficulties, helping prepare classroom discussions, publishing a school newspaper, or decorating the school with their own art work. One noteworthy project was a model of the Second Temple exhibited on Jerusalem Day in the Youth Wing of the Israel Museum. Also of interest were contributions by the gifted children to the work of their homerooms: Some children presented summaries of their individual or group studies and some helped discuss the classes' social problems (including problems generated by the very existence of the project).

Overall, a rich and interesting array of activities was created in the different schools, and after the first year in operation, some achievements and problems could be enumerated. One clear gain was the enrichment of the schools by additional educational materials, instructional equipment and library items, and by expert personnel from outside the school. These extra resources appeared to benefit the rest of the students and the educational staff. Following an activity pattern introduced for the gifted, a weekly group study day was introduced in one school for the entire student body, in which all students participated according to their interests, regardless of their homerooms. Personalized attention and guidance given by the project teachers to gifted students had an effect on the school as a while, leading to greater care for the individual student. In several instances, teachers assumed counseling roles akin to those found in open schools. In the case of one integrated school, the introduction of the project for the gifted prevented parental requests for transfer of some more able students to a nondistricted, more selective institution.

Certainly, some teachers expressed doubts about the preferred treatment given the gifted vis-à-vis the underprivileged students. These quandaries were at times reflected in the attitudes of the homeroom teachers, who hesitated to permit the gifted to join their special sessions (project children were excused from their classrooms for up to eight hours per week). Some teachers did not encourage the contributions of the project students to the homeroom activities. Others complained that the departure of the most able students was harmful to the level of studies in the classroom. It also became apparent that the boundaries of personal and pedagogic responsibility for each child were not clearly delineated between the homeroom and the special teachers. In addition, the special teachers made insufficient effort to involve the parents in the enrichment program for their children. After this initial experience it was also felt that the special teachers needed to broaden their knowledge of both contents and methods of instruction.

Social problems between the project students and the other children came to the fore in varying degrees of severity. Apparently, the gifted children found themselves in a conflict between the desire to remain "equals" with their schoolmates and the opportunity for special treatment. The impression was that the intensity of the social reaction was related to the prevalence of school norms recognizing individual differences among students. Discussions of the project in special meetings for teachers and students eased some of the tensions. The contributions of project students to their homerooms and their schools also played a role in reducing conflicts.

The tasks of school psychologists included referrals of candidates for psychological testing, follow-up of project students, counseling teachers and parents, and active participation in the planning of the programs. Most important, psychologists took up the role of consultants to principals and entire school staffs, with the aim of facilitating acceptance of the change in the organizational structure of the school and in their personal attitudes.

On the whole, the initial experience gained positive appreciation on the part of most participants from inside and outside the schools. We ourselves were so satisfied with the primary and secondary gains of the program, that myself and my associate on this project rushed to describe it in Hebrew and English papers. We truly felt that our cherished scenario of a resourceful school, the school that has the motivation and finds the means to flexibly accommodate and promote all of its students, was beginning to materialize. This sentiment prevailed throughout the successful continuation of the program in the two following years, when the individual schools gradually took up the leadership, guiding developments in accordance with their specific circumstances. On our part, we were intent on extending the program to other sections of the system, especially schools which had greater numbers of gifted students.

As was said earlier, an essential feature of the project was that each individual school make its own decisions about the shape the project takes. A school's capacity for indcpendent decision making was a rather exceptional and innovative expectation at that time, when the elementary education system was still quite centralized, and much more power an initiative rested with central authorities (the Ministry and local government) than with school managements. Principals were accustomed to be offered programs worked out by central administrations, and — at best — accept them to the extent that they appeared beneficial and relevant to their schools. Indeed, the present program had been thought out and generally shaped by outside authorities, and was then proposed to certain schools for implementation.

The chosen schools identified with the concepts from the start, and with the aid of administrative support and professional guidance gradually adopted the approach and made it, largely on their own, an integral part of their educational credo. This transformation entailed an abrupt departure from prevalent policies and practices regarding exceptional children, who were normally placed (with diagnostic recommendations) in settings outside the confines of the "regular" school. Evidently, the schools concerned were willing to accommodate the change, not only for its intrinsic value, but also because of what they may have gained in resources and reputation by displaying activities for its gifted students. Of importance was also the fact that the schools succeeded in getting their parent communities, including the parents of the gifted, aligned behind the project.

In retrospect, the developments can be viewed as a case of educational authorities endowing schools with "sponsored autonomy". The ideas and visions, the patterns of implementation, the means of support and resources, were conceived and supplied by sponsoring policy makers outside the schools. Given the common sense of schooling at the time, an undertaking of this nature could never have been conceived by a rank and file school, let alone by parents of gifted children. When eligible and interested institutions adopted the offerings they came to foster their growth autonomously, and on the way underwent other change processes.

And then, abruptly, the project came to an end. For the year of 1979/80, a school for the enrichment of the gifted was added to the City's educational system, offering special courses for one day a week to eligible students from all elementary schools. Schools were quick to have their gifted children IQ-tested and to allow them to join the new program. For our project schools, this was an offer that could not be refused, since parents insisted on enlisting their children in the new program. The four principals, bewailing the demise of an exciting endeavor, were utterly helpless in the face of the alternative attraction. We ourselves were stupefied, seeing three years of investments and attainments go to naught, and ever since have been wondering where we went wrong. For, compared to innovations successfully introduced by the School Psychological Service into the educational administration and schools, this was a singular case of failure.

It will be remembered that the Ministry's initiative of four years before was geared towards creating separate schools or classes for the gifted. Jerusalem objected to this scheme, and managed to channel the Ministry's provisions (especially teacher salaries) into what was deemed more correct for the City. However, while the project of the four schools evolved, the Ministry did not retract its original conception. A number of exchanges took place between political and educational representatives of the Jerusalem Educational Authority and the Ministry's administrative and pedagogic leadership. One formal complaint was that the cut-off scores for eligibility to the Jerusalem programs were lowered for the sake of promoting select children from underprivileged backgrounds. In a more ideological debate, the City's equalitarian policies were sharply assaulted with pointed meritorcratic arguments, which Jerusalem could not accept on either political or educational grounds.

At that period, local educational authorities had almost no say in matters of school programs, and Jerusalem's resistance proved to be an exception. It was apparently at this point that Jerusalem parents of gifted children were called into the picture, and exerted pressures on administros and supervisors. To be sure, parents' influence on educational policies was only incipient in those days, but the privileged parents in question were naturally more powerful. Principals of schools blessed with talented students also tended to concur with the Ministry's strategy to allow special classrooms for the gifted. A long series of negotiations ensued, resulting in the establishment of the said school, in which gifted pupils were to attend one day per week; for the other five days they remained in their original schools.

Even though the resolution was a compromise between the Ministry's and the City's strategies, it defeated the point of our original endeavor. Instead of trying to make the elementary school a richly meaningful place for all pupils, it reverted to the easy "default" of enabling specialized placements for segments of student populations. Parents probably turned to this alternative for all kinds of relevant and irrelevant reasons, and it is doubtful that they made educated decisions. Since no effort was invested by either party in clarifying to parents the pros and cons of alternative strategies, their decision was based on what appears to the present author to be "misguided choice".

Thus the rise of our endeavor came about by sponsored school autonomy, and its fall by misguided parental choice. Whether one agrees with its educational philosophy or not, one lesson that can be learned from this story is that in an educational system one cannot forego education of either schools or parents.

Chapter Three

SCHOOL CHOICE AS A BARGAIN IN A SECTARIAN EDUCATIONAL SYSTEM

MICHAEL CHEN
Tel Aviv University

The purpose of this chapter is to describe the established patterns of school choice by parents and to place the new demands for additional controlled or free choice in the context of the sectarian structure of Israeli education on the one hand and the conflicting demand for ethnic integration on the other hand. The chapter argues that the main consideration of most Israeli parents in selecting a school is civilizational: namely, the transmission of national identities, religious beliefs, and political orientations to the next generation. The emphasis on transmission of civilizational heritage has determined the sectarian structure of Israeli education since its beginning and accounts for the prevalent traditions of school choice. As the sectarian structure, which grew out of a bargain made long ago by declining political bodies, has become outdated, parental demands for school choice are construed as an effort to restructure the sectarian system according to the diverse composition of Israeli society at present.

The Israeli Version of the School Choice Debate

Parental choice of a school for their children is a common practice in Israel. Parents first of all choose one of the sectors of primary education at the first grade. Some of them make an additional choice of a magnet school, or a special program of study, within their sector. Towards the age of junior high school some parents choose between a neighborhood school and a boarding school (preferred by 20 percent of all parents). Choosing a senior high school is always a matter of free choice.

In spite of the wide prevalence of school choice, a heated public debate has been engendered by the introduction of new forms of school or curricular choices since the mid-1970s. Indeed, more and more choice arrangements are spreading in many communities. Shapira, Haymann, and Shavit report that:

> "The new schools of choice can be described as magnet ideology schools or magnet content-based schools, which are focusing on a particular area of study, employing selective registration procedures ... They combine a particular educational ethos with parental involvement and choice. Their particular characteristics are in potential conflict with the conservative Israeli educational system."
>
> (Shapira, Haymann and Shavit 1995: 360)

School choice has usually been formulated as the parents' right to choose a school outside of the official school district. A modified version of "controlled-choice" is now spreading in Tel Aviv, and in the religious sectors of education. It requires parents to choose one out of several similar schools from the same sector in the same district (Shapira, Haymann, and Shavit 1995). The legality and the equity of the controlled choice version has been under consideration by the Supreme Court of Justice during the last couple of years.

Elmore argues that the school choice controversy is about

substantial issues, such as finance, organization, and control. He indicates two dimensions of choice: "Should parents and students be empowered to choose among schools?" and "Should educators be empowered to organize and manage schools, to recruit and select students, and to receive public funds for providing education to those students?" (Elmore 1986: 79).

In Israel, new arrangements of school choice are considered by many as a retreat from the official policy of equity, as expressed in ethnic integration, which was expected to be accomplished by means of diverse classrooms, enforced zoning of school catchment areas, and bussing. Ethnic integration policy, as the main part of a far reaching school reform, was formulated and recommended by the Israeli Knesset in 1968, upheld by the Supreme Court and officially adopted by the Ministry of Education and many local authorities, ever since.

Gal indicates that many appeals to the Supreme Court of Justice by parents asking to register their children outside the official school zone have been rejected. Referring to the Knesset decision: "The Court determined that the authority's actions in matters of education are in the interest of society. When an authority opposes an individual... and there is a conflict, the interests of society prevail" (Gal 1995: 82). In 1980 Justice Barak commented that:

> "In conclusion we are concerned here with redistricting, which results in the petitioners' children having to attend a high school, the educational level of which is lower than the school they expected their children to attend. The consideration which led to this result is that of accomplishing integration in education... In order to accomplish this goal there is no choice at times but that children be transferred from one school to another."
>
> (Goldstein 1995: 322)

A review of the deliberations of the Israeli Supreme Court over a period of 25 years shows that it has become commonly accepted that school integration, as an operational definition of educational equity, and school choice, as an operational definition of freedom to choose, are two contradictory concepts.

A few Israeli advocates of school choice argue that choice advances school effectiveness, as it requires parents to be more involved in the school's functioning, and forces teachers and administrators to do their best to attract more students in order to survive in a competitive marketplace. Eden and Klinov even claim that the most capitalistic style of parental involvement, namely, free choice of a school in the educational market as a whole and full private tuition, leads to optimal integration, namely to maximal ethnic equity. They add:

> "In our model schools act as private firms, trying to achieve maximal profits while in practice few of them do: schools are mostly public or non-profit organizations... (But), the maximum-profit assumption does not matter much. The same claim could be proved on the assumption that schools want to produce maximum amount of human capital."
>
> (Eden & Klinov 1995: 386)

Most advocates of extended school choice, however, do not subscribe to the demands of school effectiveness, but to civilizational, cultural and political considerations: "Society is concerned with the maintenance and reproduction of cultural values. The community looks for a local definition of needs and goals, while the school (of choice) is interested in developing its unique educational identity" (Shapira, Haymann, and Shavit 1995: 369). Following a comprehensive study of the culture and curriculum of the Israeli educational system, Harrison, Elazar and Marom (1994) report great variations in the degree of responsiveness of the authorities to parental demands among the different school subsystems. While state public schools rank civilizational demands at the bottom of the list, state religious schools place them at the top. However, the rigid sectarian structure of schooling indicates the overriding importance of political and religious considerations for all.

Expectations relating to academic achievements, occupational opportunities, costs, and involvement, or the pursuit of happiness, are secondary in importance to civilizational considerations. Most parents concern themselves with the quality or effectiveness or equity of their choice only after making the main decision about the school sector.

A heated debate about choice is taking place in the United States as well. Marshall, Mitchell, and Wirt, in discussing the four most influential education values, quality, equity, efficiency, and choice, argue that: **"Choice inherently opposes all other values"** (Marshall, Mitchell, and Wirt 1989: 147).

Choice opposes quality because it leads to different definitions of quality, such as academic achievements, religious devotion, political conformity, or positive self-concept at any price. Parental involvement in curriculum planning, and choice of parochial schools, exemplify the tension between these two dominant values. The contrast between choice and equity has been evident in the frequent unwillingness of decentralized educational authorities to divert appropriate educational resources to the poor, minorities, handicapped people, and other students with special needs. Only by applying to higher levels of the legislatures, the courts, and the government, have disadvantaged groups succeeded in generating equity-based laws, leading to affirmative action. Choice and efficiency are opposing values, according to Marshall, Mitchell, and Wirt (1989), because: "Providing a range of options for decisions runs against the 'one best way'... favored by meritocrats"... and very frequently the opposing "participatory and rational forces arise from different segments of society, from lay citizens versus the professional managers of the service systems" (p. 149).

Elazar (1994) argues that choice is related to parents' "civilizational demands", namely the transmission of the cultural heritage to the children, regardless of quality, equity, or effectiveness: "This includes the transmission of the overall culture of the civilization and the treasures of its heritage. The problem becomes one of defining the content of that heritage" (Elazar 1994: 15).

The division of the educational system into several separate and autonomous administrations, such as Hebrew and Arabic, or general, religious and parochial, indicate the responsiveness of the authorities to national, religious and political interests. However, the maintenance of a sectarian educational system is considered as a useful compromise between the enthusiastic followers of rapid "nation-building" and the extremely diverse constituencies of Israel.

Elboim-Dror commented that:

> ... Structural separation and autonomy for different ideological orientations has proved to be very effective in reducing conflicts about educational policy ... As every one is for education and every ideological organization gets its share of the public fund, disagreements and public debates are very rare.
>
> (Elboim-Dror 1981, 222)

Generous allocation of educational and social services, free of charge, to all sectors of society, had been practiced by all Israeli governments. The relative advantage of the government over dependent populations enables it to bargain with diverse or opposing political sectors in many ways. Usually, the ruling parties offer different sectors substantial educational means and services in exchange for political collaboration or conformity. Frequently these kinds of bargains lead to strange outcomes, when some marginal sectors get a greater share of educational resources than the ruling sectors of society (Horowitz and Lissak 1989).

Traditional Patterns of Choice in Hebrew Education

The tendency to send children to a school in which most of the students carry on the same cultural traditions, have similar religious or ideological belief systems, possess similar cognitive modes of functioning and learning interests, and belong to a similar social class is very common among parents of all sectors[1].

However, social homogeneity of educational units exists much less in state general education than in state religious education, and it dominates the organization of parochial education.

In the orthodox independent sector, almost every religious community has a different system of schooling, which further differentiates among students of Western (Ashkenazi) and Eastern (Sephardi) origin. Habad, Gur, Beit-Ya'akov, Satmer, Netury-Karta and other Hasidic organizations maintain exclusive educational systems.

Freedom of choice of primary school between and within school sectors was very common in Jewish education since its very beginnings in the nineteenth century. The Law of State Education of 1953 restricted it somewhat by the introduction of flexible districting. But the decision of the Israeli Knessset in 1968 to reform the structure of education, to establish large scale junior high schools, and to enforce ethnic integration by bussing restricted it considerably.

The spread of various forms of school choice was made possible by the ambiguous versions of the two basic laws of education: the Law of Compulsory Education (1949) and the Law of State Education (1953). The 1949 law obligates parents to register their children at the age of five in an officially approved learning institution, and to ascertain that they study there according to regulations. The parents must make a choice of one school out of several recognized possibilities. Not only can they choose any school out of the numerous educational sectors and subsectors, they are entitled to enroll their children in a private institution, as long as the Minister of Education has recognized it as a "place of learning" (Goldstein 1995). In practice, all schools are fully financed by public funds, and education is considered to be free for all. Goldstein and Lederer (1994) conclude their analysis of the legal aspect of school choice by claiming that a great selection of public and private, elementary and secondary schools, which differ in their value orientations, cultural perspectives, curriculum, instructional methods, or administration are offered for choice to Israeli parents. As all of them are fully financed by the public: "The parents are indeed free to make a choice" (Goldstein and Lederer 1994: 119)[2].

The Sectarian Features of the New Choice Movement

Demand for two kinds of school choice by parents appeared in various sectors of the educational system after the "political upheaval" of 1977, when the right-wing parties came to power for the first time and a religious leader, Mr Z. Hammer, became the Minister of Education. One kind is the establishment of new subsectors alongside the existing sectors. The other is the emergence of new networks of cross-district magnet schools **within** the established sectors.

Chen and Addi (1995) report that following the 1977 upheaval, the efforts to establish new integrated junior high schools practically faded away. The new right-wing government replaced the centralized policy of integrated schools with a decentralized policy of "neighborhood renewal", which, symbolically at least, has diminished the advantages of centralized institutions and the services of large urban centers and enhanced the power of local government, especially of depressed and marginal communities.

With the slow decline of the centralized, urban, school integration policy, alternative cross-district schools started to emerge in many communities.

A former Minister of Education, Y. Navon, being concerned with the growing number of cross-district schools, appointed a special advisory committee to study the prevalence of the new phenomena, and to advise how to deal with it. In a letter of appointment to the head of the committee dated December 1989, Prof. Y. Kashti, the minister, estimated that about half of the students of the state religious sector study in cross-district schools, according to their parents' preferences (Kashti 1991). Shapira reports that about 150 cross-district elementary schools of different kinds operate in the state educational system.

Most of the value-oriented magnet schools were proposed and founded by organized groups of interested parents. Most of the subject-matter magnet schools, which specialize in the arts, in physical education, or in the sciences, were founded by local authorities in response to parents' demands (Shapira 1989: 151). From a legal point of view, the debate is about permitting students to be enrolled in an elementary or a junior high school, which maintains a particular program of study, and is located outside of the official school zone, as prescribed by the local authority and approved by the Ministry of Education.

New sectors

The common feature of all sectors and subsectors, old and new, is their full commitment to a particular national, religious or political mission. Even the most general state public system is committed, by Section 2 of the State Education Law of 1953, to a list of particular purposes of education. It defines the goals of education in secular terms, grants administrative and curricular autonomy to the State-Religious school system, and discriminates against the Arab population by ignoring their national, religious, cultural, and linguistic traditions.

Tali, a network of cross-district schools, which advocates the reinforcement of Jewish studies within the secular state education sector, had started to operate in the late 1970s. It is defined as an independent administrative unit, and operates in about 15 separate schools or in special spaces in comprehensive schools. It was initiated by American immigrants to Israel, who belong to the Conservative Movement. It maintains a particular curriculum, employs special teachers and administrators, and is overseen by a special supervisor. It admits all applicants from a very wide geographical region (Goldring and Zisenwine 1989).

Noam is a network of cross-district junior high schools, which started to attract many able students, mostly of Ashkenazi origin, within the state religious sector, in the early 1980s. In selecting and admitting students, the authorities consider both intellectual ability and intensity of religious life (Dagan 1993).

Another emerging sector is El-Ha'Maayan (To the Source). It operates within the independent sector, and consists of about 80 schools, which serve children of Eastern families, who support Shas, a new Orthodox party.

After 1977, Druze schools started to operate as a separate subsector within the Arab educational division, and the same applies to the Bedouin segment of the Arab population.

Cross-district magnet schools

The second kind of schools of choice are regular schools, which offer a particular program in a specific field of study (e.g., magnet schools) or offer an additional course of study, financed by the parents ("grey education"). Some of the magnet school subscribe to specific ideological orientations.

The Gonen School in Jerusalem was established in direct response to the 1977 political upheaval by activists of the Labor Party in order "to inculcate in its students the ideology and traditions of the Labor Movement". It attracts Western students from all parts of Jerusalem and Eastern students from nearby neighborhoods (Shapira 1989).

Magnet schools attract students from many remote parts of the official district, and offer special programs in the sciences, the arts, sports, or Jewish studies. As they operate under the auspices of local authorities, as "schools of particular mission", or "schools with additional programs of studies", they are financed by the government like other schools, but they also enjoy additional support from local authorities, and charge additional payments from the parents.

One of the first indications of the change of policy towards cross-district enrollment has been the establishment of special classrooms for the gifted in primary or junior high schools in many municipalities. Since 1976 parents have been entitled to enroll gifted children in magnet schools. This arrangement violated the policy of school integration in two ways. Not only did it encourage cross-district registration, it also permitted able students, mostly of Ashkenazi origin, to escape from integrated classrooms. At present, many local authorities maintain very different delivery systems for the gifted. They may exist in a separate school, a separate classroom or in an extracurricular setting, but they all attract students from great distances (Goldring, Milgram, and Chen 1989).

Another cross-district network is Kedma ("Eastward" in Hebrew), which developed within state education, admits mainly secular children of Eastern origin to several secondary schools, which pursue particular avenues to higher education.

One of the most controversial experiments in school choice is being implemented in Tel Aviv. The "controlled choice program" asks parents in disadvantaged neighborhoods to choose one out of three junior high schools: one in their own neighborhood and two others in a more affluent section of the city, either academic or comprehensive. The "controlled choice program" aims to enable disadvantaged parents to make a decision according to their personal conditions and inclinations. If they are interested in advanced studies and in directing their children toward higher education, and if the children are ready to cope with the difficult cognitive and emotional demands of the integrated classroom, they may choose to bus them to an academic-oriented high school. If they prefer the supportive and less strict climate of their neighborhood school, they are entitied to choose that.

Degani (1979) reports that selecting appropriate primary schools in or outside the poor neighborhood, according to the children's motivation and abilities, has been practiced in disadvantaged neighborhoods for many years. While local authorities strictly forbade cross district enrollment in affluent sections of the city, they practically permitted it in disadvantaged neighborhoods. Most of the disadvantaged parents are now well experienced in the challenges of the integrated classroom in the comprehensive junior high school. They are capable of evaluating by themselves who can cope with the demands of the academic program and who is unable to do so. The findings of several studies indicate that some disadvantaged students made substantial cognitive and motivational gains by attending integrated classrooms in academic schools, but that many others paid a high psychological price for their academic failure (Kfir and Chen 1985).

Community schools were another response of left-wing educational leaders to the 1977 political upheaval.

> The community school program, as formulated in 1978, aimed at two objectives: 1. Improvement of school performance by intensifying the interaction with the neighborhood's population (parents and local organizations), 2. Providing a center of social and cultural activities for the adult population.
>
> (Peres and Pasternak 1993: 12)

Although most of the 150 community schools do not require cross-district bussing, their establishment requires parental consent for additional payments and involvement. Most of the community schools operate in middle-class communities: "Community schools can succeed in places where the parents have the necessary means and the motivation to be involved in school work. Such necessary conditions are usually available in affluent small-size cities" (Peres and Pasternak 1993: 47).

The cost of maintenance of the average student in a magnet, a community, or a controlled-choice school is higher than the cost in regular schools. Parents are expected to contribute in money, time and involvement more than the average parent. However, very few parents refuse to participate because of the extra payments. The high costs of schools of choice are sometimes criticized as a violation of principles of equity, but so far the debate is theoretical.

The new controlled choice seems to violate other guiding values: equity, quality, and effectiveness. It violates the norms of equity as it costs more per student than many other programs, and it classifies young students into exclusive tracks of educational and socioeconomic levels. It violates quality as it offers to apprehensive students who are reluctant to choose a demanding school, a relatively poor course of studies, and it compromises choice of irrelevant academic environments. It violates effectiveness as it permits the choice of useless curricula, does not provide proper returns for the educational investments of many students. Nevertheless, the controlled choice program is well accepted by many, and by disadvantaged parents in particular (Chen 1995).

In conclusion, school choice by parents indicates a decisive change in public approach to the maintenance and distribution of schooling. Since 1977, an intensive search for the preservation of civilizational value by school choice characterizes Israeli parents more than the pursuit of equity, quality, or effectiveness.

The emergence of many new kinds of school choice in each sector can be viewed as a bargain between the centralized authorities, who want to enlarge their constituencies by school integration, and parents of all groups who want to transmit their particular civilizational heritage to their children.

School Choice Versus School Integration

Kashti, head of the Ministerial Committee for the Study of Cross-District Registration, is one of the opponents of extended choice. He argues that "cross-district schools indicate social tendencies toward resegregation" and "permitting parental choice of a school outside the official zone tends . . . to drive the educational system to a point of no return". In the diversified and hostile educational system of Israel, it exacerbates societal inequities and increases the gap in educational achievements between students of high and low SES, and of dominant and marginal sectors of society. More specifically, school choice privileges violate established practices of ethnic integration, by which local educational authorities decide about the place of learning of every child (Kashti 1994: 14). Yonah, a philosopher of education adds: "The likely overall result of the implementation of parental choice may be detrimental to equal educational opportunities" (Yonah 1995: 149).

The decision of the Israeli Knesset in 1968 to restructure secondary education by the introduction of junior high schools and integrated classrooms was planned and implemented by top leaders of the Labor Party, in collaboration with other liberal parties, all of them dominated by the Ashkenazi elite. The school reform and integration policy reflected the secular establishment's belief, shared by many secular left-wing and liberal thinkers of European origin, that equality in educational achievements is a necessary prerequisite for equality in socioeconomic status and political power, and that equality in socioeconomic status would blur and extinguish the importance of hostile ethnic identity. Such a decline of Eastern ethnic identity was considered a necessary condition for the preservation of the existing balance of power, and the prevention of the emergence of a hostile "Second Israel".

School restructuring for ethnic integration was imposed on the state educational system by the top leaders of several political parties. In order to close the gap in educational achievements between the ethnic groups, the Ministry of Education tried to cover exclusively all the costs of schooling and to shelter the integrated schools from parental involvement. It was assumed that by requiring some tuition fees and by encouraging parental involvement, differences among parents in educational achievements, and in ability to pay, might compromise equality of opportunity for all students, and prevent the preferential treatment of disadvantaged students (Chen 1987: 11).

The adherence of educational authorities to school integration persists in spite of widespread opposition. Its advocates argue that the attainment of very important goals has been facilitated by it. Extensive economic resources have been invested in providing proper primary and secondary education to Eastern students by the Labor Party governments. Its supporters argue that many significant achievements have been accomplished by it. At present about 85 percent of an Eastern cohort graduate from high school, and about one-third continue in various kinds of postsecondary education. The gap in standardized educational achievements between the groups has dropped from a full standard deviation in the early1970s to about half of it at present (Goldring and Addi 1989). The infiltration of educated Easterners into professional occupations is also gradually growing. They now comprise about 40 percent of the teaching force, which is just 12 percent less than the expected proportion.

Adler maintains that: "The gaps between Jews of Western and Eastern origins has decreased significantly" (Adler 1995:

263). At present neither ethnicity, nor the variables which were highly correlated with it, continue to define populations at risk. Furthermore, as school integration attempted to reduce segregation, many ethnic groups do not perceive themselves any more as marginal groups. Thanks to its achievements, there is no further need for enforced school integration as a strategy of educational fostering (Adler 1995: 258).

However, many sections of the Israeli public do not agree with the authorities that school integration is a "success story". The Knesset's resolution to get integrated was severely opposed by many segments of the Jewish population. Shmueli (1987), former director general of the Ministry of Education, notes four sections that interfered with the implementation of school integration: religious organizations, the kibbutz movement, the teachers' union, and parents' pressure groups.

Chen and Addi (1995) report that local authorities with high proportions of state religious and parochial schools and/or Eastern students (of Asian-African origin) tended to reject the Knesset recommendation "to get integrated". They conceptualize school integration as a process of educational and cultural socialization, arranged by the dominant Ashkenazi (European) elite, in order to incorporate the newcomers from Asia and Africa into its patterns of culture and social organizations. The Westerners aspired not only to maintain their European civilizational heritage, but also to transmit it to the newcomers.

In spite of some encouraging results of the school reform, the public at large tend to view school integration as an unsuccessful enterprise. Affluent parents complain that the diverse classroom delays the full educational development of their children. Ashkenazi students tended to segregate themselves in homogenous informal groups, according to their high ability groupings.

Many Eastern parents have rejected the demand to go from their familiar neighborhoods to the hostile and competitive integrated school. Even high achievers of Sephardi origin had to pay a heavy "psychological price" for the little educational advancement granted by the integrated classroom. They frequently experienced a significant drop in the level of their self-perception (Kfir and Chen 1985). The most important indication of the shortcomings of the integrative approach has been the fact that most of the bussed Sephardi students dropped out of high schools. Most of the students who were bussed from disadvantaged neighborhoods to integrated junior high school in affluent neighborhoods failed to be admitted to the high schools, and either dropped out, or moved to private or vocational schools (Tel Aviv Municipality 1992).

Sami Shitrit, a leader of Kedma, a new exclusive network of cross-district schools for secular Sephardi students, argues that under the humiliating conditions of Ashkenazi hegemony in the integrated schools, Sephardi children are unable to realize their intellectual abilities.

> "It is necessary to compensate for the wrong treatment by the discriminating authorities. Only students who are proud of their culture and history and are confident of their academic ability can succeed".
>
> (Resnik, 1996)

From the point of view of the Labor Party, the founder and carrier of the school integration policy, its main political goal has been to gain the support and vote of Easterners. It has **not** been accomplished. Between 1968 and 1977, a period of massive investments in school integration, Eastern voters discarded the sponsorship of the ruling Labor Party and started to support the right-wing Likud, and Orthodox parties. The Easterners' change of political behavior contributed to the dramatic "political upheaval" of 1977, in which the left wing lost power for the first time.

In conclusion, the era of school integration is approaching

its end, not because the Westerner in Israel (or the whites in the USA) refused to collaborate in its implementation, but because after many years it has been rejected by the target populations: the Easterns in Israel or the blacks in the USA.

In summarizing the era of school desegregation in the USA, *Newsweek* magazine uses the following headlines:

> "Separate but equal is wrong, right? From busing to voting, a new generation of politicians and thinkers, black and white, is rethinking integration."
>
> "Segregation: After years of Jim Crow — separate schools, separate worlds — the '50s and the '60s brought the dream of a racially mixed America." But,
>
> "Celebration: The old roads to integration are in disrepair, replaced by the sense that black-only institutions are a better way to go".
>
> (*Newsweek*, April 29, 1996: 34)

Parental Choice of Sectors: A Political Perspective

It has been assumed that the competition among sectors for student enrollment leads to parental empowerment and increased prevalence of school choice within sectors. In order to prevent frequent transitions of students from sector to sector, more opportunities are offered to parents. Indeed, a significant inconsistency between parents' political orientations and the school sector of their children has been indicated.

Table 1 describes the changes in the relative size of the three major sectors in the primary schools of Hebrew education during the high time of the right-wing government: 1980–91.

The data in Table 1 indicate a slow increase in the total number of students in primary schools between 1980 and 1991.

Table 1. The distribution of primary school students in 1980 and 1991, by school sector (in percentages and indices)

	No.	State Education	Religious Education	Independent Education	Total %
1980	424.173	74.2	20.1	5.7	100.0
1991	476.055	71.3	21.6	7.1	100.0
1980 = 100.0	1991 =	107.8	120.1	139.8	112.2

Source: Table 22.18 of the statistical abstract of the Israel Central Bureau of Statistics, 1992.

In index numbers, the total number grew from 100.0 to 112.2 during that period. The indices indicate a relative decline of state education (107.8). Its growth is substantially slower than the moderate growth of the state religious sector (120.1), and the rapid growth of the Independent (Orthodox) sector (139.8).

The differences in rates of growth among the different sectors may be explained in part by demographic factors, namely, a very high birthrate among the Orthodox, or by educational considerations as the growing holding power of the Orthodox sector. But they definitely indicate a state of competition between the school sectors, as a large number of parents choose a different sector for their children than the one in which they had studied by themselves (Schwarzwald, 1990).

It is pertinent to ask if parents do consider several alternatives before making the first choice of school for their children. There is some evidence, however, that many parents make their decision in two steps. First they consider the appropriate sector, according to their value system and cultural norms. Then they consider practical issues, such as the school's reputation, its student body, teachers, programs of study, proximity costs, and, most importpant, chances for admission to higher education and/or occupational status (Benyamini and Tatar 1992).

As the educational system is so clearly subdivided, most

parents have no hesitation in enrolling their children for the first time in a sector which fits their ideological and religious inclinations. However, there is a large margin of parents whose sectarian commitments are ambiguous. They have to weigh the pros and cons of each possibility and consider several alternatives. A very crude estimation of the size of the fluctuating group ranges between 10 percent and 20 percent. The estimate is based, first of all, on the differences between the number of students in the various sectors in Hebrew primary and in secondary education. As only 69 percent of the primary school students were registered in 1992–3 in state education schools, compared with 77 percent in secondary schools, a significant shift from sector to sector is assumed to be associated with the transfer to secondary schools. Furthermore, as there is a significant gap between the large numbers of students in the religious sectors (31 percent in primary schools in 1992–93) and the relatively small numbers of voters for the respective political parties (about 20 percent in the 1988 elections), once again, some inconsistencies between the parents' beliefs and choice of a sector may be indicated.

It is assumed that this group of fluctuating parents explains the concealed competition for parents' choice among the educational sectors. One form of subtle effort to attract students from another sector is demonstrated by the ways in which vocational schools present themselves to the public. In most televised advertisements of the secular state education sector, many male students appear with a cap on their heads, according to religious regulations.

This practice is especially common among the religious sectors, which try to attract as many students as possible from other sectors. The popularity of religious schools among non-observing parents is surprising. It is explained by the "spiritual transformation" of the Jewish community on the one hand, and by several practical considerations on the other hand. One practical consideration is that female graduates of state religious schools are exempt from compulsory military service. Furthermore, the military service for male students of many **Yeshivot** (religious boarding schools) is shorter and less demanding. A second consideration is ethnic integration. The state religious sector never fully subscribed to school desegregation (Schwarzwald and Amir 1989). Many Westerners, and some Easterners, probably prefer the segregated classrooms of this sector to the integrated classrooms of the state educational sector. The third advantage of these schools is the preferential treatment given to able students. They are usually allocated to separate high ability tracks and placed in orthodox **Yeshivot** at a relatively young age. It seems, however, that the practical advantages offered by the state religious sector are not enough, especially for Eastern students.

At present, about half of the high-school students in the state religious sector study in cross-district schools, either in boarding schools or in selective urban schools, which attract students from a very wide geographical area (Kashti 1991: 7). It seems that the aim of the leaders of the state religious sector in granting many parents a free choice of school is to prevent Western parents from shifting to Orthodox boarding schools (**Yeshivot**) and Eastern parents from enrolling their children in El-Hama'ayan, or in secular vocational schools.

Ayalon and Yogev (1996) show how external political pressures influenced routine processes of curricula planning in the state religious education:

> Since the 1970s, the state religious sector has faced a dwindling student population resulting from the growing secularization of Israeli society on the one hand, and the growth of alternative religious education on the other. The establishment of El-Hama'ayan of the new political party of religious Mizrachim — Shas — coupled with the dwindling political power of the National Religious Party vis-à-vis the rise of right wing parties, are among the

reasons for this deterioration in enrollment. As a response, the state religious sector increased the emphasis on religiosity.

(Ayalon & Yogev 1966: 12)

Dagan, Director-General of state religious education, recommends intensifying the religious character of primary school by extending parental choice. Such a move is necessary to satisfy parents who consider the state religious system as "too loose" on the one hand, and to satisfy other parents who seek a modern, science-oriented curriculum, on the other hand (Dagan 1993: 37).

The growing number of religious and even secular parents, who choose an orthodox school is surprising. The choice of such a segregated school implies relinquishing any chance for academic higher education as well as many occupational opportunities for both males and females. Struch and Schwartz (1989) report that students in Orthodox schools, as well as students from other sectors, consider themselves in a state of mutual conflict. All groups, including the Orthodox group itself, report mutual negative attitudes to one another, associated with strong aggressive tendencies. However, all view the Orthodox as strict observers of Jewish tradition (**Halacha**), and respect their readiness to sacrifice so much.

The behavior of parents who enroll their children in Orthodox schools indicates how important is school choice to the cultural pride and positive self-presentation of a great number of Israelis. It is hardly related to equity, effectiveness, or quality of education.

Integration versus Choice: Easterners' Flight from Ashkenazi Supremacy

Both the declining policy of school integration and the spread of school choice seem to reflect important changes in the balance of power among the three major political divisions of the Jewish community: first, Labor and the left wing, secondly, the Likud (Unity) and the right wing, and thirdly, the religious bloc. As is now well known, the Easterners' flight from the Labor Coalition and the National Religious parties towards the right wing and the Orthodox parties has changed the political make-up of Israeli society (Horowitz and Lissak 1990). As newcomers in the 1950s and 1960s, the Easterners used to go to the schools which had been assigned to them by local authorities, and vote for the ruling parties which had the power and the means to support them economically and politically. Accordingly, they voted generously several times for the Labor and the National Religious parties, which were in power for many years. The Easterners voted for them, although the secular leaders of the left wing, or the Ashkenazi leaders of the religious sector, viewed their cultural traditions as outdated (Horowitz and Lissak 1989).

From the early years of the State until the 1977 political upheaval, the distribution of the Easterners' votes resembled the distribution of the total population, with one exception: a significant preference for the National Religious Party. As this party controls, by law, the state religious sector, it attracted a very large segment of the Easterners both as students and as adult voters.

According to Schmelz, Dellapergola, and Avner (1991), in the 1965 elections, the National Religious Party got 9 percent of all votes (compared with about 4 percent at the 1992 elections), while more than 75 percent of its voters were Easterners. The state religious sectors registered then about 27 percent of all students in Jewish primary schools, most of them of Asian-African origin.

At about the same time, the coalition of the left-wing parties, with the Labor Party at its center, attracted about 51 percent of the votes, most of them Westerners, but also a substantial Easterners' vote. The Labor Party then practically controlled the Ministry of Education, and the administration of the state education sector, which attracted then about 65 percent of Hebrew primary education, including a representative proportion of Sephardi students.

During that period, both leading sectors of Hebrew education, the state and the state religious sectors, sensed that they did not properly serve Sephardic students, but they employed different strategies to cope with this shortcoming.

The state religious sector employed a strategy of educational excellence, by offering special programs of advanced religious studies to very able students of both ethnic groups, and by providing generous compensatory education to other students. It virtually separated high, moderate, and low achievers in different schools. Most of the low achievers were students of Asian-African origin, who resented very much being segregated in poor schools.

From the early 1970s many Easterners started to desert the National Religious Party and to establish independent and exclusive parties, organizations, and school networks for religious Easterners only. The results of the Easterners' flight from National Religious sponsorship are very apparent at present. First of all, most of the Sephardi graduates of the religious school system do **not** vote for the National Religious party. In the election of 1988, only 3.9 percent of the Jews voted for it, and only half of them were Easterners. Schmelz, Dellapergola, and Avner (1991) report that Sephardim with religious inclinations shifted their votes from the National Religious party either to more Orthodox parties, or to the new Likud party, which is the backbone of the right-wing coalition. In the 1988 elections, the Likud attracted 31 percent of all eligible votes, but many more Easterners (57%) than Westerners (43%). About 44 percent of all the Easterners voted for the Likud, compared with 27 percent for Labor, 5 percent for the National Religious party and 9 percent for Shas, an exclusive new party of Orthodox Easterners (Schmelz, Dellapergola, and Avner 1991: 201). The state religious sector, which attracts only 14.6 percent of the adult votes, continues to attract about 20 percent of all Hebrew primary school students. However, many of the parents, as well as many of the students, vote for other parties.

There is probably no simple solution to the crisis of the National Religious Party and its state educational sector. On the one hand, the emphasis on excellence and Ashkenazi religious hegemony drives out disadvantaged Easterners to the ethnically exclusive Shas party, and to its school sector. On the other hand, the ethos of impractical religious studies drives upwardly mobile Easterners towards the pluralistic, patriotic and semi religious Likud party (Schmelz, Dellapergola and Avner 1991: 93), and to secular vocational schools of the state education sector.

For lack of any better solution to the growing Easterners' flight, the leaders of the religious sector permit reluctant parents to do whatever they want, and to choose for their children any school or any program of study as long as they remain loyal to the rules of the system.

D. Landau, Chief Supervisor of Hebrew Studies in the state religious sector, argues that:

> Permitting parents to decide by themselves about preferable programs of study could lead to the neglect of many important cultural domains. In schools, choice does not ascertain anything. Students, in contrast to materials, refuse to be classified in a reliable way, as required by marketing.
>
> (Landau 1993: 21)

In practice, however, religious parents continue to choose a school from several possibilities inside and outside the religious sector.

The competition between the National Religious Party and other political bodies seems to be the probable explanation for its tendency to provide parents with more and more choice of school and more opportunities to be involved in school management.

The Flight from School Integration and from the Labor Party

The flow of events which led the Labor Party to reconsider its adherence to the school integration policy, and to approve more and more parental involvement and school choice, differs in many details from that of the National Religious Party. But there is a common factor in both cases. Labor, like the National Religious Party failed to attract the support of most of the Eastern voters in spite of the pursuit of school integration, and other affirmative action. It employed school choice to encourage reluctant parents to maintain their support of the official educational policy.

In the 1965 elections, before the adoption of the school integration policy, the ruling Labor Party got 37 percent of all votes (about 40 percent of the Jewish votes). In 1977, after nine years of aggressive implementation of the school-integration programs, it got only 30 percent of the votes. About 65 percent of the voters were of Western origin and only 35 percent of Eastern origin. Only a little more than a quarter of the Easterners supported Labor, which placed equity in educational opportunities at the top of its social agenda, while almost half of them supported the right wing, which tended to be indifferent in matters of education (Chen and Addi 1995).

It is not surprising that the dramatic outcome of the elections of 1977 indicate the beginnings of parental involvement in school affairs and growing demands for school choice, especially among parents and educators of the left wing. Two major changes of mind among them should be noted. One is related to the sobering idea that the centralized instruments of the state cannot serve the interests of the left wing.

The second is that many Westerners understood that real social convergence by different ethnic groups cannot be achieved by administrative actions. It can be accomplished only when both groups are ready, able, and motivated to do it.

The founders of the Gonen cross-district ideological

school, which aimed at reviving the values of the Labor movement, restricted enrollment to Easterners who were ready to accept the school's goals, and to Westerners who approved of ethnic integration. They viewed enforced bussing of reluctant parents and students of both ethnic groups as the reason for the Westerners flight to the suburbs, and of the Easterners' flight to right-wing and religious (mostly Orthodox) schools (Yael Levin, Gonen school principal, personal communication).

Similar considerations are reported by Goldring and Zisenwine. In a study of the cross-district network of Tali (Reinforced Jewish Studies) schools, which started to operate at the time of the political upheaval, most of the Eastern students admitted were able to cope with educational and social demands of the integrated framework. It turned out that Eastern parents who chose Tali schools and agreed to pay the extra costs usually registered in Tali only their more able children, who were ready to cope with the challenges of the integrated classroom. Less motivated children were referred to regular schools (Goldring and Zisenwine 1989).

If the pressures for ethnic integration could be construed as a counteraction against the common tradition of free choice, then the present growing demand for parental choice, by both Western and Eastern parents, may indicate a partial disappointment with the modest accomplishments of school reform and ethnic integration. It also indicates a growing awareness on the part of many Eastern parents that they are entitled to preserve their own cultural heritage and to reject the hegemonic tendencies of the Western traditions, which prevail in integrated schools.

Discussion: The Future of School Choice in a Diverse Democratic Society

The emergence of new kinds of educational sectors, particular schools or special programs of studies, which are offered for free choice by parents, is not unique to Israel. In describing the situation in the USA, Crowson (1992) comments:

> One of the most fascinating phenomena of the late 1980s and early 1990s has been an explosion in types of schools and school-enrollment alternatives. With roots in the "alternative school movement" of the 1960s, the concept of the "magnet school" has become enormously popular in metropolitan areas. Usually with a firm racial quota, the magnet schools tend to offer to city schools desegregation through voluntarism.
>
> (Crowson 1992: 206)

In contrast to Israel, school choice in the USA gains momentum as a reform towards consumer sovereignty, and is considered one of the important issues of the 1996 elections. It appeals to conservatives as it advances the

> "market place" assumption, which is that schools will become more accountable to their publics and certainly more attentive to student needs if forced to compete for enrollment. Schools that do not perform will "go under", as students leave to enter more productive environments.
>
> (Crowson 1992: 207)

In Israel, in contrast, the choice movements aim to help students to "become somebody", namely, to form a distinguished national, religious, political, and psychological identity. The choice movement revolutionized two predominant conceptions of the "nation-building" period. First, that being a Jewish citizen in Israel is a meaningful and satisfactory component of personal identity, and secondly, that ethnic origin is unfortunately a most important indicator of social status. Those two concepts serve as a legitimizing myth of the school integration policy, and are still strongly accepted by left-wing educators of the state school sector. The emergence of many new subsectors implies that in addition to being identified as Israelis of a certain ethnic group, students wish to declare their national affiliation, religious orientation, political inclinations, and intellectual interests by joining the most appropriate school sector. In the absence of such a sector, their parents can get organized and establish it. One of the unanticipated outcomes of the sectarian development is the gradual decline of the state education system from a forum of national authority to one of the particular sectors. This decline is not only a result of the declining number of enrollments in integrated schools and in vocational schools, the two kinds of institutions of affirmative action, which were established by left wing educational leaders. It is mainly a result of growing homogenization and of indecisive systems of value orientations.

The Easterners' flight from integrated schools and from vocational education to other parties and to other sectors of education lead to the social homogenization of the student body of state education. The percentages of Easterners, of children of poor families, of residents of development towns, is significantly lower in this sector than their percentages in the Jewish population at large. **As a homogenized section of the population, state education cannot be regarded any longer as a universalistic entity. It is only one more educational sector.**

A. Frank, principal of one of the leading boarding schools of the Kibbutz movement, the leading subsector of the ruling state education system, argues in a recent paper that:

> Our particular mission was threefold: accomplishment of Zionist-National missions; realization of a socialist way of life, and struggling for peace with the Arabs. All those missions have already been accomplished, and are carried out by bureaucratic administrations . . . The educators of the Kibbutz are now in an urgent need of a new particular program.
>
> (Frank 1995: 2)

If the Kibbutz educators are confused by the absence of a leading value system, then other educators of the state education division are indeed embarrassed.

The above survey of the developments of parental choice in Israel does not indicate any advance towards the establishment of an "educational marketplace". On the contrary, it seems to promote more sectarian divisions of the school system than ever before.

In spite of the basic contrast between the USA and Israel they have a common element, that is, the emergence of parents as a well-organized pressure group. In comparing traditional and emergent forms of parental involvement in the school site, Crowson (1992: 188–96) argues that 1) responsiveness replaces involvement guided by school authorities; 2) openness of the school to parental involvement replaces formal, administrative enclosure; 3) participative interaction in some processes of decision making replaces limited formal representation; 4) parents take part in the educational process as fellow educators instead of being passive recipients of professional authorities.

In Israel of the 1990s, parents intervene in the planning of "enrichment programs", determination of school districts, placement of students in schools, classrooms, or various forms of special education, and in the financial management of the school.

As in the USA, educational issues now rank very high in the list of issues, which are debated in election campaigns at the national and local level. Kaspi (1994) comments that: "in the last municipal elections [Nov. 1993], disagreements about educational issues emerged in most local authorities. Most of the candidates learned that educational issues attract voters' attention, and made frequent use of it" (p. 15).

The growing power of parents to choose a school sector or a school site within a sector is just one aspect of their increased involvement in education. However, it seems that educational empowerment is just one indication of broader individual freedom, which is spreading in Israel, as well as in other democracies. A common element of individual empowerment in different societies is the acknowledgment of the diversity of different civilizational heritage and the rights of self-presentation of ethnic and racial communities by the centralized authorities of the state, and by their readiness to bargain with every group about its particular rights and obligations.

In conclusion, different forms of school choice seem to be the hottest item on the agenda of educational policy and practice, but the sectarian system easily absorbs all new forms of school choice between and within the extremely divided educational structure. One cannot disagree with Elazar (1994) who, after stating that parental choice is the key value in Israeli education, comments that:

> We can surmise that the outline of Israeli elementary and secondary education has been laid out. It is one that attempts to incorporate parental choice into the existing system, although there are discussions of privatization, and one committee of the Ministry of Education recommended it. It seems that parental choice will more likely be confined to allowing parents to utilize the flexibility at their disposal under the present law, which is substantial.
> (Elazar 1994: 18)

So far, the established educational authorities have bargained quite effectively with the emerging educational sectors and subsectors. Satisfactory compromises have been usually accomplished between the excessive demands of the dissenting sectors and the establishment's demands to preserve the existing social order and to conform to the democratic norms of struggle for power and leadership. However, the flow of events is so rapid that Elazar's assumption, "that the outline of Israeli education has been laid out" seems an underestimation.

The voluntary gradual transformation of the state education system from a dominant center of comprehensive power and overriding authority, into just one of several sectors, of equal legal status and of restrained political power, is a very new phenomenon. It is very difficult to envisage the implications of this development for the future of the divisive Israeli society.

Notes

1 Officially speaking, the Israeli educational system consists of two main divisions, Hebrew and Arab education. Hebrew education comprises four sectors (namely, separate supervisory and administrative units). Arab education is divided into three sectors. Of all 697,200 primary school students in 1993, about 79.5% studied in Hebrew, and 20.5% in Arab schools.

Hebrew education is further subdivided into: state education = 56.8%, state rural (communal) education = 2.9%; state religious education = 15.9%, and independent religious (orthodox) education = 4.7% of all students.

Arab education consists of state arab education = 14.5%, state bedouin-education = 3.1% and state druze education = 2.1% of all students (Sprinzak, Bar, and Levi-Mazlou 1994, Table 1.1 and Figures 1.2 and 6.6).

2 All quotations from Hebrew sources have been translated into English by the author.

References

Adler, C. (1995). "Equality and excellence: Contradictory or complementary goals in Israeli Educational Policy". In Kahane R. (Ed.), *Educational Advancement and Distributive Justice: Between Equality and Equity*, pp. 253–266. Jerusalem: The Magnes Press.

Ayalon, H. and Yogev, A. (1996). "The alternative worldwide view of state religious high schools in Israel". *Comparative Education Review*, **40**(1), 7–27.

Benyamini, K. and Tatar, M. (1992). *Prestige, Evaluation and Choice of School by Parents.* Jerusalem: The NCJW Research Institute, The Hebrew University (in Hebrew).

Chen, M. (1987). "Education in Israel between equality and excellence: A symposium". *Skira Hodshit*, **34**(10–11), 3–23 (in Hebrew).

Chen, M. (1995). "Residents of disadvantaged neighborhoods know their children's needs". *Hed HaHinuh*, **69**(7,8), 4–6 (in Hebrew).

Chen, M. and Addi, A. (1995). "Community politics, school reform and educational achievements". In Kahane, R. (Ed.), *Educational Advancement and Distributive Justice: Between Equality and Equity*, pp. 341–57. Jerusalem: The Magnes Press.

Crowson, R. L. (1992). *School Community Relations Under Reform.* Berkeley, California: McCuchan Publication.

Dagan, M. (1993). "Attributing a particular character to religious schools by choice". In *Bama: A Publication for Supervisors*, pp. 37–40. Jerusalem: The Pedagogical Committee, The Ministry of Education) (in Hebrew).

Degani, A. (1979). *Ha'Tikva Neighborhood. A Methodological Framework for Social Renewal of a Disadvantaged Neighborhood.* Tel Aviv: Geography Department, Tel Aviv University.

Eden, B. and Klinov, R. (1995). "The effects of different financing methods on the extent of school integration". In Kahane, R. (Ed.), *Educational Advancement and Distributive Justice: Between Equality and Equity*, pp. 358–74. Jerusalem: The Mgnes Press.

Elazar, D. J. (1994). Introduction. In Harrison, J. A., Elazar, D. J. and Marom, Z. R. (Eds), *Unity and Diversity of Culture and Curriculum in the Israeli Educational System.* Jerusalem: Institute for the Study of Educational Systems.

Elboim-Dror, R. (1981). "Conflict and consensus in educational policy making in Israel". *International Joumal of Political Education*, **4**, 219–32.

Elmore, R. F. (1986). *Choice in Public Education.* Santa Monica, California: Rand Corporation.

Frank, A. (1995). "A disappearing gap: The particular nature of kibbutz education is fading". *Hinuh Acher*, **52**, 4 (in Hebrew).

Gal, N. (1995). *The Individual, the Authority and the Letter of the Law: The Supreme Court Position on Parents' Choice of School.* Jerusalem: The NCJW Research Institute for Innovation in Education, The Hebrew University (in Hebrew)

Goldring, E. B. and Addi, A. (1991). "Using meta-analysis to study policy issues: The ethnic composition of the classroom and academic achievements". *Studies in Educational Evaluation*, **15**, 231–46.

Goldring, E. B., Milgram, R. and Chen, M. (1989). "Toward formulation of educational policy for gifted students". In Ministry of Education and Culture, The Pedagogical Secretariat, *Planning Educational Policy*, pp 207–38. Jerusalem: The Publication Department (in Hebrew).

Goldring, E. B. and Zisenwine, D. (1989). "Developing Jewish identity: Parents and schools in Israel". *Jewish Education*, **57**(1), 28–34.

Goldstein, S. (1995). Liberty vs. equality: Concerning student placement in state schools in Israel: The changing views of the Supreme Court. In Kahane, R. (Ed.), *Educational Advancement and Distributive Justice: Between Equality and Equity*, pp. 290–310. Jerusalem: The Magnes Press.

Goldstein, S. and Lederer, E. (1994). "The status of parents in determining their children's education: Legal considerations". In Danilov Y. and Inbar, D. (Eds), *Choice in Education in Israel*, pp. 117–37. Jerusalem: Publication of the Ministry of Education and Culture and The Hebrew University of Jerusalem (in Hebrew).

Harrison, J. A., Elazar, D. J. and Marom, Z. R. (1994). *Unity and diversity of Culture and Curriculum in the Israeli Educational System.* Jerusalem: Institute for the Study of Educational Systems.

Horowitz, D. and Lissak, F. M. (1989). *Trouble in Utopia, The overburdened policy of Israel.* Albany, New York: State University of New York Press.

Israel, Central Bureau of Statistics (1992). *Statistical Abstract of Israel, 1992*. #43 Jerusalem: Central Bureau of Statistics.

Kashti, Y. (1991). *Cross-district Educational Frameworks: Report of the Public Committee*. Jerusalem: The Ministry of Education and Culture (in Hebrew).

Kaspi, D. (Ed.) (1994). *Self-Administration of Local Educational Authorities Between Centralization and Decentralization: A Bibliography*. Jerusalem, The Szold Institute (in Hebrew).

Kfir, D. and Chen, M. (1985). "Ethnic integration and the adolescent personality". In Ziv, A. (Ed.), *The Exceptional Age*. Tel Aviv: Papyrus (in Hebrew).

Landau, D. (1993). "Culture and education in a techno-economic age". In *Bama: A Publication for Supervisors*. Jerusalem: The Pedagogical Committee, The Ministry of Education (in Hebrew).

Marshall, C., Mitchell D. and Wirt, F. (1989). *Culture and Education Policy in the American States*. New York and London: The Falmer Press.

Peres, Y. and Pasternak, R. (1993). *Community Schools in Israel: Success or Failure*. Tel Aviv: Eitav Publishing House (in Hebrew).

Resnik, D. (1996, February 16). Kedma is a failure. *Ha'ier, Weekly Magazine*. Tel Aviv (in Hebrew).

Schmelz, U. O., Dellapergola, S. and Avner, U. (1991). *Ethnic Differences Among Israeli Jews: A New Look*. Jerusalem: The Institute of Contemporary Jewry; The Hebrew University of Jerusalem.

Schwarzwald, J. (1990). *A Research Perspective on Religious Public Education in Israel*. Ramat Gan: Bar Ilan University.

Schwarzwald, J. and Amir, Y. (1989). "Interethnic relations and education: An Israeli perspective". In Krausz E. (Ed.), *Education in a Comparative Context*, pp. 245–68. New Brunswick and Oxford: Transactions.

Shapira, R. (1989). "Schools of particular mission: Background, development and challenges". In Ministry of Education and Culture, The Pedagogical Secretariat, *Planning Educational Policy*, pp. 207–38. Jerusalem: The Publication Department (in Hebrew).

Shapira, R., Haymann, F. and Shavit, R. (1995). "Autonomy as ethos, content and commodity: An Israeli model for controlled choice of autonomous schools". In Kahane, R. (Ed.), *Educational Advancement and Distributive Justice: Between Equality and Equity*, pp. 358–74. Jerusalem: The Magnes Press.

Shmueli, E. (1987), *School Reform and its Opponents.* A paper presented before the Jerusalem Forum (in Hebrew).

Sprinzak, D., Bar, E. and Levi-Mazloum, D. (Eds) (1994). *Facts and Figures about Education and Culture in Israel.* Jerusalem: State of Israel, Ministry of Education and Culture.

Struch, N. and Schwartz, S. (1989). "Perceptions of conflict with Haredim (Orthodox Jews): Differences in basic values and antagonistic expressions". *Megamot*, **32**(1), 5–26 (in Hebrew).

Tel Aviv Municipality (1992). *Statistical Yearbook No. 32.* Tel Aviv: The Municipal Center for Social and Economic Research (in Hebrew).

Yonah, Y. (1995). "Parental choice in Israel: Rationales and practical considerations". In Kahane, R. (Ed.), *Educational Advancement and Distributive Justice: Between Equality and Equity*, pp. 132–52. Jerusalem: The Magnes Press.

Chapter Four

SCHOOL AUTONOMY AND PARENTAL CHOICE: STEPS IN LOCAL EDUCATIONAL PLANNING

FLOR HAYMANN, HAYA GOLAN, AND RINA SHAPIRA[1]
Tel Aviv University

During the last decade there has been an ongoing process of change in Israel's educational system, expressed in such matters as the call for decentralization of the system and increased participation of local authorities in planning educational policy, enhancement of the schools' autonomy, empowerment of teachers, diversification of curricula and teaching methods, and increased parental participation (Reshef 1984; Poor 1985; Kashti, Arieli, and Shapira 1991).

These trends were reinforced by the Ministry of Education report (1992), **Israel Believes in Education**, which gave expression to the need and the wish to transfer power and responsibility from the national to the local level and the school.

The changes involved in decentralization and autonomy affect all the participants in the educational process — the central authority, the local authority, the school, the supervision, the parents — and require a responsibility redefinition of the roles and authority of each of them in planning educational policy.

One domain in which we can observe the changing trends is in the role of the local authorities. Until the 1970s the local authorities in Israel were perceived as the executive arm of the central authority, lacking any real power to adapt educational patterns to the specific needs of the local population. Recently, however, local authorities have started to undertake more responsibility for education — particularly at the elementary level. This involves decision-making processes through identifying problems and finding local solutions (Ony 1988; Yogev 1990).

In this chapter we have chosen to focus on the education system in Tel Aviv-Jaffa as an example of the changes that have characterized Israel's education system over the past two decades: the move towards decentralization and diversification, the changed concept of the local authority's role, and the increasing view of the school as an independent entity with its own educational-social world.

The two main purposes of this chapter are: (1) to describe those developments in Tel Aviv-Jaffa's education system that led to the consolidation of a local educational reform plan in 1992, designed to meet the special needs of this city; (2) to discuss the educational-social implications of this program and analyze its planned implementation, especially in those aspects that are designed to renew the entire system.

After describing the existing trends of change, we will focus on three major stages in the development of Tel Aviv-Jaffa's education system: (1) early stages — milestones on the road to diversification, autonomy and choice in the system; (2) the Junior High School Reform Program — its implications, prospects and problems; (3) the Tel Aviv-Jaffa Educational Reform Plan.

In addition we present results of a parents' survey which aimed to learn about parents' attitudes towards the educational reform plan.

Trends of Change in the Israeli Education System

Like many industrialized countries, during the first three decades of its existence Israel was characterized by a high degree of centralization, designed to ensure equity in public services generally and in education particularly (Volensky 1992).

A centralized education system and uniformity in the schools were considered during that period as the appropriate means to achieve the social and educational goals of the young state, which was gathering in masses of immigrants from a wide variety of cultural and social backgrounds.

During this period autonomy was granted only to certain sections, such as religious education and kibbutz education (Kamanda 1985). The "right of diversification" in curricula, embodied in the State Education Act (1953), was rarely exercised, and the education system made no more than halfhearted attempts to cope with an increasingly varied population (waves of new immigrants that came to Israel from all around the world, especially in the 50's from Asia and North Africa) and its needs.

Finally, in the 1970s, the Ministry of Education realized that no centralized system could fulfill to entire range of needs of a pluralistic society. The processes of democratization in education, empowerment of local communities, accelerated technological developments, professional demands for organizational and structural reforms, and the academization of teaching — all these together led the Ministry to seek an innovative educational approach to address these developments. The approach adopted was decentralization of the education system and promotion of autonomy in the school (Reshef 1984; Inbar 1987; Volensky 1992).

School-based autonomy became the official policy of the Ministry of Education in the early 1980s. This called for empowerment of the school by "giving it a considerable degree of pedagogic independence, with wider powers and responsibilities in planning and implementation of teaching, educational and cultural activities" (Ministry of Education 1981). However, this often failed to go beyond the declarative level, while the schools themselves and/or the system did very little to turn it into reality.

The concept of autonomy was somewhat vague (Reshef 1985; Friedman 1987; Shapira, Green, and Danilow 1994) and lends itself to a variety of interpretations: ideological, organizational, professional, pedagogical, humanistic, community-oriented, and so forth. The Ministry of Education adopted the organizational and educational aspects of autonomy, in response to the pressure to transfer power from the center to the local authorities (Inbar 1987; Friedman 1987). The basic assumption was that pedagogical independence of schools is possible even within a centralized system.

The broader implications of school-based autonomy gradually consolidated around certain central components. The autonomous school has its own, unifying "credo"; the teaching staff jointly set the school's aims and subject matter, and are responsible for pedagogical and organizational decisions and for the curricula and evaluations that derive from the school credo; the students and parents are not "passive clients" but active partners in the school's policy making; and finally, the school's budget is independent as far as possible (Gordon 1987; Shapira, Green, and Danilow 1994).

Until the end of the 80's, while supporting the concept of autonomy, the Ministry of Education continued to maintain its supervision and control mechanisms and limited the resources allotted to schools, thus detracting from the schools' ability fully to implement autonomy (Inbar 1987; Green, Haymann, and Shapira 1991).

In the meantime, changes were occurring in the status of local authorities and in their influence on educational policy planning in the areas of their jurisdiction.

The laws and decisions of the Knesset (Israel Parliament) reflect the nature of the relations between the central and local authorities. The Ministry of Education and the local authorities are together responsible for the education system and the implementation of educational laws. The responsibility and power rest with the Minister of Education, while the local authority is the agency that implements the directives of central government (State Education Act 1953; Compulsory Education Act, 1949; Knesset Record 36, 1963). Neither the legislation nor the procedures that evolved over the years clarified the nature of the partnership and distribution of power between the authorities or how educational budgeting should be divided.

The perception of the local authority's role reflected two principles: separation of the pedagogical and administrative domains, and local government as an executive rather than a policy-making body (Elboim-Drod 1985; Ony 1988).

The tasks assigned to the local authority were as follows: construction and maintenance of school buildings; student enrollment and determination of enrollment zones; student transferal from one school to another; health services; educational counseling; supplementary education, and student insurance. The Compulsory Education Act (1949) determined varying levels of education. The state was responsible only for compulsory elementary education. At the secondary level, which was not defined as compulsory, local authorities were to take more responsibility for planning and implementation.

In the course of time it became clear that separation of the administrative and pedagogical domains was not feasible in practice, and each local authority defined its function and status in the education system in its own way. Elboim-Dror (1985) mentions a number of factors that played a role in the shift towards decentralization and the increased power of local government in educational policy making:

(1) The educational reform of the 1960s, and the political pressures accompanying its implementation, which led to an unplanned transfer

of power from the center to the local authorities in matters of zoning, establishing junior high schools, and dealing with strikes.

(2) The channeling of funds to local authorities for the purpose of establishing junior high schools, expanding vocational education, and establishing comprehensive schools.

(3) The growing involvement of the second generation in key functions in local authorities, and their placing education as a top priority.

The drastic cuts in the education budget over the 80's, and the transfer of the burden of funding educational projects and programs to local authorities and parents, helped to legitimize their increased involvement in determining the direction of development of local education systems.

These factors, along with the experience accumulated in other countries that encourage community involvement, have strengthened the belief that the transfer of power and authority from central public systems to communities and local authorities facilitates appropriate treatment of local problems (Volensky 1992).

These trends of change — the promotion of school-based autonomy and decentralization of the education system — which were envisaged from the early 1970s, acquired new validity in the 1990s and rose to the top of the educational agenda. This is reflected in a document issued by the Ministry of Education and Culture, **Israel Believes in Education** (February 1992), and in the establishment of a steering committee for "autonomous schools" (1993) peopled by senior Ministry staff and academics, to discuss issues of decentralization and autonomy, and their implementation.

The newly emerging policy views relations between the center, the community, and the school in a new perspective. It entails many problems, some of them common to the entire system and others specific to certain communities.

To illustrate the above, we will describe some milestones in the development of Tel Aviv-Jaffa's education system.

Early Stages: Milestones on the Road to Diversification, Autonomy and the beginning of Choice in Tel Aviv-Jaffa's Education System[2]

An examination of the educational system in Tel Aviv-Jaffa (one of the largest and most complex in Israel — for data see Appendix) reveals two major changes, which have characterized the education system as a whole since the 1980s. The first is the significant increase in the number of two-year-olds attending nursery school, and the second, and more outstanding, is the growing number of students in secondary education: 90 percent of the city's children who began the 7th grade in 1984–85 finished the 12th grade in 1989/90 (Dagany 1990) — though not necessarily in the same school where they started junior high.

This increase in the number of children entering secondary schools, and the rising demand for secondary education on the part of less advantaged social groups, led to a broadening and diversification of secondary education in the city, such as the establishment of comprehensive and vocational schools as alternatives to the regular, academic high school.

These developments placed various constraints on the city's education system, partly due to the city's particular demographic make-up. Tel Aviv-Jaffa, like metropolitan cities elsewhere in the world, typically consists of neighborhoods that are largely homogeneous in socioeconomic terms. The city is divided into three social-geographical areas, marked by large socioeconomic gaps: the north (the high socio-economic area), the center, and the east and south of the city (the low socio-economic area). The socioeconomic gaps find expression, among other things, in their respective percentages of disadvantaged pupils, which are significantly higher in elementary schools in south Tel Aviv than in the center and north of the city.

One of the distinguishing features of elementary education in the city is the existence of neighborhood schools, which by their nature tend to be homogeneous in terms of student population. This largely determines the kind of problems faced by the schools as well as the initiatives that have arisen in the field in an attempt to find differential solutions to the different needs of children in the city's elementary schools.

For example, the Education Authority initiated a project for enrichment and reinforcement of the core subjects (Lewy 1984) which was administrated, in 1984, in 41 state and state-religious schools, most of them in the southern and eastern parts of the city.

Another initiative designed to meet the special needs of a certain population is a project of community schools in middle-class neighborhoods which previously had no community services. This project, based on the involvement of parents and local community in processes occurring in the school in order to enhance the educational work (Harpaz 1985), was started in many Tel Aviv schools in 1986.

The establishment, in 1984, of specialized non-neighborhood schools with specific orientations (School of science, the environment and society, and School of the arts) formed a major watershed in the development of the municipal education system, setting an example of educational innovation to other schools (Shapira 1988).

This was an attempt to create a new type of school, offering a special curriculum to pupils from all parts of town. The schools were established by the Education Authority with the support of the Ministry, according to the model of the "magnet" schools in the United States. Their initiators saw them as an attempt "to improve the education system and the quality of education, to meet the needs of gifted pupils, develop their special talents, and provide an alternative attempt at voluntary integration" (internal document, Tel Aviv-Jaffa Municipality 1984).

Setting up these schools entailed considerable financial investment and their daily operation was made possible by the financial support of the Education Authority and the parents, who pay graduated school fees according to income.

These schools, from the beginning, have enjoyed a high degree of administrative and pedagogic autonomy, expressed, among other things, in setting curricula, choosing some of the teachers, determining admissions policy and managing budgets — subject to the overall aims of the education system.

One school with a special orientation had previously existed in Israel's education system. This is the A. D. Gordon School, established in the 1930s and functioning ever since as a school with an ideological-social orientation admitting students from outside its designated zone (Shapira, Haymann, and Dintzman 1990). Unlike the A. D. Gordon School, Tel Aviv's new magnet schools attempted to locate pupils through selection mechanisms that fitted their particular nature, setting administrative criteria to ensure a balance between girls and boys and between students from different parts of the city.

These schools are described as fertile educational environments (Shapira and Haymann 1991) and as models for other schools in the city and in other cities. Nevertheless, they have sparked — and continue to spark — public opposition due to what is seen as their break with some of the fundamental principles of education in Israel: integration and equal opportunities (Shapira 1988; Kashti Commission 1991). The small number of these schools, and the fact that two of them only admit children with suitable ability, creates imbalance between them and the other schools, violating the principle of equal opportunity and equating uniqueness with excellence.

Thus the advantages and disadvantages of the specialized non-neighborhood schools pointed to new possibilities in education, while raising trenchant educational and social questions, marking a new stage in the development of the education system, to which we will return later.

Parallel with the opening of the specialized non-neighborhood schools, changes began to appear in other schools, initiated by principals, teachers, or supervisors. Changes took the form of greater variety in teaching methods, curriculum development, reorganization of staff, and introduction of various educational and pedagogic projects (Shapira, Haymann, and Posner 1994). In a few cases the system also permitted students to choose between local schools. The development of autonomous schools was considered by policy makers as a means to minimise the differences between the magnet schools and the rest.

Feedback tests held in 1991 showed that Tel Aviv-Jaffa's elementary schools did better than the parallel population nationwide (Ha'Aretz December 5, 1991). The findings show schools that excel alongside schools that "fail", and the results do not always correspond with the area in which they are located. Particularly outstanding were a number of schools in the south of the city whose achievements considerably exceeded their achievements in the past.

The Tel Aviv elementary school system in the beginning of the 90's made its first steps in promoting diversification and innovation. However, these trends were still local initiatives and didn't characterise all the schools.

The Junior High School Reform: Implications, Prospects, and Problems

In secondary education, the introduction of educational reform and the establishment of junior high schools in the late 1960s generated pluralism in the education system and led to the increased involvement of the local authority in pedagogic issues.

Tel Aviv-Jaffa's junior high schools were usually set up in

affiliation with academic secondary schools. Concurrently, four-year technological schools continued to operate, admitting 9th graders. Some of the secondary schools were turned into six-year comprehensive schools. Gordon School and the magnet schools, including a new one opened in Jaffa, were permitted to function as nine-year schools. One school functioned as a four-year school and one as an independent junior high school.

Thus, in Tel Aviv-Jaffa the reform generally took on a six plus six-year structure, with some cases of nine plus three and three separate junior high schools.

The decision to establish junior high schools was based on the belief that they could provide an organizational framework for equal opportunities in education, in order to reduce gaps in achievements, and for the social integration of students from different backgrounds (Ministry of Education 1970).

Implementation of the junior high school reform in Tel Aviv-Jaffa was accompanied by difficulties and problems, due to the city's demographic structure, its north-south polarity, and to the affiliation of the junior high schools to academic secondary schools, which conducted selection procedures for entrance to the higher grades. Moreover, the municipality's decision to avoid bussing from the north of the city to the south failed to create demographic balance in the junior high schools. Some junior high schools in the south of town remained homogenous, and others were heterogeneous in varying degrees.

Another problem, shared by junior high schools elsewhere in the country, is the lack of adequate training for teachers of heterogeneous classes, and the difficulties in using teaching methods appropriate to the new situation. These problems were "solved" in Tel Aviv, as in other places, by creating tracks and grouping in the junior high schools. This brought about a segregated situation within the integration program, since these groups are based on students' achievement levels, which often parallel their social backgrounds, and thus perpetuate the social gaps instead of bridging them.

The Education Authority, preparing to contend with these problems on different levels and in different ways, invited research and surveys to locate the problems as a basis for planning and change. Recent studies on secondary education have identified several major issues, such as: the lack of curricula suitable for heterogeneous classes in junior high schools; teachers' pedagogic experience and ambivalent attitudes towards the heterogeneous class' tendency to create homogeneous classes for high and low achievers, causing social separation in the junior high schools (Report of the Institute for the Advancement of Social Integration 1988); and violence in some very heterogeneous schools (Dagany 1990). In addition, there are reports of negative attitudes towards integration among some parents and students, and of opposition by some of the population to compulsory integration and the concomitant bussing system (internal document, Tel Aviv-Jaffa Municipality 1991).

In the transition from junior high to high school many students change schools or even drop out. In 1990, only 46.5 percent of the students who had begun their studies at a specific junior high school continued until 12th grade in the same school. Of those students who had transferred to other schools, 74 percent had moved to lower levels, and 10 percent had dropped out of the system altogether between 7th and 12th grades (Dagany 1990). Most of the dropping-out was found among students from low socioeconomic areas.

The 1992 high school matriculation results in Tel Aviv-Jaffa reveal that only about 60 percent of those students who sat for the exams passed them. It should be noted that even in those schools which conduct selection procedures in the move from 9th to 10th grade the results are not optimal, as might have been expected in this type of system (from a personal conversation with the head of Tel Aviv-Jaffa's Education Authority, August 16, 1993).

Among the reasons for the problems that the secondary education system faced were: selectiveness in admission to high schools; the conception of junior high and high schools as separate frameworks; the absence of an overall concept of the six-year secondary school, and insufficient curricula appropriate for heterogeneous high school populations.

Moreover, although the structure of secondary education permits choice between tracks in comprehensive schools, and between various academic and vocational schools, the actual choice is left to the schools themselves. Since they, on their side, use rigorous selection criteria, this has led to segregation in schools and strengthened the problematic connection between students' social background and the type of education they receive (Svirsky 1990; Shapira, Haymann, and Shavit 1995).

What emerges from this analysis is a sharp shift in policy from the junior high school to the high school, amounting to a reversal. The enrollment to junior high school is dictated from above, not necessarily according to zones or to parents' and students' preference. In the high schools a different situation prevails: the choice is free and the schools themselves practice selection. In fact, the parents have free choice of high schools but this is partly blocked by the schools on educational grounds, without any social education control. These processes in the high school thus perpetuate the social inequities which the junior high school was designed to cure.

Over the years various plans for reorganization of the secondary education system have been proposed and partially implemented, including a proposal to open controlled enrollment zones (1985); changes made in the channeling method; and examination of the possibility of moving to independent junior high schools (internal document, Tel Aviv-Jaffa Municipality 1991). In the context of these experiments, graduates of five elementary schools in south Tel Aviv were given the choice between two options: a junior high school near their homes or one in the northern part of town. Students completing other elementary schools were not offered a choice and were simply referred to schools in their own zone or elsewhere.

This set in motion a process which has also been found to characterize such situations in other countries. The schools in the northern neighborhoods which have successfully implemented the reform in terms of social integration and student achievement attract students from the southern neighborhoods. Other schools which have been less successful in coping with the many problems arising out of social integration encounter resistance on the part of pupils and their parents. Simultaneously, schools in the south have gained strength and begun to attract those youngsters who prefer to attend school near their homes or avoid threatening competitive encounters at this stage in their lives.

The commissions that examined Tel Aviv-Jaffa's junior high schools (The Commission for the Advancement of Social Integration, and the Commission of Inquiry into the Status of Junior High Schools) emphasized the need to develop suitable curricula for heterogeneous classrooms and train teachers to work in such situations, alongside administrative and organizational changes.

This awareness of the need for change in teachers' working methods has led to the introduction of intervention programs to improve teaching methods in the heterogeneous classroom (learning-style-based instruction, differential teaching, adaptive teaching, etc.).

In summary, the survey of the development of Tel Aviv-Jaffa's education system indicates several major problems: achievement gaps between schools; lack of continuity between schools in the move from elementary to junior high school, and particularly from junior high to high school; separation of academic from vocational education; difficulties in accomplishing the aims of the educational reform; and the existence of different and sometimes even contradictory policies concerning school integration at the various levels: elementary, junior high and high school.

In an attempt to address these problems, various solutions were proposed, offering answers and underlining the need to consolidate an educational reform program of change based on an overall systemic conception. The following is an example.

The Tel Aviv-Jaffa Educational Reform Plan

In September 1992 the Tel Aviv-Jaffa Education Authority drew up a plan for reform of the city's education system: "There are other ways — Tel Aviv-Jaffa Believes in Education" (Tel Aviv-Jaffa Municipality, Education Authority 1992).

The purpose of this plan was to raise the standard of education and of students' achievements, and its planners believed that this could be accomplished through "the creation of a broad range of educational options, suited to the educational, cultural and social pluralism of the population, while taking into consideration the students' needs and areas of interest" (Tel Aviv-Jaffa Municipality, Education Authority, 1992: 4).

The plan covers all levels of education from preschool on, including special education, the basic assumption being that all levels of education should be treated as one whole, while paying attention to the special nature of each stage.

The proposed change was based on four main principles:

(1) School-based autonomy;
(2) Increased freedom of choice for parents and students;
(3) Integration of various groups of students;
(4) Diversification of curriculum and teaching practice.

An examination of the principles and their translation into recommendations reveals a perception of the school as the focus and fount of educational action and renewal.

The first principle, which runs like a thread through the entire program, sees the school as the main factor in making decisions regarding its aims and methods, and hence also responsible for the results. As an autonomous institution, team work, curriculum, evaluation and feedback processes and parental involvement.

The second principle relates to three topics:

- **Controlled choice of schools**: the program recommends broadening the existing enrollment zones, allowing every school in the extended zone to develop autonomously with its own orientation, and permitting students to choose between several schools, while maintaining a control mechanism to ensure balance in the student population of each school.
- **Increased flexibility** in the secondary system: it is proposed to introduce more variety in study tracks to suit students' different paces and levels (flexibility in time, choice of subject matter and teaching methods).
- **Comprehensiveness and continuity**: the school should offer as wide a variety as possible of learning frameworks, enabling all students to learn according to their abilities and realize their potential in one school. This means breaking the barriers between academic and vocational education and abolishing the track system. The principle of comprehensiveness embraces that of continuity, according to which the school has to provide for continuity of studies in the same school for every student from 7th grade through to 12th grade.

The third principle relates to the issue of integration among different groups of students (from different parts of town, new immigrants and mainstream children, Jews and Arabs, pupils in special and regular education, kindergarten and elementary school children), in order to achieve stable and durable social integration.

The fourth principle deals with the diversification of teaching and learning methods to match the pluralism in Tel Aviv-Jaffa, stressing the contents and subject matter the schools need to deal with in order to meet the needs of the students and of an advanced technologies society.

We will now turn to a consideration of the two principles which are the focus of this article: **school-based autonomy** and **controlled choice** — and their implementation, examining to what extent they can provide an answer to the needs of the education system in Tel Aviv-Jaffa.

The transfer of responsibility for implementation to the schools, changing their organizational work patterns and adjusting these to each school's special needs, is a key concept of school-based autonomy. Educational research has found it to be a basic condition for school effectiveness. There is broad consensus among researchers that certain effectiveness factors characterize successful schools. Among these are: clear definition of school aims; curricula adapted to the attainment of these aims; evaluation and feedback on educational achievements; and parents' involvement in the educational process (Edmonds 1979; Hallinger and Murphy 1986).

Along with autonomy's potential contribution to the education system, promoting school autonomy raised basic questions: what is the right balance between the aims of the school, the city and the state education system as a whole? To what extent should the school be authorized to determine priorities in resource allocation, select teachers and set curricula, and at what point does this really become "school-based management"?

Developing autonomous schools with different educational ethoses was perceived as a basic and necessary condition for fulfilling the second principle — the controlled choice of schools by students and their parents. Choice in education is one of the major issues with which education systems everywhere have been concerned in the last decade or so. Those who support free choice argue that it will help to improve the education system by improving students' achievements, furthering the professionalization of teachers and enhancing their job satisfaction, and increasing parents' involvement in their children's education (Chubb and Moe 1990; Bradford 1991).

At the same time, those who warn of the dangers of free choice if exercised without any social control. They claim that such choice often serves the interests of the stronger groups, who choose schools according to their demographic make-up and their prestige. This may harm social integration and increase inequality of educational and social opportunities for children from different groups (Bridge and Blackman 1978; Shapira 1988; Svirsky 1990; Yona 1991; Cooper 1991; Shapira, Haymann and Shavit 1995).

The principle of choice is not new to Israel and certainly not to Tel Aviv-Jaffa. Patterns of choice that have developed — between religious and nonreligious schools, between regular schools and those with special orientations — together with increased flexibility of enrollment zones and the choice between tracks in secondary education — indicate both the advantages and disadvantages of choice when it is free and uncontrolled. The middle classes today can choose the educational framework of their preference, both by choosing where to live and by applying to a selective or magnet school (Menahem *et al.* 1993).

Thus the proposal for controlled choice in the Tel Aviv educational reform plan may be seen as the institutionalization of an existing practice developing it, while being aware of its prospects and avoiding its inherent pitfalls. The program, which has aroused heated public debate due to the social dangers associated with the notion of choice, does not, in fact, as mentioned, advocate free choice but refers to controlled choice, perceiving the stage of choice as part of an overall strategy to improve the quality of education in Tel Aviv-Jaffa.

The planners of the educational reform plan saw the implementation of controlled choice, combined with the other principles (autonomy, continuity, comprehensives, new teaching methods, etc.), as a means of creating a better integrative balance at all levels of education out of choice and not by coercion, and as a way of enhancing motivation for learning, helping students to realize more of their potential (Tel Aviv-Jaffa Municipality, Education Authority 1992: 29).

On the elementary level, the program recommends a general orientation to encourage school autonomy and to conduct an initial experiment in the south of the city. In the first stage, processes of autonomy would be developed in the school, and at a later stage parents would be given the choice between a number of schools forming an extended enrollment cluster.

In secondary education, it is proposed to allow 6th grade leavers to choose between the junior high school in their zone and one or two others, while preserving the principle of integration. The program does not propose giving the schools freedom to select their pupils. Choice is to be "controlled" by central authority, so that the demographic make-up of each school is a fair representation of the population of the city as a whole.

Between the writing of this chapter and its publication, the original plan passed from the thinking stage to the operational stage. During this period the planning teams worked, both at the municipal level and the school level, to consolidate the plan and prepare the infrastructures necessary for implementation of the plan.

In secondary education the plan of action that developed differed from the original plan[3]. At the first stage 6th grade students from schools in disadvantaged areas are offered the choice of five secondary schools, three of them local and two in wealthier parts of the city. Pupils from other areas are assigned to schools by enrollment zone. An information center for parents was set up in the framework of the program and transportation is provided for the students. Control over school placement is based on several criteria in order to ensure social heterogeneity of schools (students' socioeconomic status, school achievements, gender).

At the school level a process of reorganization took place in several domains: physical conditions, diversification of learning frameworks, curricula and teaching methods, support for underachievers, promotion of positive attitudes towards social integration among students and their parents.

The implementation of the program in the first two years has revealed the following trends: more heterogeneity in the student population in the various schools, fewer students dropping out of school in the transition from junior high to high school, a high level of satisfaction among students and parents, and increased awareness concerning education.

The program is now in the third year and the policy planners are currently considering how to expand it to other areas in the city, so that it will include the entire pupil population in the city. (Evaluation reports about the implementation of the reform program in Tel Aviv were published recently by the Tel Aviv education authority (1997).).

Tel Aviv-Jaffa's Educational Reform Plan: The Parents' Point of View

The introduction of changes in education often presents difficulties, arising in part from the resistance of clients who are usually not involved in the various steps. Reforms like the one Tel Aviv has planned was crucial from the parent's point of view. This is why Tel Aviv-Jaffa's Education Authority decided to conduct a survey among the parents of 6th graders, to learn about their attitudes towards various aspects of the educational reform plan[4]. The findings were supposed to help to determine future action.

The assumption was that the question of choice was particularly relevant for this population, most of whom would move on to junior high schools the next year. The survey would also provide data about the general attitudes of this group to the educational reform plan as a whole. The survey data were processed according to four urban areas, broadly corresponding to the north, center, east and south of the city. Given the similarity found between the first two categories on the one hand and the second two, on the other, this discussion will refer to two broad areas.

The findings show that parents were generally in favor of the two basic principles of the reform plan: development of individual school orientations, and controlled choice. About 80 percent of the parents support the idea that junior high schools should develop specific educational approaches, and this reinforces the trends existing in the field. In the matter of controlled choice, about 90 percent of the parents thought that all 6th graders and their parents should be entitled to choose a junior high school from among a number offered, while preserving the principle of integration.

The findings provide important information which could guide educational policy makers in the development and application of the controlled choice model in Tel Aviv-Jaffa. For instance, the fact that no differences were found among parents in different parts of the city, who nearly all saw choice as "the natural right of parents in a democracy", justifies the development of a controlled choice model for all the city's children.

To the question: "Which junior high school would you choose today?" 40 percent answered that they did not know enough about the various junior high schools to answer the question. This finding, which occurred more or less equally in all areas of the city, is interesting in itself as it reflects lack of awareness, or even indifference towards, a matter that is central to every parent's life: their child's education. It seems that the absence of choice leads to indifference and lack of interest, and many parents are unaware of the options available to their children in secondary school. If parents have to choose a school for their children they are likely to look for information to help them make the best choice.

These results suit research findings showing that choice can make parents more active and concerned, more responsible and informed, leading them to share with the school in the responsibility for their children's education (Nathan 1987).

This finding calls for the opening of information centers for students and parents, offering comprehensive information about the schools available and helping those who have difficulty in making the best choice for their child, matching the characteristics of the school with the needs of the child.

About 60 percent of the respondents knew what junior high school they preferred. Two points stood out in the parents' responses: (1) parents from the northern part of town tended to choose a junior high school in the central to northern areas, though not necessarily the one they are referred to today; (2) among parents in the south who knew what they wanted, half preferred their children to study near home, and the other half wanted schools in the north of the city. The findings also indicate that a situation of entirely free, uncontrolled choice might lead to excess demand for certain prestigious schools in the north of town while emptying some other schools of their pupils. This calls for initiatives to improve the less popular schools so as to ensure a choice between viable alternatives.

In light of the goals of the educational reform plan (improvement of education and of students' achievement, and promotion of social integration in secondary schools), the findings underline the need to bolster schools in the south of the city in order to bring them up to the level of other schools. Quality education should be available to those who prefer a school near their homes in south Tel Aviv. Schools in the north of the city should adapt to their students' needs and offer a wider variety of learning frameworks and teaching methods, enabling all their students to finish high school and offering genuine alternatives.

Moreover, if the educational policy makers decide to change the present intake system and make all schools heterogeneous on the basis of choice, there will be a need for unconventional measures such as turning schools in the southern part of Tel Aviv into magnet schools which can attract students from the north.

An interesting fact that has emerged was that about one-third of the parents in north Tel Aviv (the high SES area), when asked about some of the "desirable" and "undesirable" characteristics of their preferred school, indicated an "integrative school" as desirable, while about a third of the parents in south Tel Aviv did not mention social integration as a desired characteristic. Furthermore, about 70 percent of the parents whose children attend schools in the south of the city attach great importance to choosing schools that stress study achievements, compared with 40 percent of the parents in the north of the city. Parents in the southern neighborhoods see parental involvement as an important factor in their choice of school, the general trend being that the farther south, the greater the wish for a school that welcomes involvement.

The two latter findings indicate that among parents in the south of the city there is a growing awareness of the school as a vehicle of social mobility which provides them with an opportunity for active involvement. It seems they increasingly feel that "If we don't do things for ourselves, we'll go on lagging behind Israeli society". These data support a trend whose first signs became manifest in the early 1970s, namely the growing strength of the local community. Parents who previously showed no interest in what happened in the school, relying on the teachers and principal, "who know best what's good for our children", now seek greater involvement, because of the centrality of education.

We may assume that this change in parents' attitude to involvement was partly influenced by the Neighborhood Renewal Project and the various committees dealing with education in neighborhoods and development towns. These raised people's awareness of the need for community involvement in general and parental involvement in education in particular.

With regard to other considerations in the choice of a school, a general tendency was found among all parents to emphasize two points: the quality of the teaching staff, and suitability to the students' needs. In answer to the question: "In what track do you want your child to study, or how important is it to you that your child studies a vocation at secondary school?", parents in the south of the city attached more importance to schools with a vocational orientation. They considered vocational education important, in contrast to parents in the north, who preferred an academic track, leading on to higher education.

Thus, in developing a model of controlled choice of schools with specific identities, it is necessary to consider the variety of trends found among groups of parents and provide opportunities to match these trends while maintaining the goals set by the system.

On the whole, the findings show the parents' readiness to adopt the fundamental principles of the program, and the information they provide may serve as a basis for the decisions of the city's educational policy makers, indicating the range of educational opportunities the system should offer, and the accompanying need for control in order to guarantee all the city's children an equal prospect for an education that meets their needs and aspirations.

Conclusion

In concluding this review of the development of educational thinking and action in Tel Aviv-Jaffa, we may point to a "double loop" learning process (Argyris 1990), analyzing needs, planning activities in response to these needs, rethinking based on examination and evaluation of the anticipated and unanticipated results, new planning and implementation.

This process gave rise to an overall reform program which was rooted in society's most central educational concepts, and constructed in the light of Tel Aviv-Jaffa's unique character and the needs of its education system.

In this chapter we examined two of the principles of the educational reform plan — school-based autonomy and controlled choice — taking into account their educational and social implications.

Our main question was: what is the educational process community should take to ensure that implementation of these principles will promise optimal achievements while avoiding damage to central social aims? Autonomy and choice play an important part in many reform programs that are designed to enhance educational quality, promote social integration, empower teachers and increase parental involvement. The process by this time showed that introducing these changes in Tel Aviv-Jaffa may help to advance the system towards the desired goals, provided that they are introduced as part of a strategy for change which also embraces some other elements.

The problems that characterize the city — social polarization, differences between schools in levels of achievement, and difficulties in accomplishing the aims of (the previous) educational reform — point to the need for careful thinking and strategic planning for change, taking all these aspects into account. Parents' attitudes to the program — as presented in the survey — further stress the need for system-wide planning to ensure the best possible combination of the needs of the individual, the community and society.

The program as implemented in practice differs, as stated, from the original plan, both in terms of structure and pace of implementation, which was in fact swifter than anticipated. The difference between the two programs is largely a consequence of the need to base the reform on principles determined at the beginning while at the same time providing an answer to social pressures from the field and developments in the schools themselves.

Analysis of the stages that led to the consolidation of the program in Tel Aviv-Jaffa reveals a number of basic conditions which are essential for the success of this kind of program:

- There must be administrative procedures to guarantee all social groups equal access to educational services and promote social integration in schools and between schools. These procedures include:
  - Student admission based only on interest, with no selection process on entrance or between levels, ensuring continuity of studies at the same school.
  - Choice of school subject to central control, to ensure that the social make-up in every school reflects the general population of the city.
  - Central enrollment procedures with clear and fair criteria for choice.
- Tools must be provided for creating educational alternatives which are seen by the students as comparable.
- The authorities must provide the public with reliable information about the schools and about the process of choice to ensure that

choice can be exercised by all parents on the basis of full and true information.

- School staff should be included in the planning and implementation processes, freely and not by coercion.
- The cooperation of all those involved in the process of change should be ensured.
- Ongoing evaluation must become an integral part of the program, in order to provide flexibility and learning.

In conclusion, we can say that this analysis of the state of education in Tel Aviv-Jaffa, in addition to what it teaches us about the city, can also be seen more generally as an illustration of the processes of consolidation of educational thinking as an answer to changing educational and social needs.

Notes

1 The authors work with the steering committee for implementation of Tel Aviv-Jaffa's Educational Reform Plan on issues of autonomy and controlled choice.

2 In the present chapter we cannot enter into a description of all the programs implemented in the city, but merely present educationally and pedagogically significant milestones that determined the system's special nature.

3 For a detailed description of the implementation of the program, see Dorfman *et al.* (1994).

4 The data are taken from a report on a parents' survey carried out by the authors (Tel Aviv University 1993, unpublished). The sample consisted of 1,200 parents from all state elementary schools in the city, one 6th grade from each school. The questionnaire was returned by 752 parents, 70 percent of the sample. Other surveys concerned the attitudes of secondary school teachers and a follow-up among teachers in the south of the city who participated in the experimental program.

Appendix: The Number of Students in Tel Aviv-Jaffa's Education System, at all Levels, in 1991/2

The data are taken from the **Statistical Yearbook**, 1991, no. 31.

In kindergartens, elementary and secondary schools there were 51,959 students. The kindergartens included 198 frameworks for 3–5-year-olds, with a total of 325 classes; state kindergartens — 329; state religious — 41; independent religious — 9; creches — 6; special kindergartens — 35.

In elementary education there were 86 schools with 26,118 students. Of these, 75 percent attended state schools, 11.5 percent state religious, 6.5 percent were in recognized education, 4.1 percent in special education and 2.9 percent in Arab schools.

Secondary education was mostly in six-year schools, comprising a junior high school, mainly academic. On completion of the educational reform program in 1988, Tel Aviv-Jaffa had 17 junior high schools with 9,829 students. Eleven of the junior high schools were state schools, three state religious, two were apprenticeship schools and one, an Arab school.

There were 42 secondary schools in 1991, 29 of which offered academic tracks and 22 vocational tracks. The total number of students in secondary education was 17,976. Fifteen of the secondary schools belonged to the municipality: seven of them academic, six comprehensive and two vocational.

References

Amir, Y. (1988) *Report of Institute for Advancement of Social Integration*. (Internal Report — Unpublished).

Argyris, C. (1990) *Overcoming Organizational Defenses: Facilitating Organizational Learning*. Boston: Allyn and Bacon.

Bradford, M. (1991) "School-performance indicators, the local residential environment, and parental choice". *Environment and Planning,* **23**, 319–332.

Bridge, R. G. and Blackman, ?. ?. (1978) "Family choice in schooling". *District Policies and the Implementation of Change*, 4. Santa Monica: Rand Corporation.

Chubb, J. E. and Moe, T. M. (1990) *Politics, Markets and American Schools*. Washington DC: Brookings Institute.

Cooper, B. (1991) "Parent choice and school involvement; Perspectives and dilemmas in the United States and Great Britain". *International Joumal of Educational Research*, **15**, 235–250.

Dagany, A. (1990). "Student mobility and dropping out in the municipal school system". *Geocartographia*, The Institute for Spatial Research. Report presented to Tel Aviv-Jaffa Municipality (in Hebrew).

Dorfman, S., Haymann, F., Shapira, R., and Shavit, R. (1994). "Local education reform: The case of Tel Aviv-Jaffa". *Journal of Education*, **1**(16), 2.

Edmonds, R. (1979). "Effective schools for the urban poor", *Educational Leadership*, **37**, 15–24.

Elboim-Dror, R. (1985). "Educational policy making in Israel". In Ackerman, V., Cannon, A., and Zucker, D. (Ed.), *Education in an Emergent Society*, pp. 116–135. Tel Aviv and Jerusalem: Hakibbutz Hameuchad (in Hebrew).

Friedman. Y. (1987). "Meanings and focus in autonomy". In Friedman, Y. (Ed.), *Autonomy in Education — Conceptual Frameworks and Implementation Processes*, pp. 32–44. Jerusalem: Szold Institute (in Hebrew).

Gordon, D. (1987). "Student autonomy and school autonomy". In Friedman, Y. (Ed.), *Autonomy in Education — Conceptual Frameworks and Implementation Processes*, pp. 45–57. Jerusalem: Szold Institute (in Hebrew).

Green, R., Haymann, F., and Shapira, R. (1991) "School autonomy and Central Control — The case of Israel". *Bildungund Etzeihung* (in German).

Hallinger, P. and Murphy, J. (1986) "The social context of effective schools". *American Journal of Education*, **94**, 328–355.

Harpaz, Y. (1985). *The Community School. Development of an Idea.* Jerusalem: Association of Cultural and Sports Centers, Joint Distribution Committee.

Inbar, D. (1987). "Is autonomy possible in a centralistic system?" In Friedman, Y. (Ed.), *Autonomy in Education — Conceptual Frameworks and Implementation Processes*, pp. 58–75. Jerusalem. Szold Institute (in Hebrew).

Kamanda, Y. (1985). "The distribution of power in educational policy". In Ackerman, V., Carmon, A., and Zucker, D. (Eds), *Education in an Emergent Society*, pp. 847–864. Tel Aviv and Jerusalem; Hakibbutz Hameuchad (in Hebrew).

Kashti Commission of Inquiry into the Status of Non-zoned Schools (1991) *Report.* Jerusalem: Ministry of Education and Culture (in Hebrew).

Kashti, Y., Arieli, M., and Shapira, R. (1991) "Educational reforms in Israel". In Cookson, P. Sadownik *Sem. International Handbook of Educational Reform,* Greenwood Press.

Lewy, A. (1984). *Student Achievement in Tel Aviv Schools*, part 2. Tel Aviv University, School of Education, The Measurement and Evaluation Research Unit (in Hebrew).

Mandler, N. (1991) "Secret report — 17 schools of advantaged pupils head the list of failures in the National Standardized test for the 4th. and 5th. grades". *Ha'Aaretz,* 5 December, p. A4 (in Hebrew).

Menachem, G., Shapiro, S., Goldring, E., and Shapira, R. (1993) "Parental choice and residential segregation". *Urban Education*, **28**(1).

Ministry of Education and Culture (1981). Director-General's circular, 8. Jerusalem: Ministry of Education and Culture (in Hebrew).

Ministry of Education and Culture (1992) *Israel Believes in Education.* Educational Planning for the 21st. Century. Jerusalem: Ministry of Education and Culture (in Hebrew).

Nathan, J. (1987) "Results of future prospects of state efforts to increase choice among schools". *Phi, Delta, Kappan*, **68**, 746–752.

Ony, D. (1988) "How an educational issue reaches the public agenda in a local authority". Unpublished M.A. thesis, Tel Aviv University, School Education (in Hebrew).

Pur, D. (1985) "Discussion: Autonomy in education". *Hed Hachinuch* **59**, 6–11 (in Hebrew).

Reshef, S. (1984). *Autonomy in Education — Meaning and Implementation* — Tel Aviv University, School of Education (in Hebrew).

Reshef, S. (1985). "Development of the concept of autonomy and how to achieve it". *Hed Hachinuch*, **59**, 18–19 (in Hebrew).

Shapira, R. (1988) *Educational and Social Uniqueness: Background and Problems*. Position paper submitted to the Ministry of Education and Culture. Tel Aviv University, School of Education, Unit for the Sociology of Education and Community (in Hebrew).

Shapira, R. and Haymann, F. (1991) "Solving educational dilemmas by parental choice: The case of Israel". A special issue of the *International Journal of Educational Research*, **15**(3–4).

Shapira, R., Green, R., and Danilow, Y. (Eds) (1994). *School Autonomy: The Lesson of Implementation*. Jerusalem: Ministry of Education (in Hebrew).

Shapira, R., Haymann, F., and Dintzman, A. (1990). "No need for nostalgia — it's here and now: A. D. Gordon School's survival in a changing society". In Shapira, R. and Kasher, A. (Eds), *Historical, Philosophical and Social Aspects of Education*. Tel Aviv University (in Hebrew).

Shapira, R., Haymann, F., and Posner, Y. (1994) "Towards school autonomy — Autonomy approaches survey in the Israeli education system". In Shapira, R., Green, R., and Danilow, Y. (Eds), *School Autonomy — The Lesson of Implementation*. Jerusalem: Ministry of Education. (in Hebrew).

Shapira, R., Haymann, F., and Shavit R. (1995). "Autonomy as ethos, content as commodity: An Israeli model for controlled choice of autonomous schools". In Kahane, R. (Ed.), *Educational Advancement and Distribute Justice: Between Equality and Equity*. Jerusalem: Magnes Press, Hebrew University.

Statistical Yearbook 1991, no. 31. Tel Aviv-Jaffa Municipality. Center for Economic and Social Research (in Hebrew).

Svirsky, S. (1990). *Education in Israel — A Region of Separate Tracks*. Tel Aviv: Breirot (in Hebrew).

Tel Aviv-Jaffa Municipality (1991). *Proposed Plan for Secondary Education in Tel Aviv-Jaffa*. Submitted by Yitzhak Cohen, Secondary Education Dept. Internal Document (in Hebrew).

Tel Aviv-Jaffa Municipality, Education Authority (1992). "There are other ways — Tel Aviv-Jaffa believes in education". *Renewal program for Tel Aviv-Jaffa's education system*, part 1 (in Hebrew).

Volensky, A. (1992). *Proposal for Change in the Education System; Decentralization, Parental Choice, Feedback*. Planning document no. 1. draft b. Jerusalem: Ministry of Education and Culture, Planning, Evaluation and Control Section (in Hebrew).

Yogev, A. (1990). "The high school of the future: Processes of change in policy planning". In *Educational policy planning*. Jerusalem: Ministry of Education and Culture, the pedagogical secretariat. (in Hebrew).

Yona, Y. (1991). "Cultural pluralism vs. cultural integration: Implications for education". *Magamot*, **34**(1): 122–138 (in Hebrew).

Chapter Five

PARENTAL CHOICE AS LEVERAGE TOWARD SOCIAL CHANGE IN EDUCATION: TWO SCHOOLS IN JERUSALEM

TAMAR HOROWITZ
Department of Education,
Ben-Gurion University of the Negev

In recent years, parental choice has become firmly established on the public agenda in Israel and in the rest of the world. This multifaceted topic brings a variety of issues into play: the capitalist versus the socialist outlook; market and economic factors in the educational system; the right of the individual to choose in a democratic state; the relationship between school and social stratification; the significance of parent and community participation in the school; innovation and change in the school; accountability of the school; and manifest versus latent curricula (Chubb and Moe 1990).

Parental choice of schools is defined as a principle according to which parents have the right to decide on which school to send their children, irrespective of residential area. The school selected from among the variety of educational institutions and programs is usually considered by such parents to be superior to those available locally.

Ellen Goldring (1991) notes that apart from the act of parental choice itself, the right to select a school entails a number of additional implications such as greater parental involvement, improved quality of education, enhanced social and ethnic interaction and, as a consequence, more equality.

More often than not, parental choice is associated with choice among magnet schools, that is, parents and the student choose a school according to their wishes and their evaluation of the child's abilities and interests. Such schools may be geared to the arts, sciences, gifted children, and so forth. Magnet schools are characterized by special programs or specific teaching methods and voluntary integration in a certain district or region. They are chosen by the child or the parents, and they accommodate students irrespective of their home district.

In Israel, debate on the topic of parental choice usually centers around the issue of magnet schools. However, the term "magnet" does not necessarily connote "advanced level" but rather the notion of schools catering for special areas of interest. Thus, in Israel, the advocates of the principle of parental choice generally argue that magnet schools should be developed but not at the expense of the principle of integration.

A number of magnet (or specialist) schools have been established in Israel in recent years. The establishment of these schools was made possible by a clause in the Compulsory Educational Law which allows parents to have a say in 25 percent of the curriculum in their children's school on the condition that any such change is supported by at least 75 percent of the parents. Thus, many of Israel's magnet schools were established by changing the format of an already existing school framework.

Four broad categories of magnet schools are to be found in Israel:

(1) Schools cultivating specific skills and contents: such as schools for arts, nature study schools, and so forth.
(2) Schools associated with religious systems: for example, the Tali schools, Reform Movement schools, the Noam school network.
(3) Schools with a social ideology: such as schools furthering the values of the Labor Movement.
(4) Schools based on special educational doctrines: for example, the open schools etc.

This chapter deals with value-oriented magnet schools as a medium for educational and social change. Two magnet schools in Jerusalem will be described: the School for Furthering the Values of the Labor Movement (hereinafter abbreviated to Labor Movement School) and the Conservative School established by the Tali school network (schools with augmented Jewish studies). The focus will be on the educational and social processes taking place within them as a result of the implementation of the principle of parental choice. The description is based on an ethnographic study conducted by the author during the 1980s. These two schools were classified by the researcher as value-oriented schools, one advocating religious values and the other social democratic values.

A value-oriented school is defined as a school that places emphasis on education toward values. According to Parson's conceptual framework (1964), a social organization has to satisfy four needs, namely, goal achievement (G), the integrative need (I), the adaptive need (A), and the need for values (L). He argues that these needs coexist and vary according to their emphasis, form, situational and temporal context. There are schools whose main emphasis is goal achievement, for example, schools for the gifted. Some place emphasis on the integrative need, for example, certain kibbutz schools. Other schools emphasize adaptation to the external environment, such as vocational schools which stress the acquisition of occupational skills in preparation for the future. Value-oriented schools include the various religious schools.

Of course, each educational system tries to satisfy all four needs: Apart from academic achievement, schools are expected to fulfill immediate social needs, to prepare students for functioning in society, and to educate them on the basis of the values of society.

Although the two schools to be described in this article are similar in that they emphasize values, they differ in that one is religious and the other is socioideological. Parsons talks of four value systems: religion, ideology, philosophy, and science. Thus, the Labor Movement School can be defined as based on ideological values and the Conservative School can be defined as based on religious values.

Ethnographic research tends to concentrate on the focal points that are thought likely to explain the social phenomena. It would seem that the focal points of magnet schools chosen by parents are:

(1) The motivation underlying the parents' choice of the school: The motivation may be to implement a certain worldview, to improve social status, to improve learning, to develop specific abilities or skills, and so forth.

(2) Curriculum: Curriculum can be viewed according to various perspectives. Are unique subjects and topics taught in the school? What is the place of extracurricular programs? What is the place of formal and hidden curricula? What role do parents play in crystallizing the curricula?

(3) Teachers' orientation: The teacher as a transmitter of knowledge; the teacher as an agent of values; the teacher as a partner in learning; the teacher as facilitator; the relationship between teachers and parents.

(4) The student in the school and in the community: The student as an individual; the student as part of a restricted collective — the class; the student as part of a broad collective (intergenerational closure) of parents and the ecological or functional community (Coleman 1984).

(5) Parent participation in the school: Supportive parents, active parents, parents as partners in school activities; parents as partners in decisions and policies (Friedman 1990).

The two schools will be described according to the above parameters.

Motivation for Choice of School

The rationale underlying the establishment of the two schools was the assumption that school can teach values. Indeed, according to their respective founders, school is the ideal framework to teach values — values in general and the particularistic values of groups.

The Labor Movement School

The Labor Movement School was founded in 1981 by a group of educators and university professionals associated with the Labor Movement who wanted to establish a school which would promulgate the universal values of justice and equality alongside the particularistic values pertaining to the heritage of the Labor Movement.

The school was established in a deprived neighborhood with 60 percent of its population disadvantaged students. The worldview of its founders is contained in the school covenant, the introduction of which reads as follows:

> Socialist Zionism envisaged the creation of a society based on the principles of equality and justice as a guarantee that national independence will be achieved. When independence was achieved, the goal of promoting identification with the state took priority over affiliation to the political factions within it . . . and, therefore, the Labor educational system, the educational instrument of the Labor Zionist Movement, was disbanded . . . The state secular school which emerged was void of any content that might be interpreted as serving to convert souls to its [Labor Movement] values. In retrospect, students, were, thereafter, exposed to a worldview representing minimalistic bourgeois consensus — conformist, reluctant to accept change, and alienated from the notion of social commitment
>
> (Horowitz 1990: 55)

The social covenant listed four main goals: (1) education toward a democratic and egalitarian society; (2) education toward Zionism and peace; (3) education toward Jewish humanism, and (4) education suited to the needs of future society and able to influence the form it will assume.

The details of the value-orientation were not spelled out exhaustively in the covenant. This was accomplished only after extended deliberation (lasting over a year) during which potential parents and potential educators discussed the issue, argued, and invited experts until they eventually succeeded in clarifying and formulating the value system that they wanted to convey to students. In other words, what we had here was a long socialization process of parents and educators and the desire to translate their sociopolitical outlook into an educational outlook.

The ideological discussions did not cease with the founding of the school. Moreover, reality often begat situations generating conflict between action and ideology. Indeed, the relation between ideology and reality appeared constantly on the agenda of all discussions, on both the parent level and the teacher and student level. For instance, on the issue of compensatory classes, the parents argued that in a school wielding the banner of equality, there was no place for compensatory classes even if it was difficult for teachers to help students with learning difficulties in the regular framework. Another topic arousing controversy was the issue of school trips to areas annexed after the 1967 war. Some of the parents saw this as a nonissue, void of ideological connotation, while others, viewing it as part of the educational program, saw it as running counter to the antioccupation position of the school. The question of achievement versus equality arose in arguing whether parents should be allowed to augment their children's studies in certain subjects such as English and mathematics beyond what is learned in class, or whether learning should only take place in class.

The Conservative School

The process of founding the Conservative School was similar to that of the Labor Movement School. A group of parents and educators, some of them Israelis who had spent extended periods in the United States, and some of them new immigrants from North America, organized themselves in the mid-1970s with the aim of establishing a school for their children in which the dominant principle would be a positive attitude to Jewish religious tradition. The group was quite varied with regard to religious outlook: Some observed each and every religious rule while others were what is called in Israeli jargon "traditional" – people with a positive attitude toward Jewish tradition without being overly observant.

In a letter to the Minister of Education in March 1975 they wrote:

> On the one hand, state-religious schools emphasize values and commitment to tradition but they tend

towards dogmatism and demand commitment to a prescribed life-style. On the other hand, the state-religious schools, while amenable to contents of modern society, deny the spirit of Judaism and its traditional components.

A group was set up with the aim of elaborating educational and value issues, thus initiating a preliminary socialization process of parents and educators. The main goals of the school were defined in the initial stages:

> The goal of the school is to cultivate a Jewish way of life based on traditional values and customs. The educational conception will be oriented to the status, practice, and cultivation of Jewish tradition in contemporary pluralistic society ... in an attempt to combine the values of Jewish tradition with humane social values in general while, at the same time, taking into account the prevailing predicaments of modern man. The pluralistic principle will be expressed, among other things, by representing the various streams of thought existing in the House of Israel as part and parcel of Jewish society. (unpublished position paper 1975)

The conspicuous absence of discord in the deliberations of the group was, perhaps, a result of the school's acceptance of the principle of pluralism. It was also clear to the founders that the school would be associated with the state-secular system and not the state-religious system.

The school opened its doors in 1976 despite opposition from the religious establishment as well as the Jerusalem Municipality and its Educational Department: The former was unwilling to accept the new model of alternative religious education and grant it legitimation because, in their view, it represented a threat to the various forms of orthodoxy, and the latter, because it was viewed as a threat to the integration policy of the city.

Yet another motivation for the establishment of the school (albeit less salient), mentioned by a number of parents, was the desire for a more progressive form of education. A large part of the founding group emanated from the liberal tradition of American schools and the regular Israeli schools appeared to them, rightly or wrongly, as old-fashioned and rigid.

Another important issue was the nature of the school's association with the Conservative Movement: The school is not officially affiliated with the Conservative Movement but a loose form of "association" exists (many of the founding group were members of the Conservative Movement). Similarly, the conservative youth movement was later the movement which received legitimation in the school. However, the philosophy of pluralism (which is also the philosophy of the Conservative Movement) prevailed as a central theme in the philosophy of the school.

In the Conservative School (as in the Labor Movement School), the debate on the ideological path to be adopted continued over a number of years: For instance, what should be the status of prayers? How should they deal with conflicts between the home and the school concerning the observance of religious rules such as ritual hand-rinsing, grace after meals, and so forth? Think tanks were convened to deliberate on the educational path to be adopted by the school and joint workshops were run for parents, teachers, and students on these and many other issues.

Curriculum

The Labor Movement School

In crystallizing its basic approach toward curriculum, especially in the junior high school, an attempt was made to arrive at a balance between the various disciplines — the arts, the sciences and the humanities. The aim was to convey the notion of one humanistic cultural whole rather than three separate "cultures", each alienated from the other. Thus, emphasis was placed on interdisciplinarity.

The school elaborated its approach to important issues related to its ideology such as the Israel-Arab conflict, integration and equality, and so forth. These topics were studied in different contexts and in various activities. The marked effect of the programs on the outlook of the students was confirmed by the researcher.

Special subjects related to the values of the school were developed: Current affairs was a central subject with students relating to values in the context of social realities by analyzing and evaluating them. The humanities were also studied in the context of the value system of the school, as far as possible. What would normally be extracurricular programs were part of the curriculum, their rationale also anchored in the value system.

The hidden curriculum was no less important than the formal curriculum. The school conveyed its values in each and every area: in sport, in art, in the humanities, and in science. The school newspaper and wall bulletins, apart from reporting school activities, gave space to students' opinions on issues relating to values. A school climate reflecting the value of "pioneering socialism in Israel" was cultivated.

Parents were partners in the design of the blueprint of the curriculum. Parental influence on the curriculum was facilitated via the curriculum committee (consisting of teachers and parents) and by the fact that many of the parents were qualified in teaching and other related professions (e.g., counselors, psychologists).

Among the programs developed by the committee was the school's unique Bar Mitzvah program which combined biological, historical, and social aspects of the subject. This successful program provided a model used by parents in developing other topics of study. Other subjects unique to the Labor Movement School were: the 1st May holiday, the writings of the Labor Movement philosopher A. D. Gordon, dramatic events in Jewish and Israeli history. However, they did not develop the full curricular program. The parents were also partners in designing and developing the extracurricular program, especially with regard to choice of youth movement. They also helped in elaborating the hidden curriculum and in promoting the climate of pioneering socialism in Israel.

Their enthusiasm was such that it was even said, with humor, that some of the parents seemed to be yearning to go back to school and to relive their youth experiences of yore. In summary, it can be fairly stated that the parents were highly involved in the conception of the curriculum and its elaboration: They developed a number of subjects, they took part in creating extracurricular programs, they helped to develop the latent curriculum and cultivate the school climate. In practice, the parents served as gate-keepers of the basic values of the school, ensuring that they would not be compromised in the changing face of reality.

It should be noted, however, that the challenge of developing the curriculum was taken on by a very select group of parents, that is, educated parents, a fair number of whom were qualified in education and similar fields. The neighborhood parents did not take part in this process, not only due to their lack of qualifications but also due to their low self-confidence. Some of the more well-to-do parents expressed disappointment that, in actual fact, they had not developed one single completely new subject pertaining to the ideology of the school, but that their main activity had been global thinking or initiating specific activities.

The Conservative School

The school's conception of Jewish Studies was stated in a memo sent out to parents in 1976:

> ... In the new program, emphasis will be placed on the study of the branches and ramifications of Jewish history.

The aim will be to transmit knowledge to the students and to verse them in the sources, on the one hand and, on the other, to guide them in coping with the values of modern society.

The premises of the school curriculum are contained in a document compiled in 1981:

> ... The underlying premise of the curriculum is commitment to Jewish tradition ... This tradition is perceived as having developed around the events experienced by the Jewish people, to which special meaning has been assigned. These perceptions range from divine control as all-pervasive and everlasting to various cultural aspects of Judaism. The school is committed to a variety of public norms selected due to their centrality in Jewish ritual (prayers, Talmud Torah, grace, etc.).
>
> The second axiom of the curriculum is openness, questioning, and investigation. The third axiom is good education and Jewish education.

The school even detailed the meaning of its basic axioms in terms of curriculum:

> (1) The school will cultivate Jewish values (the sanctity of life, responsibility to the people and culture of Israel, sensitivity to others, personal insight and consideration of others) as an integral part of the curriculum. These values will be expressed in various spheres — curriculum, prayers, social activities.
>
> (2) In religious studies, the school will aim at a historical perspective, taking into account traditional sources and knowledge accumulated over the centuries in the areas of literature, ideology, etc. In Bible study, for

instance, traditional interpretation will be blended with that of early Middle Eastern texts, in this way also gaining insight into the history of Biblical thinking. Emphasis will be placed on the worldviews of the Jewish Sages (Chazal) and in the upper grades, the differences between the writings of Chazal and the Bible, on the one hand, and between Chazal and later philosophies, on the other, will be examined. In the study of the Halacha, time will be devoted not only to the Halacha itself, but to the status of the Halacha in various periods and, as far as possible, the historical circumstances that influenced the design of the Halacha.

(3) The school will attempt to cultivate a positive attitude towards the sanctity of the Sabbath and Jewish holidays and to promote a willingness to observe them according to tradition.

(4) Prayers will be conducted in the school. The traditional prayer book will be used, allowing for omissions and additions in accord with events and the spirit of our generation.

(5) Boys will wear a skullcap for religious studies and prayer.

(6) The school will be mixed, with girls participating in prayers and religious studies.

In principle, the Jewish Studies curriculum in the Conservative School did not differ formally from that of the state-religious school. However, its emphases and mode of transmission were somewhat different. For instance, the study of Oral Law and Halacha stressed acquaintance with the world of Chazal, the dynamics of Halacha (values, beliefs, and opinions), and the development of the thinking of Chazal in the historical context. In general, the hidden message was to arrive at an awareness of the problematics of Judaism and to show that doubt has always been with us. The school employs many ways to exemplify and illustrate the world of Chazal, including drama and the personification of the Halachic sages.

The topic of prayer was especially problematic. The question was how to bring people to pray when they are nonbelievers. This issue was dealt with by emphasizing that prayer was a public act, an act of identification. The Jewish holidays, which have an important place in the curriculum, were dealt with using a multidisciplinary approach. They were studied in centers across age groups and each year the study material was extended. Holidays and religious ceremonies were prepared jointly — by parents, teachers, and students — and served as an important instrument for imparting values.

In addition to the curriculum dealing with Judaism, community programs were developed, such as road safety, the cyclic nature of life, and the "honor thy teachers" program. Parents participated in some of the community programs and some programs were initiated and produced by the parents themselves.

Extracurricular programs served as an auxiliary aid to the regular program. In practice, it was sometimes difficult to distinguish whether certain programs were part of the formal program or part of the extracurricular program.

The hidden curriculum was considered very important. Art, literature, and singing programs centered around the topic of Judaism. The school walls carried selected saying of Chazal; projects and art work related to learning contents such as Noah's Ark; songs with religious content were sung, usually adapted to the more modern versions; the school newspaper, in addition to news, printed quotations from Chazal. And the students, for their part, had the feeling of belonging to a modern, joyful religion, and not a religion that imposes nothing but limitations and restrictions.

Much thought was devoted to how to transmit the curriculum. At a study seminar, the following types of questions were raised: What to do when one explanation is preferred over another? What about double messages? The solution to such questions was formulated by one of the founders, as follows: "... Attention should be drawn to the fact that any given answer represents only one of several possibilities... The student should be provided with a variety of opportunities for choice and the chance to justify any such choice".

The principles guiding teaching in the school (1984) were versed in the spirit of Chazal:

(1) From the sources: studying, the student, and the teacher;
(2) But his delight is in the law of the Lord (Psalms 1:2);
(3) Whoso keepeth the fig tree shall eat the fruit thereof (Proverbs 27:18);
(4) And he weigheth the waters by measure (Job 28:25).

The Teacher in the School

The Labor Movement School

The teacher in the Labor Movement School was, first and foremost, an educator and translator of values rather than a facilitator providing access to knowledge. As it says in the social covenant:

> The teaching staff, guided by the school principal, will translate and implement the educational and social policies as determined by the school board. Action-learning will serve as one of the main means of implementing the major goals of the school. The work of the school will serve as a democratization model for the educational system, drawing its strength from the school board, the teaching staff, and the student community.
>
> (Horowitz 1990: 294)

The teacher's role in the school was not restricted but perceived in the widest possible terms, as stated in the covenant: "The teaching staff will take on additional roles far beyond those usually expected of teachers".

Since the teacher is perceived as a socializer, considerable attention was paid to selecting suitable teachers. Most of the teachers were supporters of the Labor Movement or close to it in their outlook. Some teachers employed in the original school even resigned when it became the Labor Movement School, "rather than stay on and compromise their personal beliefs", to use the words of one of them. The teachers also attended a series of enrichment courses on topics that were central to the outlook of the school, such as democracy, human rights, tolerance, and prejudice.

The teacher-student relationship was different from that generally pertaining in other schools, this being confirmed by the students. Friendly egalitarian relations prevailed and some of the teachers modelled their role on that of the youth movement leader. Personal example was emphasized. In practice, however, familiarity and closeness to teachers proved problematic and sometimes led to discipline problems. Egalitarianism was seen by the teachers as the main source of lack of discipline and unwillingness to help in class. The teachers perceived themselves as part of an inclusive system of socialization of the new generation; they tended to treat the parents as equals but this, too, sometimes led to tensions between parent and teacher and, in crisis situations, the question arose as to who should be the one to decide. Theoretically, the school took precedence but, in practice, this was not necessarily the case. This often put teachers in an uncomfortable position in their relations with parents.

The Conservative School

The teacher in the Conservative School was also perceived as an educator and as a translator of values. Since the school advocated pluralism, there was no urgent need for teachers to have fully crystallized opinions and outlooks beyond a basic minimum. Teachers were expected to set a personal example and to be ready and willing to cope with basic values. Some of the teachers came to the school from the state-secular system and some from the state-religious system. Some were religious, some traditional, and some completely secular.

In the early years, the value system to be conveyed was not entirely clear and this is why some teachers argued that interpersonal values should be stressed. Nevertheless, slowly but surely, a staffroom believing in and advocating the values of the school emerged. One of the most important instruments for the socialization of the teachers was the workshop. Both teachers and parents participated in workshops, the aim of which was mainly to study new content relating to the values of the school.

There were not too many discipline problems in the school. The pattern of relations between the teachers and students was basically conventional but, maybe, somewhat more relaxed than most Israeli schools.

Measures were adopted to elevate the status of the teacher. A "Teachers' Day" ceremony was held every year on Lag B'Omer, the day commemorating the martyrdom of the rabbinical teachers during the Revolt of Bar Kochba in the second century AD.

Students in the School and in the Community

The Labor Movement School

The student was seen as a member of the community, and the community as consisting of several layers: classmates, schoolmates, the youth movement, parents, and the wider community. So we have here what Coleman called "intergenerational closure" (Coleman 1984). Students were expected to be oriented to each of these layers and not to relate only to their classmates, for instance. In the same vein, parents were expected to relate to wider circles and not just to their own child's class. A number of mechanisms were created to bridge the various layers pertaining to the school itself. Workshops were arranged for parents and teachers, for teachers among themselves, and for teachers and students. There were many informal joint activities: trips, sing-alongs, and discussion groups.

The youth movement was seen as pertaining to the community and complementing the home rather than as a protest mechanism. After extended deliberations, the parents eventually adopted the youth movement recommended by the school. However, over the years, it became clear that an autonomous youth culture had emerged in school, especially among the more well-to-do population.

Classroom composition was constantly varied, depending on the subject, to provide the child with maximal opportunities of meeting as many other students as possible and, at the same time, to establish more permanent ties in the home-base class. Classroom composition was based on sociometric research carried out by one of the parents.

The need to open up the community layers was especially important in view of the fact that the school contained at least two populations representing two cultures, Middle Eastern culture and Western culture. One of the parents claimed that there were actually four cultures in the school: first, a pure neighborhood culture consisting entirely of Middle Eastern origin groups who were traditional, with an anti-Western lifestyle orientation, and somewhat isolated; second, a neighborhood culture, consisting mainly of Middle Eastern groups but with a modern Western lifestyle orientation and striving to be absorbed into Israeli society as quickly as possible; third, parents of Western European origin, but with empathy for the Middle Eastern population and eager for them to advance themselves; and fourth, pure Western origin parents displaying little in the way of empathy and with a somewhat condescending attitude toward the Middle Eastern population. In the opinion of this parent, a real community was feasible only between those displaying genuine empathy to Middle Eastern groups and Middle Eastern populations open to Western culture. The encounter between these two populations, however, was ridden with tension related to such issues as: religiousness versus secularism; traditional versus Western orientations; strictness versus permissiveness; patronization versus low status; manual labor as a value versus manual labor as a loathed necessity; achievement orientation versus competitiveness. Thus, although a degree of community consciousness was achieved, the feeling of community was not shared by all. On the other hand, on the student level, the encounter was more or less taken for granted and perceived as quite natural. Thus, despite some signs of progress, both parents and teachers realized that there was still a long way to go before the desired sense of community would emerge.

The Conservative School

The Conservative School joined the official organization of community schools a few years after its establishment. One of the founders claimed that the school represented an ideal community model in that the students, the parents, and (albeit to a lesser extent) the teachers were perceived as part of the community.

An attempt was made to keep conflict between the school and parents down to a minimum, the most critical area of potential conflict being the observance of religious rules and maintaining a traditional lifestyle at home versus the climate of the school. The school did not require commitment on the part of the parents to maintain a religious lifestyle: "This is not a package deal", in the words of one of the parents. Nevertheless, it appears that two-thirds of the parents attended synagogue, most of them, the Conservative synagogue. Parents were, however, asked to show respect for the school and ensure that boys wear skullcaps for prayers, practice ritual hand-rinsing, and say grace after meals.

The school attempted to build up consensus in as congenial a manner as possible, for instance, by inviting parents to help prepare and to participate in all workshops dealing with the Jewish holidays alongside the teachers and children. Talks were delivered by outside lecturers, holiday customs and recipes, as well as quizzes, were presented and enjoyed by one and all. The aims of the workshops were defined as follows: "Preparing educational programs; strengthening the values advocated by the school, and providing a joint learning experience for parents and their children".

The school extended its community action to the surrounding area, and even cooperated with other schools. The school was oriented to the needs of elderly, frail and disabled people, with the students offering help to them. Many joint trips were held. Every year a drama production was staged in which as many as one hundred parents and children participated.

Despite the wide range of community activity available in the school framework, the source of the youth culture that emerged was not the school. The orientation of the children was seen to be mainly determined by the home, and most of their leisure time was spent with parents. However, the children generally described their school life as very happy.

A pertinent social and community issue in this context is why children from the neighboring deprived areas did not attend the school. One of the arguments against the school when it was established was that there could be no justification for setting up a magnet school which does not have social integration on its agenda. The school management and the parents did try to recruit children from local deprived neighborhoods but were not successful. It appears that the ideology of the school was unattractive to the predominantly Middle Eastern origin population which generally adhered to more orthodox or traditional religious ideologies. The Anglo-Saxon tone of the school was foreign to them. In addition, they were wary of innovation: as one such parent put it: "In this kind of school there is likely to be a lot of educational experimentation. I cannot afford to offer my child for experiments and, therefore, I prefer a regular school".

Parent Participation in the School

The Labor Movement School

The parent-school relationship was perceived as follows by one of the activists at the time of the school's establishment: "I envisaged a forum of equals: teachers and parents sharing in the task of running the school both in the area of policy and in the area of implementation, resources, knowledge, philosophy, planning, programs, and doctrine". The social covenant states that:

A different type of involvement in the educational process will be expected of parents:

(a) Active involvement as joint educators in the educational process of their children.

(b) Creating a model for children based on involvement, social leadership and responsibility, and sharing the burden, each according to his or her talents... Parents will be expected to participate in the educational programs, take an active role in lessons, help prepare programs, take part in social and cultural projects, assist in school maintenance... However, parental involvement will not imply interference in the autonomous role of the teacher or the school principal. It will be for the teaching staff under the school principal to translate and implement the educational and social policy developed by the school board.

The school board consisted of parents, teachers, and community representatives. Similarly, committees were set up which dealt with four main areas:

(1) Ideological committees: a thinking forum, curriculum committee, and the Tzomech committee, encounter committee.
(2) Social committees: commitment committee, class committee, the children's community committee and the youth movement committee.
(3) Expressive social committees: environment and home committee, school trip committee, summer school committee.
(4) Technical committees: finances committee, municipality relations committee.

These committees have been active throughout the years.

The system of frameworks set up provided parents with a basis for wide action in the sphere of ideological translation and development. Some of the committees bore a substantial influence on what took place in the school. The ideological committees were especially influential. However, they were characterized by constant tension, much of which can be attributed to the discrepancy so often reigning between practice and ideology. Nevertheless, it was in these committees that the real debate on the aims of the school took place. In the social committees tensions often emerged between the parents from the deprived neighborhoods and the more well-to-do parents. These committees focused on bridging the gaps between the two populations, thereby crystallizing the policy on de facto integration. In general, the well-to-do parents, many qualified in the area of education, were dominant on the committees, parents from the more modest backgrounds preferring to be active on the class level.

All in all, it can be fairly stated that parents were very active, both on the ideational level and in the day-to-day activities of the school. At a certain stage parents constituted 10 percent of participants in the formal curriculum, and this on a voluntary basis: They helped students with learning problems, they taught subjects in which they were specialized, they initiated parties and trips, they assisted in finding funds. It was possible to recruit parents for any need that arose in the school, not only for one-time activities, but for long-term commitment too.

However, over the years, involvement waned and the founding group started showing signs of fatigue. In the middle of the ninth year of the school there was a serious discussion around the following issues: parent participation versus parental involvement; parents inside the classroom or outside it; founder and pioneer parents versus parents who had joined a fully fledged framework; discipline in the school — the responsibility of the parents or the school?

The Conservative School

Parent participation in the Conservative School embraced a very wide range of parents. The parents defined the goals of the school and were involved in their implementation. However, a select elite group, consisting of the founding body and management, predominated.

The school provided clear guidelines on parental participation, management and teachers having the prerogative in almost every area. The model defining the status of parental involvement could be described as "participation without interference". Nevertheless, parents were active partners in many lessons, in developing subject matter, in special events, in preparing parties, in recruiting funds. Parents were expected to take part in the parents' committees set up by the school: the enrichment committee, drama committee, library committee, funding committee, the education committee, a committee of committees in which the values of the school were elaborated, and the clean-house committee.

The parents published the school newspaper once a week, which, apart from the day-today activities of the school, contained biblical and Halachic commentaries.

The influence of parents was, therefore, quite substantial in the ideological area and in creating a congenial enthusiastic community climate. Indeed, conflicts between the school and parents were few and far between and, if they did occur, they centered around personal differences rather than matters of principle. This could explain the absence of erosion in the level of parent participation in the activities of the school.

Summary and Conclusions

These two schools serve to exemplify how the act of parental choice leads to intensive involvement, identification, and readiness for total change in the spirit of the values that the parents wish to impart to their children. In both schools, much effort was devoted to the translation of the schools' values into day-to-day life. This process led to an experimental orientation, a desire for change, and sense of activity beyond the specific values of the school. Parental participation clearly provided both the leverage and impetus for social and educational change in both these schools.

Based on the analysis of these two schools, we constructed a paradigm of interaction between type of school and the act of choice. The assumption of the paradigm is that parental involvement influences school processes, but the type of parental participation and the motivation of parents is dictated by the type of school. The typification of the achievement-oriented school, special subject schools, and the open school is based on their declared rationale and the impressions of the author of this chapter.

References

Chubb, J. E. and Moe, T. M. (1990). *Politics, Markets, and American Schools*. Washington, DC: Brookings Institute.

Coleman, J. S. (1984). Schools and the Communities They Serve. Paper presented at a conference in celebration of the 140th anniversary of the founding of the School of Education of the State University of New York at Albany.

Friedman, Y. (1990). "*The community school: practice and theory*". Magnes, Jerusalem: 1991 (in Hebrew).

Goldring, E. (1991). "Participants in an organizational framework". *International Journal of Educational Research*, **15**, 215–228.

Horowitz, T. (1990). From Neighborhood Community to Ideological Community. Jerusalem: Szold Institute (in Hebrew).

Parsons, T. (1964). *The Social System*. New York: Free Press.

Paradigm of Type of School and Parental Choice

Type of school	Motivation of choice	Student in a wider setting	Perspective as teacher's role	Curriculum
Ideological values	Implementation of worldview	Part of community	Translates values & teaches	Formal & hiden
Religious values	Implementation of worldview	Part of community	Teaches & translates values	Formal & hidden
Achievement oriented	Improving level of learning	The student as individual in class setting	Transmits & enables access to knowledge	Mainly formal
Specialist (nature, music . . .)	Cultivating skill	The student as an individual	Transmits & enables access to knowledge	Mainly formal
Open education	Cultivating thinking patterns	Student as a part of learning community	Facilitator	Formal

Chapter Six

EDUCATIONAL OPPORTUNITIES, PARENTAL CHOICE, COMMUNITY STRUCTURE, AND MOBILITY: THE CASE OF THE ARAB COMMUNITY OF JAFFA

ORIT ICHILOV AND ANDRÉ ELIAS MAZAWI
School of Education, Tel Aviv University

The purpose of the present case-study is to explore the structure of educational opportunities which are available to the Arab community of Jaffa, and to evaluate their social and educational implications and outcomes. The issues which concern us include: What educational options are available to the Arab community of Jaffa? How do they affect mobility chances and equality in educational opportunities among various groups? Data was drawn from a variety of sources. These included information collected by the municipality of Tel-Aviv-Jaffa, the Central Bureau of Statistics, and Project Renewal. In addition, interviews were conducted with all school principals in Jaffa, concerning educational policy and school routines. We also relied on data gathered over the years concerning the educational system of Jaffa, and as part of a recently completed life-history of one of the church schools (Ichilov & Mazawi 1996).

Educational Opportunities of a Minority Group

The network of Arab public schools has drastically expanded since the establishment of Israel in 1948, and while the Arab population in Israel tripled the number of students increased fifteen times (Ackerman *et al.* 1985). This change has been associated with the growing importance of "cultural capital" for social mobility, in comparison with other types of social resources such as land and familial connections (Al-Haj 1989; Smooha 1982; Al-Haj & Rosenfeld 1988).

Studies of the social and educational consequences of these developments focused almost exclusively on the issue of equality of educational opportunities for the Arab minority in comparison with those available to the Jewish majority. These studies revealed inequality between Arabs and Jews concerning access to higher education (Mazawi 1995), level of educational achievements, curriculum, etc. (Svirsky 1990; Al-Haj 1991; Shavit 1990; Peres *et al.* 1968; Mar'i 1978; Bashi & Shash 1991; Mazawi 1994).

By and large, these studies overlooked the fact that the Arab society is internally differentiated by religion, urban or agricultural life styles, and socioeconomic status. Consequently, the issue of the degree to which various sub-groups have equally benefited from the expansion of Arab state education won little attention. Yet there is evidence to suggest the existence of inequality of educational opportunities within Arab society. For example, Christian Arabs have greater access to educational resources than Muslims and Druze, as measured by numbers of students, dropout rates, matriculation certificates and achievement scores (Ministry of Education and Culture 1992).

Differences among sub-groups are often related to place of residence which is an important factor determining accessibility to education among Arabs in Israel (Semyonov & Levin-Epstein 1992). Arab settlements differ from one another. First, by the level of heterogeneity of their population with regard to religion and SES. The distinction among Muslims, Christians, and Druze often overlaps with SES, and Christians are on the average, socioeconomically more established than Muslims and Druze (Al-Haj 1985; Kraus & Hodge 1990; Kraus 1988; Shavit 1992). Christians, more often than others, tend to reside in cities and urban centers, where they enjoy access to church schools in addition to public schools (Shavit 1992; Mazawi 1989). The size of settlement determines its municipal status which, in turn, is related to the amount of resources allocated by the central government, and to the degree of autonomy that municipal authorities enjoy. Some Arab settlements can, therefore, invest more than others in education (Al-Haj & Rosenfeld 1988; Shavit 1992).

What follows is a case-study which explores the issue of equality of educational opportunities within the Arab community of Jaffa. We shall proceed by analysing the socio-economic and cultural characteristics of the Arab community of Jaffa. Next, educational networks in Jaffa, and their social and educational outcomes will be examined.

The Arab Community of Jaffa: History and Background

During the post-1948 years of Israeli statehood, Jaffa was transformed from an Arab city into a predominantly Jewish one. It was deprived of its autonomous municipal status which had been in effect until 1948, and following a short period of Israeli military administration was annexed to Tel-Aviv (Mazawi 1989; Mazawi and Khoury-Machool 1991).

Jaffa underwent radical demographic changes. Following the 1948 Jewish-Arab hostilities, only 3,500 remained of a Palestinian Arab population that was previously estimated to have numbered 70,000 people (Morris 1987). At the same time, huge numbers of Jewish immigrants from Europe, the

Middle East, and North Africa poured into Israel and settled in Jaffa as well. Jaffa, which had been an important sea port and international trade center during the Ottoman and British rule in Palestine, became a dilapidated suburb of Tel-Aviv (Avitsur 1972; Mazawi and Khoury-Machool 1991). Over the

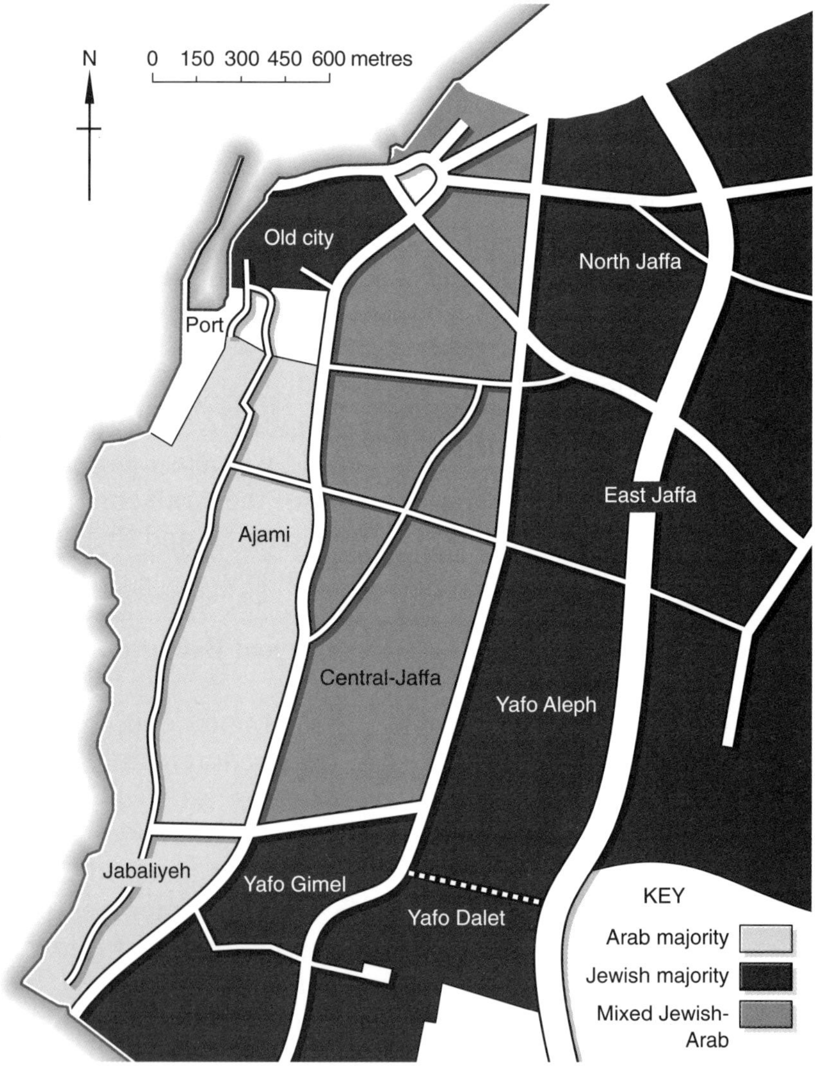

Map 6.1. The spatial distribution of ethnic groups in Jaffa.

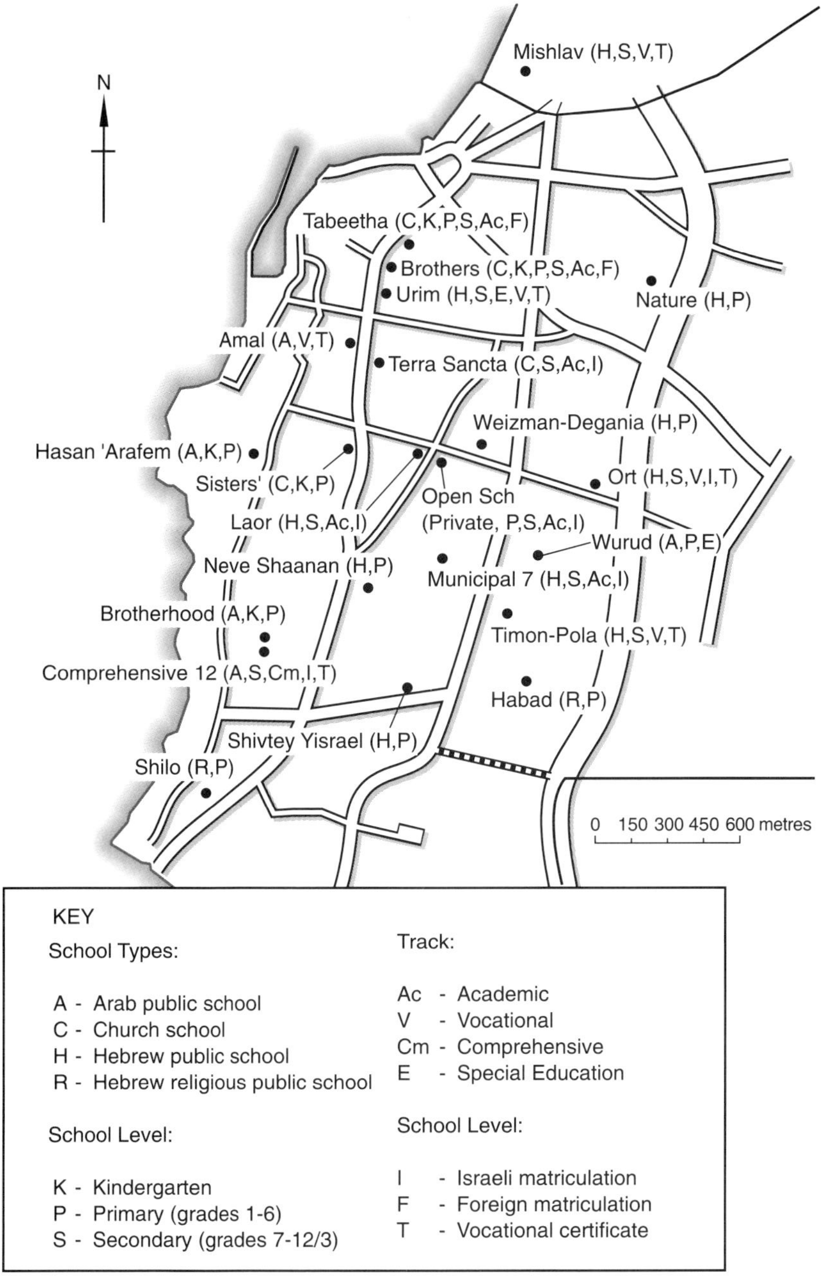

Map 6.2. The spatial distribution of educational institutions in Jaffa.

years, the Jewish immigrants of the 1950s gradually moved to more affluent neighborhoods, and it was mainly lower class Jews and Arabs who stayed on. The restoration of Old Jaffa brought in a colony of Jewish artists, and there are plans for building more residential neighborhoods for affluent Jews. Two-thirds of the Arab population resides in the extremely poor neighborhoods — 'Ajami and Jabaliyeh — which are predominantly Arab. These neighborhoods were graded 4 on a socioeconomic scale for ranking communities which ranged between 0 = lowest and 100 = highest (Municipality of Tel-Aviv-Jaffa 1986). In 1991, the average net monthly income of 47 percent of the residents did not exceed 1,000 NIS (New Israeli Shekels) (the equivalent of about 400 US dollars), an additional 47 percent earned NIS 1,001–2,000, 4 percent earned NIS 2,001–3,000, and only 2 percent earned above NIS 3,000. This indicates that the income of the majority of Arabs in Jaffa is well below the average monthly income of salaried workers — NIS 2,656 — and that many of them live below what has been defined as "the poverty line", that is, an income that does not exceed 50 percent of the average salary (Project Renewal for 'Ajami and Central-Jaffa 1991a).

The Arab community of Jaffa grew rapidly, and between 1972 and 1987 nearly tripled in size. Between 1961 and 1972 yearly growth was 3.5 percent, attributed mainly to high birthrates. In addition, the Arab population of Jaffa was reinforced by internal immigration of Arabs in Israel from villages and towns, a movement that after 1967 was joined by families from the West Bank and the Gaza Strip. Concerning socioeconomic status, the Christians of Jaffa are generally better off than the Muslims, but are not better off in comparison with Christians in other parts of Israel (Municipality of Tel-Aviv-Jaffa/Ministry of Housing and Project Renewal for 'Ajami and Central-Jaffa 1989).

The Arab community in Jaffa is heterogeneous with regard to religious affiliation. The majority (75 percent) are Suni Muslims, and the rest are Christians who belong to a great variety of denominations: Greek Orthodox, Greek Catholic, Roman Catholic, Maronite, Coptic, as well as a tiny Armenian (Christian Orthodox) community.

On average, the Arab community of Jaffa is fairly young: In 1991, 42.5 percent were under 20 years of age, 34 percent were in the 0–4 age group, and 8.5 percent were 15–19 years old (Central Bureau of Statistics 1992: 63). The fact that a sizeable part of the Arab community is of school age accentuates the centrality of the issue of schooling and choice in Jaffa.

In spite of being crowded into dilapidated and poor neighborhoods, only 5 percent of the residents of 'Ajami and Jabeliyeh expressed a desire to leave, while 43 percent considered it "a place to live in", and 52 percent regarded it as "a place to improve" (Project Renewal for 'Ajami and Central-Jaffa 1991b). These attitudes may reflect the preference of members of a national minority group within Israeli society to stay together rather than being dispersed. Consequently the option of moving out of Jaffa is rarely taken. It seems that geographical mobility is not a realistic consideration in the search for better schools.

The educational needs of the poor, small, and religiously heterogeneous Arab community of Jaffa are great. The school-age population is considerable. There is a growing demand for high quality educational facilities to support social integration and upward mobility, to reflect the great diversity of religious affiliations, and to inculcate national identity and culture.

Educational Institutions in Jaffa

The Arab community of Jaffa, it appears, enjoys a great variety of educational institutions to choose from, ranging from pre-school facilities to high school. In addition to Arab public schools, there is a network of private church schools. Jewish public schools also enroll a certain percentage of Arab students. Students can moreover commute daily to schools in adjacent Jewish neighborhoods inside and outside the municipality of Tel-Aviv-Jaffa. Although the latter option is becoming increasingly attractive, as will be discussed later, only a handful of students take advantage of it.

Arab public schools

The Arab state school system consists of two primary schools, and three postprimary schools. The language of instruction in all of these is Arabic, and there are no tuition fees.

One of the two elementary schools bears the name of a local Palestinian Arab businessman, Hassan 'Arafeh, who founded the school in 1936. Ukhuweh (Brotherhood), the other elementary school, was founded in 1980. Both schools consist of grades 1–6, and Hassan 'Arafeh includes, in addition, kindergarten classes.

All postprimary schools were established during the last two decades. These include a junior high school, grades 7–9, and a comprehensive high school, grades 10–12, offering both academic and vocational programs, and preparing students for the Israeli matriculation examinations (Bagrut) in Arabic. Both schools were established in the late 1970s. An additional vocational high school, grades 9–12, which belongs to the 'Amal (Labor) network, was established in the 1980s. It does not prepare students for the Bagrut examinations. The school offers mainly vocational training for low-prestige occupations such as carpentry, mechanics, sewing, and hairdressing.

Hebrew public schools

In Jaffa there are several Hebrew primary schools: Weizman-Degania, and Neve Shaanan. In addition, there is one special education institution: Kidma. Since there was no Arab special education school, this school has been serving the Arab population as well. The adequacy of this arrangement has been doubted because Hebrew instruction presents an extra burden for special education students who are not fluent in Hebrew. A special education school for Arab students opened recently (1992–93). There is also a postprimary school which includes junior high and high school classes (grades 7–12), and prepares students for the Hebrew matriculation examinations. In addition there is a vocational school, grades 9–12, which belongs to the ORT worldwide network. The school offers high quality academic and vocational programs, and prepares students for the matriculation examinations and for the attainment of vocational diplomas.

About 10 percent of all Arab students in Jaffa attend Hebrew public schools.

Private parochial schools

Israeli legislation granted Muslim and Christian parochial schools the status of "unofficial and unrecognized educational institutions". There are about 40 "unofficial" Christian institutions serving the Arab society in Israel: 28 elementary and 12 secondary schools. These schools enjoy great autonomy. The Ministry of Education and Culture may inspect the schools, but has no authority to request changes in the curriculum, administration, or educational personnel. Over the years the majority of these schools voluntarily accepted state control and became recognized schools. They receive some public assistance, and prepare their students for the Israeli matriculation examinations. The language of instruction is usually Arabic, but English and French are also studied. These schools enroll non-Christian students as well. In 1979 only 56 percent of their students were Christians, 42 percent were Muslims, and the rest was made up by Druze, Jewish, and others (Ackerman *et al.* 1985). In the school year of 1990–91, 6.7 percent of all Arab students in Israel attended private non-official schools, mostly church institutions, from kindergarten to the postprimary stage (Central Bureau of

Statistics 1992: 640, Table 22.19). However, in cities that have such schools available — like Jaffa, Haifa, Nazareth, Ramleh, Lod, and Acre — about half of the Arab students attend church schools (Ichilov and Mazawi 1996).

Unlike public schools, which are free of charge, private schools in Jaffa charge tuition fees. This may be a great burden, especially on families with a number of school-age children.

There are four church schools which date back to the pre-1948 period. Saint Joseph is a Catholic primary school run by the Sisters of Saint Joseph of the Apparition, and includes kindergarten through sixth grade classes. The language of instruction is Arabic. Terra Sancta is a Catholic high school, grades 7–12, that prepares students for the Israeli matriculation examinations in Arabic. It is run by a Franciscan order which is subordinated to the Catholic institution Custodia di Terra Sancta, and the language of instruction is Arabic. The other two schools consist of kindergarten through high school classes. Saint Joseph College (or the Brothers' College) is a French-Catholic school run by the Institute of the Brothers of the Christian Schools, the largest male teachers Catholic brotherhood. The language of instruction is French, and students are prepared for the French baccalaureate. Tabeetha is a Protestant school run by the Church of Scotland. The language of instruction is English, and students are prepared for the British matriculation.

Though the variety of educational institutions is impressive, the extent to which the schools in fact cater to the needs of the Arab community of Jaffa remains to be seen.

Choosing a School in Jaffa

Choosing a school for one's children is not a simple matter in Jaffa. It depends on parents' choice and on the selection criteria employed by the schools.

Parents must take into consideration the options for

further education that each school provides, as well as its curriculum, language of instruction, reputation and record, and the existence and amount of tuition fees. Parents must also think about the possibilities to transfer from a particular school to other schools, should such need arise. An important structural feature of educational systems is the extent to which alternatives are provided for disaffected students. This is so because the option of transferring schools or of "dropping down" may reduce dropping out (Lee & Burkan 1992).

Deciding between a private or a public school depends primarily on financial considerations. Tuition fees for one child per month constitute about 15 percent of an average monthly income. This is a burden that only well-to-do families can afford for one child, let alone for more than one. One must also take into consideration the fact that families pay taxes which should provide adequate educational facilities.

Parents also consider the educational reputation of the schools. Table 6.1 shows how adult residents of 'Ajami evaluate private and public educational institutions in Jaffa and indicates that the private primary and secondary schools were rated as high or very high by the majority of 'Ajami residents (56 percent and 64 percent respectively). State schools were rated much lower, with the exception of Hassan 'Arafeh which was rated by 38 percent in the category of high and very high. One may conclude that parents who enroll their children in public schools do so mainly because they cannot afford to send them to the better private schools. During the Ottoman period, and more so under British rule, the expression "community schools" (**Madares Ahliya**) was used to characterize the better church schools, while public schools were described as pursuing the mere dissemination of ignorance (**Siyasat Tajhil**).

The above evaluation of schools was supported by interviews with all school principals[1]. Principals of public schools reported overcrowded classrooms, inadequate

Table 6.1. Evaluation of educational institutions by Adult 'Ajami residents (percents)

	Very low	Low	Medium	High	Very high	Don't know	Do you have children in this school? Yes	No
Kindergarten:								
Private	2	4	13	18	11	52	15	85
Municipal	2	7	15	19	2	55	11	89
Primary Schools:								
Hasan 'Arafeh	4	6	18	27	11	34	15	85
Brotherhood	3	10	22	17	7	41	18	82
Church	1	4	9	33	23	30	25	75
Post Primary:								
Public Compreh.	2	6	25	20	5	42	19	81
'Amal vocational	6	13	13	14	3	51	7	93
Church	2	2	6	44	20	26	20	80

Source: Project Renewal for 'Ajami and Central-Jaffa (1991a). Appendixes: 23.

housing facilities, a majority of low SES students, families with problems, low parental involvement, and low quality and commitment of teachers. Principals of private schools reported a majority of middle- and upper-class students, and of children of the diplomatic personnel of various embassies. They seemed satisfied with the educational staff and housing facilities.

Arab parents who wish to plan a long-term educational course for their children face difficult decisions (Fig. 6.1).

As early as the kindergarten level parents must make a choice among community, public or church facilities. This choice is greatly determined by parents' ability to pay tuition. This early choice may determine students' next steps in the educational system. Data indicate that children who attend community and church kindergarten tend to enroll later in church schools, while those who study in public kindergarten (Arab or Hebrew) tend to go on to public schools (Menahem 1990). The socioeconomic divide between those who can and those who cannot afford to pay tuition is already visible.

The language of instruction in the community and public kindergartens is either Hebrew or Arabic. Church kindergartens which are part of Saint Joseph College and Tabeetha offer French and English programs, respectively. Since these institutions offer a complete course of education from kindergarten through high school, parents may wish to retain their children there. Kindergarten children whose parents want them to pursue their studies in a primary school with a different language of instruction will be at a disadvantage already at this early stage. The costs incurred in transferring from one school to another at more advanced stages of the educational process are even greater since curricula differ greatly among schools. Consequently, it is virtually impossible for disaffected students to "drop down" or to transfer to a similar or a more selective school.

Private schools are selective while public schools are

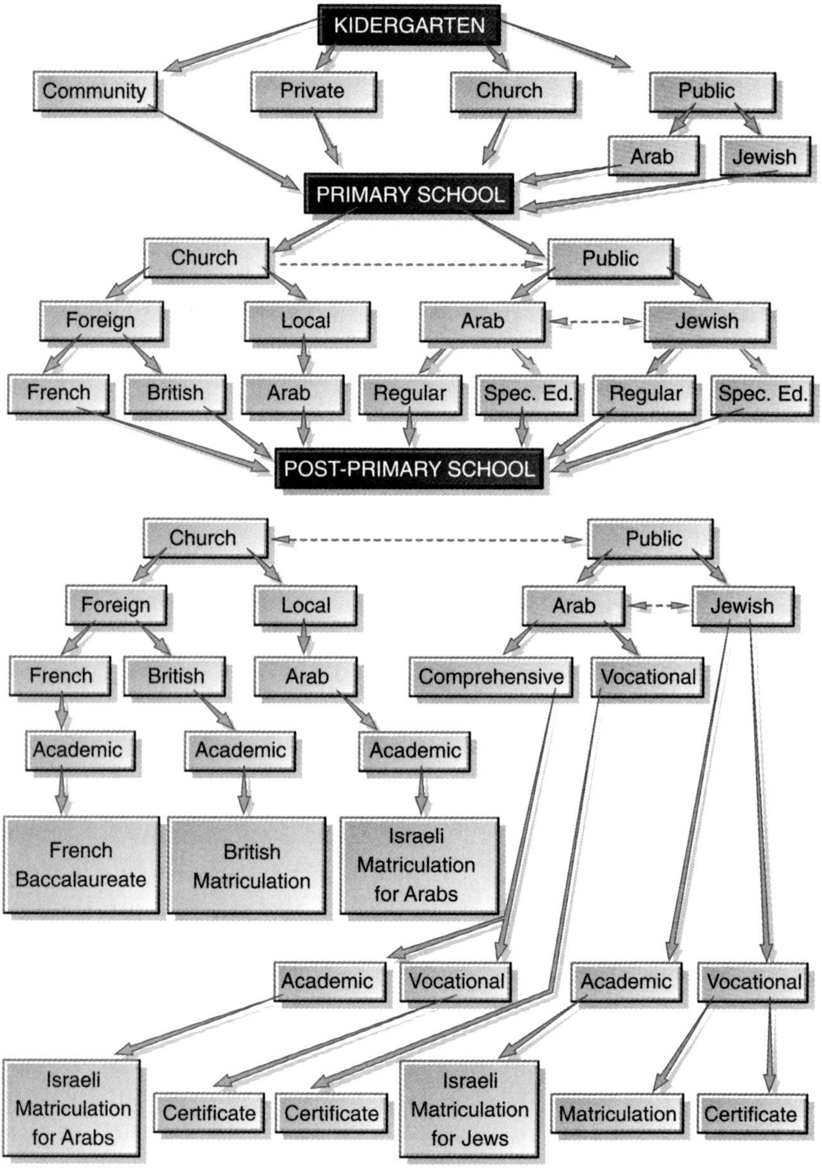

Fig. 6.1. The structure of educational choice in Jaffa.

obliged to admit all applicants. Academic record and/or screening examinations are widely used, and determine entry to the private schools. In some church schools, notably the primary division of Terra Sancta, preference is given to applicants whose siblings or relatives are already enrolled or were so in the past.

Patterns of School Enrollment

Data concerning the actual school distribution of Arab students is available for the 1980s only.

Table 6.2 reveals that Arab students are fairly evenly distributed between Arab public schools and church schools. The student population increased by 35.2 percent between 1980 and 1988; however the growth has not been uniform for all schools. Hebrew public schools witnessed the greatest increase in Arab student enrollment, an increase twice the rate of the overall growth of the student population (74.7 percent). Church schools come next with an increase of 41.5 percent. Arab public schools grew the least, by only 23.6 percent. Furthermore, it should be noted that while there has been an increase from 42.1 percent in 1985 to 44.6 percent in 1988, Arab public schools actually experienced a generally decreased enrollment from 48.7 percent in 1980 to 44.6 percent in 1988.

In Table 6.3 changes in enrollment rates are presented for the major primary and secondary schools serving the Arab community of Jaffa.

The general trend in Table 6.3 seems further to indicate growing enrollment in church schools, and decreasing enrollment in public schools. Church schools, which are considered academically superior, are in great demand. Consequently, they had already reached a point of saturation in 1986, and could no longer physically expand to supply Jaffa's growing demand for educational services. In an effort to avoid Arab public schools, parents who could not afford to pay tuition

Table 6.2. Distribution of Arab pupils in Jaffa — 1980, 1985, and 1988

	School Year				
	1980		1985		1988
Arab public schools:	1005 (48.7%)		1047 (42.1%)		1242 (44.6%)
Between years change		4.2%		18.6%	
Total period change			23.6%		
Church schools:	914 (44.3%)		1189 (47.8%)		1293 (46.3%)
Between years change		30.1%		8.7%	
Total period change			41.5%		
Hebrew public schools:	146 (7.0%)		252 (10.1%)		255 (9.1%)
Between years change		72.6%		1.2%	
Total period change			74.7%		
Total: All schools	2065 (100%)		2488 (100%)		2790 (100%)
Between years change		20.5%		12.1%	
Total period change			35.2%		

Note: Figures include pupils enrolled in special education classes, vocational tracks and apprenticeship vocational programs.

Table 6.3. Changes in enrollment rates of public and church schools

	Primary and Junior high school levels				
	State school		Church schools		
	Hassan-'Arafeh	Brotherhood	Sisters'	Brothers'	Tabeetha
School year 1980	350	221	311	269	78
School year 1985	257	363	293	436	68
School year 1988	248	328	311	421	91
Period	−29.1%	48.4%	0.0%	56.5%	16.7%

	Secondary level			
	Public schools	Church schools		
	Comprehensive	Terra Sancta	Brothers'	Tabeetha
School year 1980	434	196	49	11
School year 1985	427	204	124	64
School year 1988	415	239	139	92
Period	−4.4%	21.9%	183.7%	736.4%

began to send their children to Hebrew public schools, and enrollment in them grew by 4 percent between 1980 and 1985.

Given the plurality of curricula and languages of instruction, and the fact that Arabic is spoken at home, the integration of Arab students in Hebrew public schools proceeds far from smoothly. In order to facilitate the integration of Arab students in the Hebrew public schools, the League for the Arabs of Jaffa, a local Arab body founded in 1979, requested from the Ministry of Education and Culture that Arab students be instructed in Arabic. The Ministry's Director General angrily responded that Arab students would not be instructed in Arabic within Hebrew public schools. He blamed the Arab community for the bad reputation of Arab public schools saying that "the Arab community of Jaffa prefers the selective church schools. [Consequently] the Arab public schools educate dropouts of these [church] schools who arrive illiterate in all languages". He urged the League for the Arabs of Jaffa to do its utmost to improve the image and educational standards of Arab public schools in Jaffa.

Consequences for the Arab Community of Jaffa

This study clearly indicates that Jaffa does not have a coordinated educational system. Instead, there are three different, largely incommensurate, systems which operate alongside each other. We shall now proceed to examine the social and educational consequences of this intricate conglomeration of educational institutions for the Arabs of Jaffa.

Distribution of educational resources

Several parameters of educational outcomes were used to examine, where data permitted, how Arabs in Jaffa fare compared to Jews in Jaffa, and compared with Jews and Arabs in Israel.

In 1988, the median number of years of schooling in Jaffa

was 7.6 for Muslims, 8.4 for Christians, and 8.8 for Jews (Economic Consulting and Planning Ltd. 1988: Table 21; Israel Land Authority *et al.* 1990: 19). This falls below the national median of 8.7 years of schooling for Arabs and 11.8 for Jews in Israel in 1989 (Central Bureau of Statistics 1991, pp. 604–605). High school attendance (students aged 14–17) went up from 52 percent in 1983–84 to 58 percent in 1989–90 (Municipality of Tel-Aviv-Jaffa 1986 p. 124, section 3.7; Municipality of Tel-Aviv-Jaffa 1989). However, in 1989 the Arabs of Jaffa were below the national level of the Arab society in general, which was 62.8 percent (Central Bureau of Statistics 1991: 618, Table 22.13).

In 1983–84 29.6 percent of the 15–17 age group were out of school and out of the labor market (Municipality of Tel-Aviv-Jaffa 1986: 124). These youngsters, who are labeled "cut off/unattached youth" (**Noar Menutak**) often become delinquents. In 1991 their percentage among the Arabs of Jaffa went down to 17 percent (Project Renewal 1991a, 1991b), lower than the 20.9 percent of the Arab community in Israel (Central Bureau of Statistics 1992, p. 367, Table 12.22). In 1991, the percentage of unattached youth among Jewish residents of Jaffa was 11 percent (Project Renewal 1991a), and nationwide it was 7.1 percent (Central Bureau of Statistics 1992: 237, Table 12.22).

Concerning the attainment of a matriculation certificate, the Arab community of Jaffa did better than the nationwide record of the Arab school system in 1988–89. Of all 12th graders 87.2 percent were entered for the matriculation examination, as compared to only 78.3 percent among Arabs, and 73.6 percent among Jews nationwide. Of those, 57.1 percent received a matriculation certificate (only 39.6 percent among Arabs). Arab recipients of a matriculation certificate in Jaffa formed 47.5 percent of all 12th graders, while the comparable percentage for Arabs was 31 percent. The majority of recipients of a matriculation certificate graduated from Terra Sancta, a private church school that prepares students for the Israeli matriculation examinations in Arabic. The percentage of matriculation certificate among Arabs in Jaffa, however, is still below the average of Jews (Central Bureau of Statistics Ministry of Education and Culture 1992). The percentage of Arab 12th graders in vocational tracks is significantly lower compared with Jews in Tel-Aviv-Jaffa, Arabs, and Jews. The percentages were 15.3 percent, 28.4 percent, 23.1 percent, and, 59.8 percent respectively (Central Bureau of Statistics/Ministry of Education and Culture 1992). The lack of adequate vocational schools in the Arab society means that youngsters who fail in the academic school have no schooling alternatives (Haidar 1985). They are at risk of becoming unattached youth (Hilou 1992).

Cultural and national identity

Schools in Jaffa have a divisive influence on the cultural and national identity of the Arab community of Jaffa. The diversity of schools fosters increased segmentation in a small community which is already heterogeneous by religious affiliation. Tabeetha school inculcates culturally inclined British Yafis[2], the Brothers' College — French Yafis; Terra Sancta — Christian Yafis; the Hebrew schools — Hebrew Yafis; the Arab public schools — Israeli-Arab Yafis. This situation further complicates the already complex issue of the identity of Arabs within a Jewish state. It renders the concept "Arab community of Jaffa", to a very large extent, meaningless, and makes it difficult to involve people in common community projects, thus fragmenting the local society into distinct social and cultural groupings.

Stratification

The educational system is stratified by SES. There seems to be a vicious circle: low-income parents enroll their children mainly in Arab public schools. The social and financial returns from schooling for these students are low in terms of prestige, job opportunities, and income. This puts them back in the socioeconomic echelon of their parents. The returns from schooling in church schools, which enroll mainly middle- and upper-middle-class students, are high. This perpetuates the existing class structure in Jaffa, splitting the Arab community along socioeconomic lines as well.

The Politics of Education

Conceiving a comprehensive educational master plan for Jaffa is an extremely complex undertaking. The Ministry of Education and Culture and the municipal authorities are in charge of public schools only. Church schools, which enroll about 50 percent of the Arab students in Jaffa, are outside the direct jurisdiction of Israeli educational legislation. Their independent status stems from diplomatic agreements, and is zealously guarded by church officials. The only exception is Terra Sancta, which voluntarily became a "recognized unofficial" school[3]. The school adopted the Israeli curriculum for Arabs, and prepares students for the matriculation examinations in Arabic. This status guarantees some financial state support, while retaining the autonomy of the school. The principal, who usually is also the parish priest, is mainly accountable to the Franciscan Order. Terra Sancta is a selective school which gives priority to the admission of Christian students. It is clear what the purpose of educational policy in Jaffa should be. Jaffa needs a "learning culture" in which parents and teachers have high expectations of their children's educational achievements, whose education systems are designed to provide opportunities and motivation for learners of all abilities, and in which the labor market, and society in general, reward those who do well in education. However, how to achieve this ambitious end remains highly complicated and unclear. It has been recently suggested that the introduction of a voucher system into Jaffa might prompt the improvement of Arab public schools, and their ability to compete with the selective church schools. Assuming, however, that vouchers could be used for both public and church schools, this would probably encourage parents who cannot afford to do so at present, to enroll their children in church schools. Consequently, the demand for church schools, which are already overcrowded, would increase. Past experience shows that it is likely that church schools would consequently become more selective rather than expand due to lack of space and funds. Should church schools prefer Christian over Muslims students, as some of them do, the schools would also become more homogeneous with regard to socioeconomic status. This is so because the SES of Christians is generally higher than that of Muslims, and church schools would also prefer to enroll better-off students, whether Muslims or Christians. Under such a constellation, Arab public schools would enroll mainly lower class Muslim students. It is unlikely that competition among Jaffa's schools, which would encourage Arab state schools to improve radically, would ensue. On the other hand, closing Arab public schools down is equally unfeasible, both from a political and legal point of view.

It is possible that the enrollment of Arab students in Hebrew public schools would somewhat increase, inside or outside of Tel-Aviv-Jaffa municipality. The recent increased enrollment of Arab students in primary Hebrew public schools (during 1995–96) has already encountered the objection of Jewish parents. A massive enrollment of Arab students would pose even greater difficulties. Further, Jewish ultra-orthodox circles, which in the past vehemently opposed the enrollment of Jewish children in church schools, as well as contacts among Jewish and Arab students, would very likely resist the idea. This option is also unattractive to Arab parents who wish that their children would become familiar with Arab culture and language in school. Should vouchers be applicable in Arab public schools only, the chances are that parents would continue to avoid these schools unless they improve drastically.

Several conclusions concerning the issue of equal access to high quality educational facilities within a heterogeneous minority group become inevitable. In order to more fully comprehend how schooling is related to stratification within a minority group, it is not sufficient to compare it with the majority. The global inequality between the Jewish majority and Arab minority is accompanied by inequality intensified within Arab society. Differential educational opportunities within a minority group not only reproduce the SES structure of that community, but also affect the chances of members to be integrated into the broader society. In Jaffa, graduates of church schools clearly have an advantage over graduates of Arab public schools. They are better equipped to be upwardly mobile within the Jaffa community and within the broader labor market in general. This study also demonstrated that a plurality of curricula and uncoordinated school networks is not necessarily an asset and may rather become a liability under specific conditions. It could interfere with the inculcation of national and community identities, and with the ability of students to move within the educational system and transfer between schools.

Notes

1 Letter of Dr Shimshon Shoshani, Director General of the Ministry of Education and Culture, to the Secretary of the League for the Arabs of Jaffa, September 29, 1986. Archives of the League for the Arabs of Jaffa, Education File.

2 **Yafi** in Arabic means "a native of Jaffa".

3 In 1996, as this paper was being reviewed, the Brothers' College became an unofficial recognized school. Yet it preserved its existing curriculum.

References

Ackerman, W. *et al.* (1985). *Education in an emerging society*. Tel-Aviv and Jerusalem: Ha'Kibbutz Ha'Meuchad and Van Leer (in Hebrew).

Al-Haj, M. (1985). Ethnic relations in an Arab town in Israel. In A. Weingrod (Ed.), *Studies in Israeli ethnicity*, pp. 105–132. New York: Gordon & Breach Science Publishers.

Al-Haj, M. (1991). *Education and social change among the Arabs in Israel.* Tel-Aviv: International Center for Peace in the Middle East.

Al-Haj, M. and Rosenfeld, H. (1988). *Arab local government in Israel.* Tel-Aviv: International Center for Peace in the Middle East.

Anderson, D. (1988). Values, religion, social class and the choice of private school in Australia. *International Journal of Educational Research*, **12**(4), 351–373.

Astin, A. (1992). Educational choice. Its appeal may be illusory. *Sociology of Education*, **65**(4), 255–60.

Avitsur, S. (1972). *The port of Jaffa in its height and in its decline.* Tel-Aviv: Avshalom Institute, Malo Press (in Hebrew).

Ballion, R. (1986). Le choix du *collège*: le comportement "*éclairé*" des familles. *Revue Française de Sociologie*, **27**(4), 719–734.

Bashi, Y. and Shash, Z. (1991). Results of achievement test in reading comprehension and arithmetic in grades 4 and 5. Jerusalem: Van Leer (in Hebrew).

Central Bureau of Statistics (1991). *Statistical Abstract 42.* Jerusalem: Government Printing Press.

Central Bureau of Statistics (1992). *Statistical Abstract 43.* Jerusalem: Government Printing Press.

Central Bureau of Statistics/Ministry of Education and Culture (1992). *Twelfth grade pupils, examinees and entitled to matriculation certificates by locality 1988–89*. Special Series No. 907, Jerusalem.

Coleman, J. S. (1992). Some points on choice in education. *Sociology of Education*, **65**(4), 260–262.

Cookson, P. (1991). Private schooling and equity: Dilemmas of choice. *Education and Urban Society*, **23**(2), 185–199.

Economic Consulting and Planning Ltd. (1988). *Occupational survey in southern Jaffa and Jabaliyeh neighborhoods of Jaffa.* Tel-Aviv, April–June.

Haidar, A. (1985). *Technical-vocational education in the Arab sector in Israel.* Tel-Aviv: The International Center for Peace in the Middle-East.

Hilou, N. (1992). *The effects of environmental, family and friendship factors on juvenile delinquency in Jaffa.* Unpublished MA Thesis, Tel-Aviv University, School of Social Work (in Hebrew).

Ichilov, O. and Mazawi, A. (1996). *Between state and church: Life History of a French-Catholic school in Jaffa.* Frankfurt am Main: Peter-Lang.

Israel Land Authority; Ministry of Housing; Halamich Government Co.; Municipality of Tel-Aviv-Jaffa; The Jewish Agency (1990). *Planning and budgeting guidelines 1990–91. Tel-Aviv-Jaffa:* Jaffa Renewal Project.

Kraus, V. (1988). The opportunity structure of young Israeli Arabs. In J. Hofman (Ed.), *Jewish-Arab relations in Israel. A quest in human understanding*, pp. 67–91. Bristol, Indiana: Wyndham Hall Press.

Kraus, V. and Hodge, R. (1990). *Promises in the promised land. Mobility and inequality in Israel.* New York: Greenwood Press.

Lee, V. and Burkan, D. (1992). Transferring high schools: An alternative to dropping out? *American Journal of Education*, **100**(4), 420–454.

Mar'i, S. (1978). *Arab education in Israel.* New York: Syracuse University Press.

Mazawi, A. (1989). Geographical expansion of Jaffa and construction styles: Past and present, *Al-Jadeed*, **38**(11–12), 47–56 (in Arabic).

Mazawi, A. (1994). Palestinian Arabs in Israel: Educational expansion, social mobility and political control. *Compare: A Journal of Comparative Education*, **24**(3), 277–284.

Mazawi, A. (1995). University education, credentialism and social stratification among Palestinian Arabs in Israel. *Higher Education*, **29**, 351–368.

Mazawi, A. and Khoury-Machool, M. (1991). Spatial policy in Jaffa: 1948–1990. In H. Luski (Ed.), *City and Utopia*, pp. 62–74. Tel-Aviv: Israel Publishing Co. (in Hebrew).

Menahem, G. (1990). *Project renewal in 'Ajami and Lev-Yafo: An evaluation research.* Research report No. 7, The Institute for Social Research, The Department of Sociology and Anthropology, Tel-Aviv University.

Ministry of Education and Culture (1992). *The educational system through numbers*. Jerusalem: Senior Section for Economics and Budgets, Department of Economics and Statistics (in Hebrew).

Morris, B. (1987). *The birth of the Palestinian refugee problem, 1947–1949*. Cambridge: Cambridge University Press.

Municipality of Tel-Aviv-Jaffa (1986). *Neighborhoods and areas in Tel-Aviv-Jaffa: Physical, demographic and socioeconomic features (1983–1984)*. Tel-Aviv: Center for Social and Economic Research (in Hebrew).

Municipality of Tel-Aviv-Jaffa/Ministry of Housing and Project Renewal for 'AJami and Central-Jaffa (1989). *Planning guidelines for 'AJami and Central-Jaffa*. Tel-Aviv: Tel-Aviv-Jaffa Municipality (in Hebrew).

Peres, Y. *et al.* (1968). National education of Arab youth in Israel: A comparison of curricula. *Megamot*, **15**(1), 26–36 (in Hebrew).

Project Renewal for 'Ajami and Central-Jaffa (1991a). *Attitudes of Jaffa's residents — Neighborhood rehabilitation survey*. Tel-Aviv: Martens and Hoffman.

Project Renewal for 'Ajami and Central-Jaffa (1991b). *Survey of neighborhood's rehabilitation: 'Ajami*. Tel-Aviv: Martens and Hoffman.

Semyonov, M. and Lewin-Epstein, N. (1992). Community of residence, community of employment and income returns. *Israel Social Science Research*, **7**(1–2), 15–27.

Shavit, Y. (1990). Segregation, tracking, and the educational attainment of minorities: Arabs and Oriental Jews in Israel. *American Sociological Review*, **55**(1), 115–126.

Shavit, Y. (1992). Arabs in the Israeli economy: A study of the enclave hypothesis. *Israel Social Science Research*, **7**(1–2): 45–66.

State of New Jersey, Department of Education (1988). *Public school choice: National trends and initiatives*. Trenton: State of New-Jersey.

Svirski, S. (1990). *Education in Israel: Schooling for inequality*. Tel-Aviv: Breirot (in Hebrew).

Tibawi, A. L. (1956). *Arab education in mandatory Palestine: A study of three decades of British administration*. London: Luzac and Co.

Chapter Seven

THE FUNCTION OF EVALUATION IN SCHOOL AUTONOMY

DAVID NEVO
Tel Aviv University

Assumptions

Many definitions and characterizations of school autonomy have been suggested, although no consensus has been reached regarding one single definition of "school autonomy". However, various propositions have been put forward that are useful in thinking about school autonomy and clarifying its meaning. The following comments reflect a perception of school autonomy, which evolved during work with "autonomous schools", and serves as a basis for a conception of evaluation of school autonomy.

Since the early 1970s the Israeli educational system has shown a tendency to decentralize and to empower schools, teachers and local educational communities. This was expressed in encouraging local initiatives for innovation by teachers and principals, showing increased flexibility in implementing the national curriculum at the school level, allowing parents to determine a considerable part of the curriculum in their children's schools, and strengthening the interaction between the school and the community.

Decentralization was first of all an attempt to serve local needs and legitimize heterogeneity, but in a way it was also a reflection of the limited success achieved by central educational authorities in affecting school improvement, and an expression of the expectation of more effective school improvement through local initiatives. An autonomous school was implicitly expected to be an innovating organization, involved continuously in self assessment, restructuring, experimentation, and improvement. The assumption was that individual schools could be more sensitive than the central authority to the needs of their students and could find the best ways to serve such needs effectively.

Eventually, a significant amount of authority was granted to the schools to set priorities, plan educational activities, restructure curriculum, and revise their organizational structure. In 1981 the Ministry of Education actually issued a "call" for school autonomy, encouraging elementary schools to become autonomous and providing them with incentives to do so. But granting authority to the school never meant releasing it from the supervision and control of the Ministry of Education and its central agencies. The school could initiate, plan, and redesign, but its activities still had to be "approved" by the school supervisor or "coordinated" with some other central agency. And in this country everybody knows that when you say autonomy, you mean autonomy and not necessarily independence — not only in the field of education.

The rhetoric of school autonomy in Israel emphasized the various aspects of authority; much less was said about responsibility. Although there must have been some expectation for a balance between authority and responsibility, it was not clear what the boundaries of the school's responsibility were, nor was it clear to what extent such boundaries would change for schools increasing their authority through school autonomy. However, together with the trends of decentralization and school autonomy, the educational system seemed to be reiterating its demand for accountability by implementing national and regional assessment programs and by establishing regional evaluation teams to conduct external evaluations of schools around the country.

School autonomy in the Israeli educational system was also associated with a trend toward professionalization and an increase in teacher participation in school decision making. The need for professionalization of teachers and principals has been suggested as a major factor of school autonomy by researchers from the academe (Reshef 1987; Inbar 1987) and in official documents of the Ministry of Education (Peled 1976; Ministry of Education 1979). A more professional rather than bureaucratic conception of teaching and principalship seems to be underlying the trend toward school autonomy. This conception was expressed by raising the standards of professional training of school principals, upgrading teacher training through extensive in-service training, and encouraging teacher participation in school-level decision making.

Thus the following assumptions served as a basis for the attempt to develop an evaluation approach which would serve the needs of school autonomy:

(1) The educational system is undergoing a trend toward decentralization and empowerment of schools and local communities, but also retaining a significant amount of central control and supervision of schools by the Ministry of Education.

(2) The autonomous school is expected to be an innovative school, involved continuously in self assessment, restructuring, and improvement.

(3) Autonomy grants the school the authority to change and innovate, but it also puts the bur' of responsibility for the consequences

innovations on the school. Thus school autonomy and the demand for accountability should be complementary rather than contradictory elements in an attempt to improve the school.

(4) The professionalization of teachers and other school educators is a major component of school autonomy and an important factor of school improvement.

Such assumptions call for a wide perspective of evaluation and an evaluation system capable of serving formative and summative functions, simultaneously providing ongoing feedback for improvement and professional development, and cogent evidence for accountability. Most important, such an evaluation system should have the potential to furnish solid ground for a dialog between the school and the external audiences interested in its improvement. I believe that in our work with autonomous schools we have succeeded in approaching such an evaluation system. I would like to present it here for further review and critique. It will be presented at two levels of abstraction. At the first level its actual implementation will be described. The second level will return to its underlying principles.

Implementation

Developing school evaluation in the so-called "autonomous schools" is part of a more general interest of the Israeli Ministry of Education and some local and regional education authorities in an evaluation project which combines internal and external evaluation. The major goal of the project is to help schools develop their own evaluation systems, combining internal and external evaluation in a complementary rather than a contradictory way, in order to serve decision making within as well as outside of the school. This project of school evaluation looks at "school-based evaluation" not as a synonym for "internal evaluation" nor as an antonym to "external evaluation", but as a combination of both. Its operational plan includes four major stages.

The **first stage** provides regional in-service training workshops for principals and teachers from schools interested in developing internal school evaluation mechanisms. The workshops are conducted on a weekly or bi-weekly basis and provide 70–80 hours of basic training in program evaluation, testing, data collection procedures, and data analysis.

In the **second stage** of the project, participating schools establish internal evaluation teams composed of three or four teachers who attended the training workshops. The decision to establish an internal evaluation team is usually made by the school towards the end of the first stage of the project. A school that has established an evaluation team selects its first evaluation object (a school project, a program of instruction, or some domain of activity within the school) to practice its internal evaluation capability with the technical assistance provided by an external tutor who works individually with the school on a weekly or bi-weekly basis.

In the **third stage** the internal evaluation team is institutionalized as a permanent component of the school. The composition of the team changes periodically so that all interested teachers have a chance to participate actively in the school evaluation team. Arrangements have to be made to provide the team with some technical assistance, usually much less extensive than the assistance provided in the previous stage.

In the **fourth stage**, which may be implemented concurrently with the third stage, the school is ready for external evaluation by a national or regional educational authority. The assumption is that in the future some kind of external evaluation will be mandatory in all schools across the country regardless of whether or not they established an internal evaluation team. However, schools that decide to develop internal evaluation teams will not be evaluated externally until such time as the internal evaluation capacity of the school has become operational.

Hundreds of teachers and school principals in Israel have gone through the first stage of the project and tens of schools have established internal evaluation teams or are in the process of establishing such teams.

Underlying Principles

Five years of work with schools in developing their evaluation capabilities, as well as reflections on the concept of evaluation in education, have enriched our practical knowledge regarding the implementation of school evaluation. Several guidelines and principles can now be delineated. But I would first like to present our basic perception of the concept of evaluation and the way it is presented to the schools we work with.

Teachers and school administrators have various perceptions of the meaning of evaluation and its significance in school (Nevo and Goldblat 1988). Some of the resistance which educators sometimes show towards evaluation seems to be a result of misconceptions. Therefore, it is very important to clarify the meaning of evaluation and develop a common understanding of its role in the school.

In working with teachers and school administrators we have learned that the distinction between description and judgment as major components of evaluation is very significant for teachers, principals, and other educators within the school. Therefore, to draw attention to the essence of evaluation and to create a basis for communication between all parties involved in school evaluation, **educational evaluation was defined as a systematic activity using information to describe educational objects and judge their merit or worth**. This definition focuses on the need to make a careful distinction between two elements that are very different in their nature and function, but which are both nevertheless incorporated into one activity called evaluation. It is in line with some definitions that have been suggested in the evaluation literature (Stake 1967; Guba and Lincoln 1981; Nevo 1983) and was found especially useful in discussing evaluation perceptions with educational practitioners.

This definition had three important implications for our work with schools developing their evaluation mechanisms. First, it suggested that evaluation can not be performed on an intuitive basis but rather is a **systematic** process. This implies that principles, methods, and instruments have to be applied to evaluate events and objects within the school. Thus teachers and administrators who wish to be involved in evaluation must acquire a minimal level of technical evaluation skills. They must learn how to design an evaluation, how to collect and analyze data, and what technical assistance to employ in these areas.

Second, applying technical evaluation knowledge and skills can significantly enhance the objectivity of the **description** component in an evaluation, although no description can ever be totally objective.

Third, while subjectivity plays a greater role in **judgment** than it does in description, subjectivity need not imply arbitrariness. Systematic evaluation attempts to avoid arbitrary judgments by providing as much objective description as possible, and by making judgment procedures explicit and open to discussion. Subjective components can never be totally eliminated when making judgments. Since judgments are based on criteria reflecting local needs, attitudes, and values, they can never be totally detached from individual preferences.

Teachers and school principals find the distinction between description and judgment useful because it points out the legitimate role of each in school evaluation, and sets some realistic parameters to the role of technical assistance in description and in judgement.

Having clarified what is meant by evaluation I will now

describe the major guidelines and principles for school evaluation. They are as follows:

1 Students and their achievements should not be the only object of school evaluation.

Teachers tend to equate evaluation with testing, and use student assessment mainly for the narrow purpose of grading (Stiggins and Faires-Conklin 1992). The use of assessment to evaluate instruction, and the notion of program evaluation within the school, seems quite new to schools. It is not self-evident that any entity within the school can be an evaluation object, provided that there is some justified need to describe its nature and/or assess its merit. Typical objects of school evaluation can be programs, projects, instructional materials, teachers and other school personnel, students, or the school as a whole. Such a perception has to be presented, discussed, and adopted during the process of developing a school-based evaluation mechanism.

2 When evaluating a program, a project or any other evaluation object within the school, not only outcomes or impacts should be examined.

This does not mean that outcomes and impacts are not important evaluation variables, but a detailed description of an educational activity, and a fair judgment of its merit cannot be limited to the narrow scope of outcome measures, such as test scores. To express the full scope of an educational object, evaluation should focus on at least four types of information: goals, strategies, implementation, as well as outcomes or impacts.

This perception is not new in program evaluation literature (Stake 1967; Stufflebeam *et al.* 1971; Nevo 1983), but is not well-known in schools nor is it in line with the common-sense perception of lay people, who equate quality with performance or achievement (Nevo and Goldblat 1988).

As has been noted, focusing evaluation only on outcomes is strongly conveyed to the school through national and regional assessment programs which evaluate schools and teachers solely on the basis of student achievement. But when teachers and principals are presented with the idea that the quality of their schools can be determined (and improved) not only by looking at students' achievements, but also by examining the quality of their educational goals, curricula, pedagogical approaches, educational programs, and instructional facilities, they find evaluation much more useful to their schools, and a much more challenging component in their professional work[1].

3 School evaluation has to serve both the formative and the summative function of evaluation, providing information for planning and improvement, as well as for certification and accountability.

While the concept of "formative evaluation" stresses the importance of evaluation being constructive and useful, in certain cases it has also created an unjustified excuse to avoid demands for accountability by suggesting that formative evaluation is an alternative to summative evaluation. There is no doubt that evaluation should be constructive to pedagogy and administration within the school, but it should also be useful in assisting the school to demonstrate its merit to educational authorities, parents, and the general public. The tendency toward avoiding summative evaluation is especially disturbing when it occurs in an innovative or experimental school, which uses formative evaluation to improve its operation, but seems to forget the need for summative evaluation to demonstrate its merit as a viable alternative to a conventional school.

4 There is no meaningful way to judge the overall quality of a school (or a teacher or a student or a program) by one single criterion, or even by accepted multiple criteria, nor is there any real need to do so.

This does not mean that school evaluation should provide only descriptive data and refrain from judgments. Evaluation should provide **judgmental statements** based on diverse criteria related to various aspects or domains of the school; however, it should refrain from an overall unidimensional assessment. Criteria for evaluation should be determined within the educational and social context of the school and in accordance with the needs of its target population. As noted above, though choosing evaluation criteria involves a considerable degree of subjectivity it is not arbitrary. To be subjective rather than arbitrary, the following aspects should be considered as a basis for judging a school evaluation object: The extent to which it (a) responds to identified needs of actual and potential target populations of the school; (b) achieves national goals, ideals, or social values; (c) meets agreed-upon standards and norms; (d) is outdoing alternative objects; and (e) achieves stated (but also important) goals of the evaluation object.

Actually, overall assessments are not necessary. For formative evaluation, they are too general to provide any insightful direction for improvement. For summative evaluation, making an overall assessment is actually like making the decision. What can decision makers be expected to do after being told what the **best** thing for them is? And if what has been determined is the best thing to do, but not necessarily the best thing for **them**, then why should **they** be interested in such an overall assessment?

5 The internal evaluation needs of a school can best be served by a team of teachers and other educators, for whom evaluation is only part of their job definition, supported by appropriate training and external technical assistance.

The school evaluation team is composed of what Scriven (1967) called "amateur" evaluators, referring to those whose professional training is not in evaluation, and for whom evaluation represents only a small part of their job description. However, we are talking about amateur **evaluators** rather than amateur **teachers**. The strength of such an amateur internal evaluation team is in its educational and pedagogical background, its familiarity with the educational and social context of the school, and its not posing a threat to the school faculty and administration. The major weakness of such a team lies in its lack of evaluation skills and experience, but this can be remedied by appropriate training and technical assistance. An alternative could be to employ a part time or full time internal evaluator, who would be solely responsible for conducting the internal evaluation of the school. This alternative seems less desirable for several reasons. First, the evaluator may be a potential threat to teachers and therefore not trusted by them. Second, teachers will have less chance to experience evaluation and to make it a part of their daily work. Third, it might take the evaluator a long time to acquire a good understanding of the school context and its specific needs. Fourth, even if funds were available for hiring an evaluator for each school, such evaluators would not be immediately available in most educational systems.

In the current project, teachers serve on the school evaluation team on a rotational basis for a period of one to three years. They usually enjoy being members of the evaluation team, and put serious effort into this work. In many schools, teachers competed to become members of the team. Teachers also reported that what they learned from their work with the evaluation team had a significant impact on their teaching and their professional perceptions and behavior.

As mentioned earlier, this approach perceives the teacher as a professional teacher but as an amateur evaluator. This is in line with the conception of evaluation, which assumes that self assessment and renewal are basic characteristics of professional behavior. Involving teachers in evaluation is also crucial if they are to be perceived as equal partners in a dialogue between the school and the educational system intended to improve the school.

6 **To conduct meaningful evaluation within the school it is necessary to mobilize alternative tools and methods of inquiry from the behavioral sciences and related fields of study and adapt them to the needs of the school and the capabilities of "amateur" evaluators.**

It has to be clear to the school staff that there is no such thing as "data-free evaluation", and evaluation is not an antonym for measurement. Measurement and data collection lie at the heart of evaluation and no sound description or judgment of any evaluation object can be obtained without them. To promise schools that successful evaluation can be achieved without some kind of systematic data collection and analysis is misleading in the worst way.

School evaluators do, however, need evaluation tools that are **simple to operate** but result in **sophisticated performance**. Unfortunately, the state of the art of educational measurement is somewhat lacking in such instruments. Even in achievement testing, probably the most developed area in educational measurement, simple tools are not available and relatively sophisticated methods are used to address narrow domains of educational achievements. The result is that most of the more advanced measurement methods never reach the school (Stiggins and Faires–Conklin 1992).

7 **Learning by doing is still the best way to learn how to do evaluation.**

School professionals have always complained that when university professors or professional evaluators try to teach them how to do evaluation, their presentations are too theoretical and abstract, and they never provide them with practical tools they can really use. Confronted with such complaints, I have tried over the years to be less theoretical and abstract when interacting with practitioners. I have tried various "cook book" approaches, developing simple principles, translating them into guidelines and procedures, and always providing the pertinent instruments (e.g., questionnaires, tests, rating scales, check lists, outlines, etc.) followed by illustrative cases of implementation. But even then the complaints did not stop. The reaction to the more specific prescriptions was: "We understand you, but how do you actually do it?”. It was clear that the solution to the problem was not to be more specific, but to introduce the rule of learning by doing. Working with schools now begins with an introductory evaluation course for principals and teachers, discussing basic concepts in school evaluation, measurement and data analysis. The purpose of this course is not to teach how to do evaluation, but what school evaluation is all about, and to help them decide if they really want to learn how to do it and develop an evaluation mechanism in their schools. If they decide that they do want to learn how to do evaluation, they move, as soon as possible (even before the end of the introductory course), back into their schools and start doing their own evaluation supported by technical assistance provided by an external professional evaluator.

8 In school-based evaluation, internal evaluation is a prior condition to useful external evaluation.

The shortcomings of external school evaluations have been discussed elsewhere (Madaus 1985; Nevo 1994), and their impact on schools have been demonstrated in various studies (Glasman and Nevo 1988; Corbett and Wilson 1991; Herman and Golan 1992). Our approach to school-based evaluation calls for a combination of internal and external evaluation. The various functions of evaluation within the school should be served by internal and external evaluations, but the existence of internal evaluation within the school is a prior condition to any constructive utilization of external evaluation. As has been noted, internal evaluators are more appropriate to serve the function of formative evaluation, and external evaluators better serve the function of summative evaluation. But a school that does not have an internal mechanism for its self-evaluation will have difficulties in developing positive attitudes towards evaluation, and will lack the self-confidence necessary for a constructive dialogue between the school and the body conducting the external evaluation. In such cases, evaluation becomes a source of accusations and defensiveness rather than a basis for a dialogue between internal and external decision-makers.

Clearly those who are mainly interested in formative evaluation should focus on internal evaluation. I would like to contend here that those who are interested in summative evaluation should also encourage schools to develop internal evaluation mechanisms, not instead of external evaluation, but as a prior condition for effective external evaluation.

Conclusion

The combination of internal and external evaluation brings us to the essence of evaluation, and to the basic characteristic of our proposition for school evaluation in the autonomous school or in other schools of similar nature. If evaluation is to play an important role in school autonomy, it cannot be used only in its external capacity to tell the school how good it is, or how bad it is, or what is wrong with it, and what is best for it to do. Nor should evaluation be used exclusively in its internal capacity to provide the school with an excuse to avoid accountability or encourage it to keep stewing in its own juice.

A school is too complex to run by remote control from the outside. But schooling is also too important to leave solely in the hands of teachers and school administrators. Nobody knows what the best way to run a school is, and there are no simple answers to any of the important questions about teaching, learning, class management, or school administration. Much effort is required to identify crucial problems within the school and to develop thoughtful solutions to such problems. All parties associated with the school should be involved in such an effort in a constructive and cooperative way. And evaluation can play a crucial role in such a cooperative effort by facilitating a dialogue among all those who are interested in the improvement of the school or have a stake in its success. This might be the only role evaluation should play in a democratic society, and in educational institutions, such as the autonomous schools, committed to a professional rather than a bureaucratic conception of teaching and schooling.

Note

1 See also Nevo (1995).

References

Corbett, H. D. and Wilson, B. L. (1991). *Testing, Reform, and Rebellion.* Norwood, New Jersey: Ablex.

Glasman, N. S. and Nevo, D. (1988). *Evaluation in Decision Making: The Case of School Administration.* Boston: Kluwer Academic Publishers.

Guba, E. G. and Lincoln Y. S. (1981). *Effective Evaluation.* San Francisco: Jossey-Bass.

Herman, J. L. and Golan, S. (1992). *Effects of Standardized Testing on Teachers and Learning — Another Look.* Los Angeles: University of California, Center for the Study of Evaluation.

Inbar, D. (1987). Is autonomy possible in a centralized educational system? In I. Friedman (Ed.), *Autonomy in Education,* pp. 58–78. Jerusalem: Szold Institute (in Hebrew).

Madaus, G. (1985). "Test scores as administrative mechanisms in educational policy". *Phi Delta Kappan*, **66**(9), 611–617.

Ministry of Education (1979). *Report of the State Committee on the Status of the Teacher*. Jerusalem: Ministry of Education.

Nevo, D. (1983). "The conceptualization of educational evaluation: An analytical review of the literature". *Review of Educational Research*, **53**(1), 117–128.

Nevo, D. (1994). Combining internal and external evaluation: A case for school-based evaluation. *Studies in Educational Evaluation*, **20**(1), 87–98.

Nevo, D. (1995). *School-based evaluation: A dialogue for school improvement*. Oxford: Pergamon.

Nevo, D. and Goldblat, E. (1988) *Evaluation Perceptions and School Assessment in Schools of Active Learning: A research report*. Tel Aviv: School ol Education, Tel-Aviv University (in Hebrew).

Peled, E. (1976). *Education in the Eighties*. Jerusalem: Ministry of Education.

Reshef, S. (1987). "School autonomy: A new era in state education". In I. Friedman (Ed.), *Autonomy in education*, pp. 13–31. Jerusalem: Szold Institute (in Hebrew).

Scriven, M. (1967). The methodology of evaluation. In Stake, R. E. (Ed.), *Curriculum Evaluation. AERA Monograph Series on Curriculum Evaluation 1*. Chicago: Rand McNally.

Stake, R. E. (1967). The countenance of educational evaluation. *Teachers College Record*, **68**, 523–40.

Stiggins, R. J. and Faires Colkin, N. (1992). *In Teachers' Hands: Investigating the Practice of Classroom Assessment*. New York: SUNY Press.

Stufflebeam, D. L., Foley, W. J., Gephart, W. J., Guba, E. G., Hammond, R. L., Merriman, H. O. and Provus, M. M. (1971). *Educational Evaluation and Decision Making*. Itasca, Illinois: Peacock.

Chapter Eight

AUTONOMY AND CHOICE AS SCHOOL STRATEGIES FOR PERIPHERAL COMMUNITIES IN ISRAEL

ABRAHAM YOGEV
School of Education, Tel Aviv University

Introduction

Protagonists of school choice have by now made a clear-cut distinction between the state-wide sweeping strategy of choice and what we may name the community-based strategy of controlled choice and school autonomy. The state-wide strategy has been experimented by various American states, which have opened up their school system for parental choice (The Carnegie Foundation 1992; Cookson 1994), and by the policy of grant-maintained schools in England, which allowed schools to opt-out of their local educational authorities and open up for student choice (Chubb and Moe 1992; Power *et al.* 1994). This strategy, directly based on the market-competition ideology (Chubb and Moe 1990), has usually resulted in greater diversity between "good" and "bad" schools and in larger school segregation between privileged and disadvantaged groups of students. It has been harshly criticized on these grounds in both Great Britain and the United States (e.g., The Carnegie Foundation 1992; Ball 1993; Carl 1994).

In contrast, the community-based strategy is particularly aimed at autonomous school restructuring in disadvantaged neighborhoods. It uses a controlled choice program, which regulates school choice by balancing student composition within school, as part of an overall strategy of school renewal. This comprehensive strategy includes autonomous staff development, the development of school ethos and school-based curriculum, and a greater involvement in school of students and parents, as well as other community representatives. The strategy thus encourages the development of a network of autonomous schools of choice in the involved community.

The Carnegie Foundation report on school choice (1992) was the first to recognize the success of this strategy by focusing on the experiences of three such school restructuring programs in the US: the junior high school choice programs in Montclair, New Jersey; Cambridge, Massachusetts, and East Harlem, New York. The report claims that the achievements of ethnic minorities in these school choice programs may be attributed to other factors than school choice itself. Yet, since choice was one component of a complex strategy employed in these cases, it is obviously useless to separate the contribution of specific components to the final successful results. Later reports on community-based school choice programs (e.g., Cookson 1994; Wells 1994) have emphasized their catering to students' individual needs in addition to their contribution to ethnic desegregation and to minority achievements.

However, one of the problems of the community-based strategy has been its implementation in central rather than peripheral geographic regions. Indeed, such choice programs have been experimented in distressed neighborhoods, but only in large or medium size established cities which are not geographically remote from political and cultural centers. The reasons for that fact are diverse, ranging from demographic reasons: it is easier to employ choice programs in highly populated areas — to political and cultural ones: it is easier to draw the attention of leaders of central communities to school autonomy and choice issues, parents get involved more easily, and principals and teachers are more open to school reform.

Nonetheless, the relevance of the community-based strategy to peripheral regions is almost self-evident. Regional disparities between geographic centers and peripheries on a variety of social welfare indicators were extensively studied during the 1980s, especially with respect to developing countries, and were attributed to the political and cultural subordination of peripheral populations (Efrat 1983; Slater 1980). The control of peripheral regions by national centers has also been noticeable in the educational sphere. As noted by Chau (1985, p. 4241), "any failing behind in the development of education is quickly resented by regional communities as an expression of lack of interest on the part of the national authorities and is liable to arouse regional feelings which could raise a threat to national unity". However, educational disparities between regional centers and peripheries were mainly assessed with respect to the supply of general educational services (Carron and Chau 1980a, b) due to various measurement problems and the lack of more specific data (Chau 1985).

It is reasonable to expect such disparities to be reflected in school ethos and student treatment within school. In other words, it is quite possible that peripheral schools, being more dependent on the national center for their daily operation and supply of teaching and educational resources, cater much less to students' specific needs than schools in the geographic center. If this is indeed the case, then the community-based strategy of school autonomy and controlled choice may help in the renewal of peripheral schools. This is especially so since this strategy is known to advance ethnic and cultural minority students, who are usually concentrated in peripheral regions. Yet, as said above, the strategy has not been frequently implemented in peripheral regions. Such regions were usually included in state-wide choice programs. Peripheral schools in the United States usually suffered as a result of their inclusion in state-wide choice programs due to differences among parents in opting for choice and to the lack of bussing programs supporting the choice policy. Poor schools in peripheral areas did not disappear as a result of the choice policy, but rather became poorer due to budgetary cuts resulting from the students opting out (The Carnegie Foundation 1992).

This paper examines the relevance of the community-based strategy of school autonomy and controlled choice for schools in the peripheral regions of Israel — the Southern and Northern administrative districts of the education system. As a result of the official collection of specific data on school operations — curricular and extra-curricular activities offered by primary schools within administrative districts — we are able to compare specific characteristics of schools in the central and peripheral regions of Israel. The comparison concentrates on three related questions: (1) To what extent may peripheral schools be described as catering less to student needs than schools located in the central districts of the system? (2) Can such differences be attributed to school peripheriality, or do they result from additional relevant factors, such as student composition? (3) Are peripheral schools more dependent on the national center for their daily operations, and to what extent does such dependence influence their services to the students? Given the answers to these three questions, we shall discuss the prospects and problems involved in implementing the community-based strategy for enhancing the peripheral schools.

School Choice, Center and Periphery: The Case of Israel

School choice programs, which started in Israel in the early 1990s, stemmed out of prolonged interest in school autonomy (Reshef 1984; Shapira *et al.* 1994). Subsequently, the proposals of school choice programs rested on the community-based model from the early start, and combined school autonomy with controlled choice strategies (see Shapira *et al.* 1991). However, the debate over choice policies in Israel has surpassed the local level, and turned into a heated ideological debate between choice protagonists and supporters of the "school integration" policy (see Friedmann 1994; Shapira *et al.* 1995). The latter policy, advocating school desegregation on the basis of fixed catchment areas, has been continuously implemented in Israeli education since the early 1970s despite the growing criticism of its relevance to disadvantaged students (Yogev 1989b; Swirski 1990). The debate has necessitated the creation of a special commission for school choice by the Israeli Ministry of Education and Culture (1994). The commission's report, adopted by the Ministry, suggested a solution which may be regarded as a compromise. On the one hand, it advocated the community-based strategy of school choice, and allowed interested municipalities and local communities to establish choice policies, as long as they are aimed at school autonomy rather than "market competition" **per-se**. On the other hand, it required that the community schools meet certain "equivalence criteria" before their opening up for student choice. Unfortunately, some of these criteria may be better established as a result of autonomy and choice rather than prior to the start of such school processes[1].

This compromise notwithstanding, the heated debate over school choice has probably been an additional reason for the fact that choice policies in Israel, as elsewhere, have until now been mainly implemented in the central regions, primarily in large cities which have autonomous educational administrations. The first was the Tel Aviv-Jaffa municipality (1992), which has regarded school choice as an essential part of the renewal program of its school system. It used school autonomy programs to open up for controlled choice elementary and junior high schools in selected neighborhoods, mainly in the Southern and Eastern poorer parts of the city. The municipality now contemplates the implementation of the policy in the entire city. Jerusalem has recently followed suit and has also implemented it in less-developed neighborhoods. Other local communities which have considered choice strategies, but have not yet implemented them, are also situated in the central regions of the country rather than in peripheral ones.

In the geographical sense, center and periphery in Israel are basically determined by the division of the country into six administrative districts, already determined during the pre-statehood British mandatory period. The four central districts are Tel Aviv, Center, Jerusalem and Haifa. The two regions traditionally considered peripheral are the North (consisting of the Lower and Upper Galilee and the Golan Heights) and the South (spreading between Beer Sheba and Eilat). Together, these last two districts contain over 85 percent of Israel's geographic area, but somewhat less than 30 percent of its population. Demographically, the peripheral districts are not merely less populated than the central ones. Their inhabitants also belong more often to disadvantaged groups (Arabs and Jews of Asian-African origin), and they are more dispersed. While the large cities are located in the central districts, the peripheral ones consist of mid-sized veteran cities (like Beer-Sheba in the South, Tiberias and Nazareth in the North), small development towns established since the 1960s for the purpose of population dispersion, and various types of rural settlements (villages, cooperative villages, and **kibbutzim**).

Several studies conducted during the 1980s have shown that the Northern and Southern districts fall behind the four central ones on a variety of economic and social indicators of welfare, and therefore may be considered a periphery in the socioeconomic sense (Shachar and Lipshitz 1980; Efrat 1984; Lipshitz 1984). Efrat (1984, pp. 35–43) further attributes this peripheriality to the economic, social and political dependence of the Northern and Southern districts on the center. Indeed, the Israeli government has made several attempts in recent decades to close the gaps between central and peripheral regions, but to no avail. Some of these attempts have only increased the dependence of the periphery. For example, the Neighborhood Rehabilitation policy of the 1980s, extensively implemented in the periphery in order to increase living standards on the basis of inhabitants' participation, was only partially successful. While the inhabitants became locally involved in the program, the decentralized activities were subordinated to a rather centralized decision-making process (Carmon 1989). Similarly, development towns which have recently established, by their own initiative, residential projects aimed at attracting an outside population of higher social strata (e.g., Ayalon *et al.* 1993), are located in the central regions and not in the periphery.

The dependence on the political and cultural center is also noticeable in education. Schools in the peripheral districts are the major benefactor of the educational welfare program of the Ministry of Education and Culture. This program, based on a compensatory ideology (Adler and Sever 1988), has since the 1970s promoted special educational and cultural programs for disadvantaged students. However, the projects are determined and pre-arranged by the Ministry and leave little room for school initiative (Yogev 1989a). Several mayors and school principals of development towns in the North, facing their gloomy results in the standardized achievement tests delivered by the Ministry in the early 1990s to all primary school students, have pointed out their budgetary dependence on the Ministry as an important reason for their poor achievements. But in order to fully comprehend the picture,

Table 8.1. Percentage of Jewish primary schools conducting various activities in 1989 by central and peripheral districts ($N = 1040$)

	District						
	Tel Aviv	Haifa	Jerusalem	Central	Northern	Southern	F-test
Activity Frameworks							
Educational counseling	77.4	37.1	68.9	65.6	51.1	35.3	23.27***
Tutorial projects	81.2	71.2	75.5	74.8	48.9	65.3	11.36***
Slow-learner classes	52.2	63.6	28.3	57.2	35.2	47.4	10.29***
Ability grouping	38.2	55.3	68.2	47.6	51.7	64.7	7.08***
Homework help	46.8	56.1	51.9	50.8	65.9	57.9	3.31**
Study center	68.8	61.3	70.0	70.0	63.6	54.7	2.85*
Curriculum enrichment[1]	69.9	59.1	69.8	62.4	64.8	66.8	1.22
Teaching Programs and Methods							
Practice & diagnosis through computer[1]	56.5	32.6	26.4	18.4	37.5	36.8	15.58***
Activity — classroom	77.4	50.0	52.8	78.4	60.8	53.7	13.53***
Individual instruction	51.1	33.3	41.5	56.0	47.7	52.6	4.28***
Teaching aided by computer	39.2	57.6	38.7	53.2	44.3	47.4	3.49**
Other programs for teaching by computer	21.0	34.8	30.2	33.6	38.6	30.5	3.05**
Instrumental enrichment	22.6	29.5	17.9	25.6	33.5	20.5	2.78*
Reading center	36.0	29.5	23.6	33.2	31.8	32.6	1.18
Improving reading comprehension	70.4	65.9	73.6	71.2	64.2	71.6	0.99

Informal Educational Services							
Field school trips[1]	32.8	38.6	67.9	57.6	69.9	73.2	21.92***
Cultural activities offered by the "Cultural Basket" program[1]	35.5	31.1	62.3	56.8	71.6	66.8	20.05***
After-school student center	47.8	53.0	50.9	44.8	35.8	22.1	9.88***
Other cultural activities	86.0	78.0	64.2	68.8	61.4	66.3	7.48***
School bulletin	72.0	78.8	58.5	74.4	67.6	70.5	2.92*
Clubs							
Foreign language	36.0	24.2	47.2	30.8	18.7	18.4	8.92***
Painting	46.2	48.5	63.2	46.0	39.8	29.5	7.23***
Sciences	37.1	43.9	49.1	36.8	33.5	21.1	6.27***
Drama	35.5	28.8	44.3	36.0	31.8	20.0	4.83***
Folkdances, eurythmics	54.8	62.9	60.4	55.6	45.5	43.2	4.02***
Sport	60.2	72.7	62.3	63.2	60.2	48.9	4.01***
Sculpture, plastic arts, photography	31.7	31.8	34.0	29.2	23.3	17.9	3.20**
Computer (incl. computer games	42.5	39.4	46.2	33.6	49.4	37.4	2.81*
Handicrafts	45.2	53.0	60.4	48.4	46.6	40.0	2.70*
Choir, orchestra, music	54.8	57.6	62.3	57.2	61.9	43.2	1.33

[1]These activities were sponsored in 1989 by the educational welfare program and/or by the neighbourhood rehabilitation program.

*$p < 0.05$.

**$p < 0.01$.

***$p < 0.001$.

one should certainly examine more closely the features of schools in the peripheral districts.

School Features in the Periphery versus the Center

The Israeli Central Bureau of Statistics (1991) collected during the 1989 school year specific information on primary school activities in the Jewish sector for the Ministry of Education and Culture[2]. The activities survey was based on mail questionnaires sent to all school principals and returned by 91 percent — 1040 schools. The principals were asked to check the activities conducted by their school out of a list of 57 educational and welfare activities, divided into four groups: counseling services and student health care; curricular frames of activity; teaching programs and methods; informal education and extra-curricular activities.

In the present analysis we included 30 out of the original 57 activities, mainly excluding those activities on which school variations were small[3]. The analysis is divided into three parts: the comparison of school activities across the six districts; the effect of peripheriality versus other factors on school activities, and the contribution of dependence on the center to school differences.

School activities across districts

As shown in Table 8.1, the 30 activities analyzed are equally divided between formal ones (activity frameworks and teaching programs and methods) and informal education services, including extra-curricular club activities. This way, our analysis relates to the comprehensive universe of school activities.

We compared the peripheral and central districts across each of the activity groups. The first group of activity frameworks refers to structured formal activities of the school, and reflects its institutional treatment of students. It appears that schools in the Northern and Southern districts initiate support frameworks for students to a lesser extent than schools in the four central districts. The former are significantly lower on sponsoring educational counseling, tutorial projects, or classes for slow learners. On the other hand, the peripheral schools — especially in comparison to schools in the Tel Aviv district — are more rigorous on ability grouping of students, and offer their students more organized help in homework tasks. These distinctions are especially important in view of the concentration of disadvantaged students in the peripheral districts. Despite that concentration, the peripheral schools are less innovative in adopting new activity frameworks relevant to weak students, such as tutorial projects or slow-learner classes. They tend to stick to traditional frameworks, such as ability grouping and homework assistance sessions.

The second group of activities, referring to teaching programs and methods, reveals less acute differences between schools in the center and the periphery, and some variations among the central districts themselves. Regarding practice and diagnosis through computer, the peripheral schools do not fall behind the schools of most central districts. However, such programs of computer use are sponsored by the educational welfare policy of the Ministry, which are heavily used in the periphery, and therefore should not be necessarily attributed to school initiative. This also explains the relatively wide usage of other computer teaching programs in the peripheral schools. The-gap between peripheral and central schools on the realm of curricular programs is especially noticeable with respect to the use of activity classrooms. The establishment of activity classrooms requires school initiative and continuous staff cooperation, and is therefore more widespread in some of the central districts.

The differences in school initiative and dependence on the center between the central and peripheral schools are especially acute with respect to extra-curricular school activities. Such differences are noticeable in both the third group of activities, referring to informal educational services, and in the fourth group of club activities. Primary schools in the Northern and Southern districts are especially been on field trips and cultural activities offered by the "Cultural Basket" Program. These two extra-curricular activities are pre-arranged by the Ministry's educational welfare program, and include pre-set school tours and visits to particular dramatic and musical performances. On the other hand, the schools in most of the central regions participate more often in cultural activities outside the "Cultural Basket" program and, more importantly, establish more often student centers for enrichment programs. Indeed, the list of school club activities clearly demonstrates that all extra-curricular club activities are practised less often by peripheral schools, especially the Southern ones[4]. This is true not only about club activities offering exclusive cultural capital — such as foreign language or painting clubs — but also with respect to more standard enrichment activities, such as sports or folkdancing.

Peripheriality as a determinant of school activities

The analysis so far shows that schools in the peripheral regions tend toward a more traditional treatment of their students, are more dependent on the political center for their activities, and demonstrate a low initiative drive in developing independent and innovative school activities, both curricular and extra-curricular, Nonetheless, we should recall that in the peripheral regions there is a concentration of disadvantaged student groups (Israeli Ministry of Education and Culture 1993, p. 61). There are also large size variations among specific peripheral localities, resulting in school size differences which may account for school activity variations. It, therefore, remains unclear whether the differences revealed in school activities are attributable to the geographic peripheriality of the schools or to other related factors. The assessment of the significance of geographic peripheriality in this respect requires a multivariate analysis which includes the relevant control variables.

We subsequently performed a discriminant analysis of school activities by district which included three control variables on the school level[5]: (1) school size — number of students; (2) percentage of school students defined as "disadvantaged" according to the Ministry's "welfare index", which provides financial support to schools on the basis of their ethnic and socioeconomic composition of students (see Yogev and Ayalon 1996, pp. 117–119); and (3) school affiliation with the state religious versus state secular sector. The state religious school sector, which is particularly active in the geographic periphery, is known for its more traditional treatment of students despite the fact that both sectors belong to the public school system (Yogev and Chen 1985).

The schools of the six districts were discriminated along 12 selected activities — the six curricular and six extra-curricular activities which revealed the most significant differences among school districts in Table 8.1 above[6]. Table 8.2 presents, for comparison purposes, two discriminant functions, one without the three control variables and the other including them. The comparison allows us to assess the effect of the control variables on the original activity discrimination among school districts. For both analyses we present the results for the first major function, excluding the additional minor functions which distinguished between particular districts according to specific activities.

The first function, excluding the control variables, actually repeats the major findings of Table 8.1 in a more concise way. The function is strong, as indicated by the high canonical correlation and the low value of the Wilks' Lambda coefficient[7]. The centroids, which represent the central values of the districts on the discriminant function, indicate that the function clearly distinguishes between the central and peripheral districts according to school activities. Actually, the

Table 8.2. Discriminant analyses of selected activities in Jewish primary schools according to district: first function results

Variables and standardized coefficients of the discriminant function	Analysis excluding controls	Analysis including controls
School Activities		
Educational counseling	0.386	0.356
Tutorial projects	0.283	0.268
Slow-learner classes	0.098	0.056
Ability grouping	−0.345	−0.289
Practice & diagnosis through computer	0.216	0.247
Activity classroom	0.241	0.202
Field school trips	−0.537	−0.438
Cultural activities offered by the "Cultural Basked" program	−0.462	−0.384
After-school student center	0.200	0.177
Other cultural activities	0.163	0.092
Foreign language club	0.111	0.144
Painting club	−0.005	−0.057
Control Variables		
School size (number of students)		0.234
Percent disadvantaged students in school		−0.364
School sector: state religious		−0.208
Wilks' lambda	0.470	0.410
p (x^2 test)	<0.001	<0.001
Canonical correlation	0.561	0.584
Centroids of Districts		
Tel Aviv	1.096	1.110
Haifa	0.296	0.174
Jerusalem	−0.297	−0.014
Central	0.148	0.235
Northern	−0.702	−0.798
Southern	−0.837	−0.958

clear-cut distinction is between schools of the Tel Aviv district and those of the peripheral Northern and Southern ones, with the rest of the central districts (Haifa, Jerusalem and Central) positioned in between. The function coefficients for school activities, interpreted in conjunction with the direction of centroid values, show what we already know by now: The peripheral schools are particularly strong on ability grouping and cultural activities offered by the educational welfare or neighborhood rehabilitation programs. They are weaker than central schools on providing counseling services to students and, to a lesser extent, in initiating innovative school programs such as tutorial projects or activity classrooms.

The inclusion of the three control variables in the analysis does not significantly alter the results. Indeed, their inclusion strengthens the function, particularly due to the effect of the percentage of disadvantaged students in school. Yet the standardized coefficients for school activities do not change much. This indicates that the differences in school activities between schools in the periphery and the center are indeed attributable to their geographic peripheriality rather than to differences between center and periphery in school size, student composition or sectorial school affiliation.

Dependence on the government in peripheral versus central schools

Throughout our analysis the peripheral schools appeared to be more dependent on government initiative and sponsorship of particular activities. Considering school autonomy as a possible alternative for peripheral communities, we should concentrate more closely on this dependence issue. In order to assess the significance of this dependence for school activities we further compared, within the peripheral and the central regions, the activities of schools which participate in the educational welfare and/or neighborhood rehabilitation programs with those of the non-participant schools. The comparisons, for the same 12 selected activities of Table 8.2 above, are presented in Table 8.3. As the bottom figures in Table 8.3 indicate, the proportion of schools benefiting from one or both government programs in 1989 was about the same in the central and peripheral districts, ranging between 42 and 44 percent in the peripheral and central districts, respectively. The empirical issue, therefore, is not whether the peripheral schools are more subjected to government dependence than central schools, but rather the effect of this dependence on differences among school activities within each of the regions[8]. Therefore, Table 8.3 compares the activities of participant and non-participant schools within each of the two regions.

As Table 8.3 shows, the differences in school activities between participating and non-participating schools are significant in both central and peripheral regions. However, these differences tend to be larger in the peripheral region, as indicated by the higher *t*-test values. Furthermore, it is the comparison across specific activities which is most revealing in terms of the differential effects of government-dependence in the periphery versus the center.

Considering first the curricular activities, it appears that the dependent schools in the periphery offer their students lesser services of educational counseling than non-dependent schools, while the picture is reversed in the central districts. There, the schools participating in the government programs are those offering more educational counseling. Furthermore, three-quarters of the dependent schools in the periphery use ability grouping. This proportion is much higher than the ones for either the non-dependent schools in the periphery or for any of the two types of schools in the center. In addition, the tendency of the peripheral dependent schools to avoid innovative school practices (tutorial projects, slow-learner classes, and activity classroom), in comparison to the non-dependent schools, is much higher than the comparable differences between participant and non-participant schools in the central region. In contrast, they offer computer practice

Table 8.3. Percentage of Jewish primary schools in the central and peripheral districts[1] conducting selecting activities — according to school participation in the educational welfare or the neighborhood rehabilitation programs

	Central districts ($N = 675$)			Peripheral districts ($N = 366$)		
	Participants	Non-participants	t	Participants	Non-participants	t
School Activities						
Educational counseling	71.8	65.4	1.65*	39.6	58.5	−3.62***
Tutorial projects	78.2	75.0	0.88	68.2	49.5	3.62***
Slow-learner classes	63.6	48.3	−3.70***	56.5	30.2	5.22***
Ability grouping	58.2	44.9	3.22***	75.3	46.2	5.81***
Practice & diagnosis through computer	42.2	28.8	3.43***	50.6	27.4	4.67***
Activity classroom	65.0	70.1	−1.30	42.8	67.4	−4.83***
Field school trips	61.2	42.9	4.42***	79.9	65.6	3.02***
Cultural activities offered by the "Cultural Basket" program	55.3	42.9	2.99**	79.9	61.3	3.86***
After-school student center	42.7	50.8	−1.95*	25.3	31.1	−1.21
Other cultural activities	67.0	78.2	−3.11**	57.1	68.9	−2.32*
Foreign language club	25.7	37.0	−2.86**	21.4	16.5	1.19
Painting club	35.4	55.1	−4.78***	22.7	42.9	−4.10***
Number of schools	206	468		154	212	

[1]Central districts include Tel Aviv, Haifa, Jerusalem and Center. Peripheral districts are North and South.
$*p < 0.05$; $**p < 0.01$; $***p < 0.001$.

programs, sponsored by the Ministry, more than both non-dependent schools in their region and the two types of schools in the center.

The differential effects of government dependence on the periphery versus the center are less clear-cut regarding the extra-curricular activities. Indeed, the tendency to use the "Cultural Basket" program is the highest among the participant schools in the periphery. So is their use of centrally organized field trips, but the difference between them and the non-dependent peripheral schools, though statistically significant, is lower than the comparable difference in the central region, simply because of the higher tendency of all peripheral schools to use the educational welfare program. For similar reasons, the differences between participant and non-participant schools of the central region with respect to student centers, club activities, and independent participation in cultural activities, are higher than those among dependent and non-dependent schools in the periphery. It thus appears that the dependence on initiatives of the central government is especially reflected in the curricular activities of the peripheral schools.

Prospects and Problems of Implementing School Autonomy and Controlled Choice in the Peripheral Communities

Our analysis demonstrates three basic facts about primary schools in the peripheral districts of Israel. First, these schools are different from those located in the central regions. They tend toward traditional and strict treatment of their students, grouping them by ability while not supplying them with enough counseling services. They use traditional curricular practices and avoid innovative programs requiring initiative and involvement of school staff. They also lack extra-curricular enrichment programs widely practised in the center, unless such programs are sponsored and financed by the central government. Secondly, it is their peripheriality that determines their features rather than other related factors such as school size, sector, and student composition. Third, and not least in importance, we have shown that those peripheral schools dependent on government programs demonstrate their peripheral characteristics more extremely, especially with respect to the curricular activities.

Putting these findings together, we might say that schools in the periphery cater much less to student needs. They tend to be traditional, dull, and in general provide less interesting educational experiences than the central schools. Given these features, it is not surprising that their students, time and again, tend to fail more frequently in national standardized achievement tests. The educational disadvantages of the primary schools are detrimental to the future educational career of their students. Only 42 percent of high school students in the Northern and Southern districts study in the academic matriculation track, versus 56 percent of the high school students in the center. And the percentage of high school graduates obtaining the matriculation diploma in the peripheral districts is 34 versus close to a half of their counterparts in the central region (Israeli Ministry of Education and Culture 1993, pp. 62–63)[9].

What, then, should be done in order to change the schools in the periphery? Government policies during the 1980s advocated large educational investments in the periphery through centralized compensatory programs, such as the neighborhood rehabilitation and educational welfare programs. These policies coincided with popular notions of the 1980s regarding gap reduction in education between political centers and remote peripheries (Carron and Chau 1980; Chau 1985). However, our analysis clearly demonstrates that, at least in the Israeli case, making the periphery more dependent on the central government for educational initiatives turns the dependent schools into worse entities than peripheral schools which do not become dependent on the political center. The answer therefore does not lie in increasing the dependence of the periphery through compensatory programs. Now, almost two decades later, recommendations should follow the opposite direction: inducing change within the peripheral schools by increasing school autonomy and complementing it with controlled school choice programs. In general, it seems that only such strategies may improve the ways peripheral schools cater to student needs and turn into a place of interest for individual students.

However, the problems of implementing such strategies in the relatively remote periphery should not go unnoticed. First of all, it is clear that the strategies of school autonomy and controlled choice could produce the desired results only by implementation of the community-based model, and not as a comprehensive periphery-wide strategy. As demonstrated by the state-wide school choice policy in the United States, remote schools only suffer and become worse as a result of widely open choice and autonomy programs. They simply cannot compete for students and budgets, and yet cannot be closed down. The community-based implementation of school autonomy and controlled choice in the periphery however raises some crucial problems, which may be indicated by the Israeli case of the Northern and Southern districts.

First of all, the local communities in the peripheral districts vary by type and size to a larger extent than central communities do. As mentioned, they consist of mid-sized veteran cities and towns, smaller development towns, and a variety of rural settlements differing in ideology, social composition, and economic prosperity. Thus, we face problematic variations in the ability to motivate local communities to participate in innovative school reforms, as well as some factual problems regarding the ability of some communities to actually implement such reforms. For example, the introduction of controlled school choice requires, prior to other things, the existence of several schools within the same community. In many of the development towns or the rural settlements in the peripheral districts there is only one primary or secondary school or, as quite often is the case, two public schools belonging to the two separate sectors, the secular and the religious. Therefore, in order to induce school autonomy and controlled choice one must first create local educational authorities in specific limited regions of the periphery on the basis of cooperation among several communities. This requires not only voluntary cooperation among local communities, but also a determined initiative and efforts by the Ministry, as well as the allocation of appropriate funds for relevant expenses, such as school bussing.

But above and beyond these practical problems, there are questions related to some general characteristics of the educational services in the geographic periphery. The induction of school autonomy requires an educated corps of teachers who are capable of learning and implementing complex pedagogical principles. One crucial problem of the Northern and Southern districts in Israel has been the shortage of teacher supply, resulting in the employment of a less qualified, sometimes even unqualified, teaching force. This is a well known problem of developing countries (International Labor Organization 1996), which is sometimes ignored in industrialized states despite its prevalence in remote peripheral areas. A comparison of primary school teachers in the peripheral and central districts of the Israeli system reveals that the percentage of academically educated teachers in the periphery is 15 versus 25 to 30 percent in Jerusalem or Tel Aviv. Unqualified teachers constitute about 10 percent of the teaching force in the periphery versus 5–8 percent of primary school teachers in the central districts (Israeli Ministry of Education and Culture 1993, p. 64). It is therefore obvious that any attempt to introduce school autonomy in the periphery is pre-conditioned by an improvement of the teaching corps. This again requires concentrated efforts on the part of the Ministry of Education, as well as the relevant local communities, and the allocation of appropriate funds for teacher attraction and recruitment.

It therefore seems, in conclusion, that the decentralization of educational authority and the empowerment of local schools in the geographic periphery requires a well coordinated cooperation between central authorities and interested local communities. However, in contrast to the ongoing patterns of dependence relations between periphery and center, such coordinated efforts should concentrate on elevating dependency and inducing school autonomy by empowering the local peripheral communities.

Notes

1 These equivalence criteria served, for instance, a recent claim submitted to the Supreme Court of Justice against the school choice program of Tel Aviv. The claim for the abolishment of the municipal choice program rests on several grounds, among them the "non-equivalence" of the schools opened up for choice. The Court has rejected this claim.

2 Similar information was collected in 1996, but the data are not yet available.

3 I would like to thank Mrs. Ronit Hanegbi for her help in data analysis for this paper. We included in the analysis only those activities conducted in 25–75 percent of the schools. Also excluded were services usually provided on municipal rather than school basis, such as psychological and health services, and teaching programs and methods which, according to response distribution, were not clearly defined. For example, it was unclear if principals regarded the activity of "educational or achievement assessment" in specific or general terms.

4 This finding coincides with earlier studies showing that the Southern district in Israel, from various welfare perspectives, may be considered more peripheral than the North (e.g., Efrat 1984, pp. 35–43).

5 The control variables were measured on school level rather than community level for two reasons: our units of analysis are schools rather than local communities, and in the

central urban districts there are little or no community variations.

6 An attempt to include all 30 activities in the discriminant analysis proved unsuccessful due to specific inter-regional variations and to the mere dichotomous coding of activities' existence in school. The dichotomous coding lowered potential correlations among school activities.

7 The canonical correlation is between all 12 school activities put together and a group of five dummy variables of the school districts. Wilks' Lambda (ranging from 0 to 1) represents the discriminant power of the function. The lower Wilks' Lambda is, the better the discrimination among districts according to the function.

8 It is quite possible that the peripheral schools have financially benefited more than the central ones from their participation in the government programs. Our data, however, refers only to the fact of program participation, and in this respect there is no significant difference in school distribution between the two regions.

9 A recent evaluation study of an intervention project in Northern and Southern high schools (Lewy *et al.* 1996) further illuminates the picture regarding the dependence of peripheral schools on the center. The study evaluated the effects of fostering the academic achievements of high school students in Mathematics and English (the two most problematic high school subjects) through the intervention of central agencies and universities. It found that the intervention has had no actual effect on the matriculation achievements of the students in the peripheral districts, and recommended giving more autonomy to individual schools in the periphery in deciding on their mode of involvement in such fostering projects.

References

Adler, C. and Sever, R. (1988). "Educational fostering in Israel — goals, strategies and target-groups". *Megamot*, **31**, 5–15 (in Hebrew).

Ayalon, H., Ben-Rafael, E. and Yogev, A. (1993). *Community in Transition: Mobility, Integration and Conflict.* Westport, CT: Greenwood.

Ball, S. J. (1993). "Education markets, choice and social class: The market as a class strategy in the UK and the USA". *British Journal of Sociology of Education*, **14**, 3–20.

Carl, J. (1994). "Parental choice as national policy in England and the United States". *Comparative Education Review*, **38**, 294–322.

Carmon, N. (1989). *Neighborhood Rehabilitation in Israel — Evaluation of Outcomes.* Haifa: Neaman Institute, the Technion (in Hebrew).

Carron, G. and Chau, T. N. (Eds) (1980a). *Regional Disparities in Educational Development. — A Controversial Issue.* Paris: UNESCO.

Carron, G. and Chau, T. N. (Eds) (1980b). *Regional Disparities in Educational Development: Diagnosis and Policies for Reduction.* Paris: UNESCO.

Chau, T. N. (1985). "Regional disparities and educational planning". In: T. Husen and T. N. Postlethwaite (Eds), *International Encyclopedia of Education*, pp. 4241–4244. Oxford: Pergamon.

Chubb, J. E. and Moe, T. M. (1990). *Politics, Markets, and America's Schools.* Washington, DC: The Brookings Institution.

Chubb, J. E. and Moe, T. M. (1992). *A Lesson in School Reform from Great Britain.* Washington, DC: The Brookings Institution.

Cookson, P. W., Jr. (1994). *School Choice: The Struggle for the Soul of American Education.* New Haven and London: Yale University Press.

Efrat, E. (1983). *Geography of Welfare, Social Gap and Inequality.* Tel Aviv: Achiasaf (in Hebrew).

Efrat, E. (1984). *Geography and Politics in Israel.* Tel Aviv: Achiasaf (in Hebrew).

Friedman, I. A. (Ed.) (1994). *Parental Choice of Schools. Megamot* special issue 2–3 (in Hebrew).

International Labor Organization (1996). *Impact of Structural Adjustment on the Employment and Training of Teachers.* Geneva: JMEP/1996/II.

Israeli Central Bureau of Statistics (1991). *Survey of Education and Welfare in Primary Schools 1988/89*. Jerusalem: Series of Education and Culture Statistics 194.

Israeli Ministry of Education and Culture (1993). *The Education System in the Mirror of Numbers 1993*. Jerusalem: Administration of Economics and Budgets, Department of Economics and Statistics (in Hebrew).

Israeli Ministry of Education and Culture (1994). *Report of the Commission for School Choice*. Jerusalem: Pedagogical Secretariat (in Hebrew).

Lewy, A., Libman, Z., Zelikovitz, G., and Levy-Keren, M. (1996). *Summative Evaluation of an Intervention Project in Twenty-Four High Schools*. Jerusalem and Tel Aviv: The Jewish Agency for Israel and Tel Aviv University (in Hebrew).

Lipshitz, G. (1984). "Regional differences in the development of economic welfare in Israel". *Rivon Lekakala*, **123**, 365–380 (in Hebrew).

Power, S., Halpin, D., and Fitz, J. (1994). "Underpinning choice and diversity? The grant-maintained schools policy in context". In S. Tomlinson (Ed.), *Educational Reform and its Consequences*, pp. 26–40. London: IPPR/Rivers Oram Press.

Reshef, S. (1984). *School Autonomy: Significance and Implementation*. Tel Aviv University, School of Education, mimeo (in Hebrew).

Shachar, A. and Lipshitz, G. (1980). "Regional inequalities in Israel". *Environment and Planning*, **13**, 463–473.

Shapira, R., Goldring, E., Haymann, F., and Shavit, R. (1991). *School Autonomy in Joined Catchment Areas — Controlled School Choice in Israel*. Tel Aviv University: School of Education, Research Unit on Sociology of Education and the Community, mimeo (in Hebrew).

Shapira, R., Green, R., and Danilov, Y. (Eds) (1994). *School Autonomy — the Implementation and its Lesson*. Jerusalem: Ministry of Education and Culture, Pedagogical Secretariat (in Hebrew).

Shapira, R., Haymann, F., and Shavit, R. (1995). "Autonomy as ethos, content as commodity: An Israeli model for controlled choice of autonomous schools". In R. Kahane (Ed.),

Educational Advancement and Distributive Justice: Between Equality and Equity, pp. 358–374. Jerusalem: Magnes.

Slater, D. (1980). *Underdevelopment and Spatial Inequality*. Oxford: Pergamon.

Swirski, S. (1990). *Education in Israel: Schooling for Inequality*. Tel Aviv: Breirot (in Hebrew).

Tel Aviv-Jaffa Municipality (1992). "*It is Also Possible Otherwise — Tel Aviv-Jaffa Believes in Education*": *Revival of the Education System in the City of Tel Aviv-Jaffa*. Tel Aviv: Municipal Administration of Education and Culture, mimeo (in Hebrew).

The Carnegie Foundation for the Advancement of Teaching (1992). *School Choice: A Special Report*. Princeton, NJ: Author.

Wells, A. S. (1993). *Time to Choose: America at the Crossroads of School Choice Policy*. New York: Hill and Wang.

Yogev, A. (1989a). "Educational policy in Israel for the advancement of disadvantaged groups". In *Planning Educational Policy: Position Papers and Decisions of the Permanent Committee of the Pedagogical Secretariat*, pp. 175–205. Jerusalem: Ministry of Education and Culture, Pedagogical Secretariat (in Hebrew).

Yogev, A. (1989b). "From school reform to ethnic integration in Israeli schools: Social myths and educational policy". In A. Yogev and S. Tomlinson (Eds), *International Perspectives on Education and Society*, vol. 1, pp. 63–79. Greenwich, CT.: JAI Press.

Yogev, A. and H. Ayalon. (1996). "Between policy and research considerations: Constructing a welfare index for Arab schools in Israel". *Israel Social Science Research*, **11**, 115–142.

Yogev, A. and M. Chen. (1985). "Sponsorship as school charter: Educational mobility in religious versus secular schools in Israel". *Intemational Review of Modern Sociology*, **15**, 117–130.

Chapter Nine

POLITICS, MARKETS AND EQUALITY IN SCHOOLS

JOHN E. CHUBB* AND TERRY M. MOE**
The Brookings Institution, The Edison Project
***Stanford University, USA*

The United States faces two major education problems. The first is that the average level of academic achievement of American children is appallingly mediocre. The second is that the distribution of achievement is highly unequal, varying substantially along class and racial lines. According to the National Assessment of Educational Progress, most American children leave high school having mastered only junior high school material. However, some children finish high school having mastered considerably more than this while others never complete high school or learn alarmingly less. The SAT scores of college-bound blacks, for example, trail the scores of college-bound whites by a total of nearly 200 points. American young people are achieving less than their counterparts virtually everywhere in the industrial world, making it hard for the United States to compete. As many as a fourth of American children are achieving so lime they will struggle even to scrape by.

Both of these problems — call them problems of efficiency and equality — deserve the country's most serious attention.

And in a recent book (Chubb & Moe 1990), we proposed an ambitious solution to both of them. We outlined a new system of public education based on market principles and implemented through a mechanism that is now usually called school choice. In essence, parents would be entitled to send their children to any school of their choice, at public expense, and schools would be encouraged to improve through competition with other schools for enrollment.

The reasoning behind this proposal was derived from our analysis of public and private education. Public schools are failing, we argued, because of the institutions that govern them. Driven by politics, institutions of direct democratic control — school boards and district offices, state legislatures, the US Congress, and their respective education departments — work to overwhelm public schools with bureaucracy and rob public schools of the autonomy that is the foundation for effective organization and performance. Market institutions, which control private schools, do the opposite. Bureaucracy has little place in a competitive environment that places a premium on responding to family needs and organising to produce superior educational results. If the United States wants different — and better — public schools, it will need to control them with different institutions. The institutions that control public schools today, directly and from the top down, should have their authority severely limited, and new institutions that would control schools indirectly and from the bottom up — market institutions of choice and competition — should operate in their place.

The benefits of institutional reform, we argued, would include not only greater effectiveness or efficiency but greater educational equality as well. Urban schools, attended disproportionately by politics and bureaucracy, are the schools most likely to change in an environment of school autonomy and parental choice. Flush with resources once consumed by oversized district offices and powers once exercised by unknowledgeable and distant authorities, urban schools should find their opportunities for innovation especially enhanced. Because of the density and diversity of city populations, urban families should find their selection of schools especially wide. Poor families would also gain effective access to a right — the right to choose their children's schools — that more affluent families have always exercised through residential decisions (or by opting for private schools), but that poor families have never been able to afford to use.

Still, we argued, equality requires more support than markets alone can provide. We therefore recommended an institutional structure in which market forces are carefully regulated. Our proposal, after all, was for a **public** system of educational choice — fully funded by the government and operated according to ground rules established by the government. Among them: Parents may not top off their public scholarship with private funds and thereby choose a more expensive education. Schools may not completely control their admissions, as every school is required to accept some students it would not admit voluntarily. And all parents, but especially poor parents, are to be assisted by public authorities in finding, evaluating, and applying to schools. We proposed, we thought, a system that would revolutionize public education and increase educational opportunity to all.

To judge from what our critics have said, however, our thinking was at least half wrong. Yes, a system of choice might reduce bureaucracy, enhance school autonomy, and strengthen the role of parents. It might even stimulate enough reorganization, innovation, and rededication within schools to nudge the average level of achievement across the nation upward. But whatever it might do in these regards, it would certainly do so at too high a price, for it would certainly exacerbate inequality.

In a system of school choice, critics argue, some parents would inevitably be better choosers than others. Schools would inevitably prefer to admit the easy-to-educate over children who misbehave or have learning disabilities. Children with advantages, of parenting or of nature, would fill up the good schools while disadvantaged children would be left behind in bad schools made even worse by the departure of the ablest students. White families would, of course, flee black and Hispanic families, and some black families would choose Afrocentric schools with dubious academic curricula. The rich would get richer, the poor would get poorer and the shameful days of school segregation would return.

Of course, we disagree with this assessment. But in fairness to our critics, we must admit that our book was principally about efficiency or effectiveness, about the roots of poor performance in schools. We were concerned with issues of equality too, and we tried to address them adequately. Still, our analysis focused on efficiency. It traced the effect of school on children — all children — to school organization, to the administrative environment that shapes the organization, and finally, to the institutions that determine how schools will be administered or controlled. Our empirical analysis did not examine in any depth or detail the distribution of student achievement or the causes of inequality. Perhaps most importantly, our analysis did not compare educational equality under political and market conditions.

Now we would like to do just that. Specifically, we want to understand better what schools and school systems do to ameliorate inequality or to bring it about, and we want to consider whether political or market institutions are likely to differ on these scores. To help us, we have been analyzing a new longitudinal study of schools, students, and families — The National Educational Longitudinal Study, 1988 (or NELS: 88) — which follows a national sample of approximately 25,000 students in roughly 1,000 schools from eighth grade in 1988 through tenth grade in 1990. The study uses many of the same survey items as **High School and Beyond** and the **Administrator and Teacher Survey**, which provided the data for **Politics, Markets and America's Schools**.

NELS: 88 is therefore especially appropriate for this follow-up research.

Here we want to summarize in a nontechnical way the results of our first cut into these new data and our first effort to think comprehensively about educational equality. (The technical results are available on request.) We divide our discussion into two logical parts: equality within schools and equality between schools. When you get right down to it, educational inequality has two basic sources in schooling — the different treatment of children within schools and the different treatment of children by different schools. Either way, if the treatment differs in quality, educational inequality will be produced. Political and market institutions, we find, are not similarly prone to produce such differences. Suffice it to say that this analysis finds little evidence that political institutions are better at promoting educational equality than market institutions are. Indeed, the evidence consistently goes the other way.

Some Equality Standards

Before we consider how alternative institutions affect educational equality, we need to be clear what we mean by equality. We need to establish some equality standards against which institutions can be judged. Of course, this is no straightforward task. Equality can be defined in many different ways, and alternative conceptions of equality are regularly debated by democratic theorists, economists, and politicians — to name only some of the main contestants. We shall not venture any comprehensive definition of this complex concept but rather a set of standards that, taken together, form a loose hierarchy that builds from weaker forms of equality to stronger.

At the base of this hierarchy is the standard of **non-discrimination**: Students should not be denied access to schools or programs within schools because of their race, ethnic background, sex, or economic class. This does not mean that all schools and programs will have representative distributions of blacks and whites, rich and poor, and so forth. This is not a standard of affirmative action. The standard simply says that race, sex, and class shall not be explicit or — more likely — implicit criteria for access to education. For example, a black child might still be excluded from an accelerated math program because his or her test scores are low, but would not be excluded because, regardless of test scores, schools believe black children do not excel at math: The latter would constitute discrimination. In the analysis that follows, we focus on race and class discrimination because these are the lines along which American society is most sharply and permanently divided educationally and, in some consequence, politically and economically. Of course, racial discrimination is also unconstitutional in all public schools and in all private schools that want tax-exempt status — which means all but a handful.

Next up the hierarchy is the standard of **separate but equal**. In a nondiscriminating world, children of different racial, ethnic, and economic backgrounds are nevertheless going to end up in different schools and programs. Children attend public schools in their neighborhoods and private schools of their choosing; they are not assigned to schools randomly. Within schools, children choose different classes and programs. In addition, some qualify for gifted and talented programs, some for remedial programs, and some for nothing special at all. If children from different backgrounds are to have the chance to make equal educational gains, the disparate schools, programs, and classes in which they find themselves must provide equally good educational experiences. This means that the heavily minority urban school and the predominantly white suburban school must be places of similar quality, by whatever measures are relevant. It also means that the gifted and talented program, overpopulated with rich students, and the remedial program, home to so many poor students, must be equally well-suited to helping their respective groups of children work up to their potential.

A tougher standard than separate but equal is what we call **equal opportunity**. If we envision a society in which children from different racial and economic backgrounds do not differ radically in high school and college graduation rates or in academic achievement, it may be inadequate merely to accept well-qualified black children in college prep programs — for their test scores are too often too low — or to ensure that the remedial classes poor children take are truly excellent — for they will never be as good as the advanced placement classes rich children take. If we are serious about equalizing educational outcomes, it may be necessary to expand the availability of the best educational opportunities and to ensure maximum feasible participation in them. This means that more children, of all backgrounds, will take high level classes and follow the college track through high school. It also means that students of differing backgrounds — racial, economic, **and** academic — will attend classes together. Schools will be less likely to segregate student by ability, since the opportunities are inherently limited for those placed in slow groups. Schools will be more likely to operate heterogeneous classes where swifter and slower children work side by side or even together. This arrangement not only expands educational opportunity, it increases the interaction of children from different backgrounds. It satisfies a standard that might also be called equal, not separate.

Satisfying even a tough opportunity standard, however, does not guarantee any reduction in the education gap that divides American society. The next standard begins to do this. Following Amy Gutmann (1987), we call it the **quality threshold**. This standard of equality says that schools and school systems will ensure that the education of no child falls below a threshold necessary to participate effectively as a citizen. What this means precisely is open to debate. But what it means generally is that every child will leave school able to read, write, calculate solve moderately complex problems, think about public affairs, and hold a job sufficient to avoid public dependence. It means a good high school education. This standard does not mean that there will not be huge gaps in ultimate educational attainment; some students will inevitably want more education than others — a PhD, an MD, a JD — and not all students will be equally able at academic subjects. But this standard is about more than opportunity. It judges schools by how its weakest or most disadvantaged students actually turn out.

The final and toughest standard is also about outcomes, but about outcomes for all. We call it **equal outcomes**. This standard says that if we are truly concerned about the disparity in academic achievement that follows class and racial lines, we ought to evaluate schools and school systems by the job they do in reducing this disparity. Take two schools that begin with similar distributions of children — equal numbers of slow, average, and bright children in each school. Then look at the distribution of academic achievement within each school when the children graduate. The school that is doing the better job for educational equality is the one with the tighter distribution, with the narrower gap between bright and slow. That school is providing more equal outcomes.

This is not to say that that school is providing the kind of education that is best for solving America's education problem. It is important to bear in mind that the problem is twofold, a problem of equality and of efficiency. A school that provides equal outcomes could conceivably do so by holding down the achievement of the brightest students. Its distribution of achievement might then be narrow, but its average level of achievement might be quite low. The same problem might occur if a school placed too many poorly-prepared children in an advanced classroom; well-prepared students might be held back. In the analysis that follows we will be asking in various ways how schools measure up to our five standards of equality, but we will also be paying attention to their average level of performance — their effectiveness or efficiency.

Equality within Schools

Critics of school choice have focused most of their attention on the disparities a marketplace might create from one school to the next. Yet critics of educational inequality more generally have focused as much or more attention on the disparities that exist within individual schools[1]. A school can create or exacerbate inequality in many ways. It can discriminate against children of certain races or ethnic backgrounds in assigning children to instructional groups or curriculum tracks. It can assign children to classes in a nondiscriminatory fashion — based on test scores, for example — yet provide instruction of widely varying quality to different tracks or groups. Even if organized objectively, groups or tracks may separate children of different class or racial backgrounds so thoroughly that a school can become effectively segregated internally. In addition, a school may so limit the size of its high group or college track, that all but the brightest children are denied educational opportunity.

Grouping and tracking are widespread practices in American education, so it is only reasonable to ask whether they contribute to the wide variation in student performance. As critics of these practices often point out, there is more variation in student achievement within American schools than there is on average between schools. It is no wonder, then, that the study of educational inequality has produced more research on grouping and tracking than on disparities between schools[2]. We, too, think that group and tracking practices deserve careful attention. But we think the existing research on the subject is deficient in one major respect. It has scarcely considered the roots of the practices to which so many critics strongly object. Could it be that the patterns of grouping and tracking found in public schools today are rooted in the institutional structure of which the schools are a part? Could it be that grouping and tracking, like other qualities of school organization that we have studied before, are institutionally determined?

That is our suspicion. But existing research provides virtually no clues as to whether we are right. The problem is that research into the effects of tracking and grouping looks almost exclusively at public schools, which are all governed by institutions of the same basic form. If we want to know whether these important organizational practices are influenced by how schools are governed or controlled, we must also consider how schools that are controlled differently deal with student diversity. From an institutional standpoint, the schools that differ most from public schools are private schools. Despite their enormous diversity — from secular independent schools connected to no larger entity to religious schools that are sometimes part of sizable systems — private schools differ from public schools fundamentally. Private schools are not controlled directly, from the top down, by democratic authorities; they are controlled indirectly, from the bottom up, by market forces.

In the analysis that follows, we compare the grouping and tracking practices of public and private schools. Such a comparison cannot tell us everything we might need to know to predict how a system of school choice would affect inequality within schools. Private education today serves only 12 percent of America's children and provides only a quarter of the country's schools. If all children and schools participated in a public educational marketplace, matters might be somewhat different than they are in private education now. Nevertheless, the best evidence we can possibly have of how schools would manage student diversity in a marketplace is to compare schools that are doing so now to schools that are managing subject to political control.

Diversity within schools

Generally speaking, schools are more likely to separate children into ability groups or curriculum tracks if the children in the school differ substantially in their prior achievement or capacity to learn. Schools run a greater risk of reinforcing or worsening inequality if their students are unequal to begin with. It is generally easier for schools to integrate children if the differences among them are not enormous.

Public schools might be expected to face more student diversity than private schools do because public schools typically enroll all children in a neighborhood, regardless of background, while private schools enroll children interested in education of a particular type. This expectation must be tempered, however, by the realization that choice operates in the public sector, too. Families base their residential choices on, among other things, the quality of neighborhood public schools and the cost of housing. This choice process sorts children into public schools by economic class and educational taste, much as an analogous choice process sorts children into private schools. It is an empirical question, then, whether public schools or private schools face more internal diversity.

In our analysis we measured the diversity that typical public schools and private schools must manage[3]. We calculated the average standard deviation of public and private eighth graders on a number of dimensions of educational inequality. What we found is that the average public school does face more diversity than the average private school, but not dramatically more. On average, student bodies are 10–15 percent less diverse in private schools than in public schools. This means, for example, that in a public school, a lower-class student is slightly more likely to be around a middle-class student, or that a student living with two married parents is somewhat more likely to interact with a student living with a single mother. The biggest difference between public and private schools is in parent income, where there is 25 percent less diversity in private schools, presumably because nearly every parent in the private school must be able to afford the same tuition — high, low, or in between. A publicly funded system of school choice might well reduce this source of homogeneity because ability-to-pay would not limit access to schools.

Interestingly, the dimension along which there is nearly the least difference between public and private schools is academic achievement. Students were given standardized tests in four subjects: reading, mathematics, history, and science. In our analysis we considered the average standard deviations within schools of math scores — because math is where ability grouping most often takes pace — and combined scores. Basically, private schools are only slightly more homogeneous academically than public schools. Pending further analysis, we cannot say whether private schools have less internal inequality in student achievement because they admit slightly more homogeneous student bodies or because they organize themselves better to reduce academic inequality. The important point at this juncture is that private schools, by virtue of the control they exercise over admission, do not face substantially less diversity than public schools face. Residential sorting in the public sector is evidently rather powerful because public and private schools have similar student bodies, as measured by a number of gauges of diversity.

This similarity can be seen in yet another way. We examined the distribution of public and private schools are almost completely white — no more than 2 percent are black. There are more private schools than public schools in this category, to be sure. But when we consider schools that are more racially integrated — say, enrolling 3 percent black or more — public and private schools are similarly distributed. Public–private differences in racial diversity, like differences in other forms of school diversity, are simply not very large.

Ability grouping

The primary way in which schools currently manage student diversity is to group children homogeneously by ability. This practice is said to facilitate instruction, as teachers can concentrate on material at one level of difficulty without worrying that the material is too hard for some or too unchallenging for others. Homogeneous grouping is also criticized for locking children into achievement tracks that diverge unnecessarily. Children in the slower groups in elementary school are said never to catch up with the children in the higher groups because the instruction in the slower groups is simply not as ambitious or effective as that in the higher groups. Children in the slower groups are said to feel stigmatized by their placement. Rarely, critics say, do these children ever escape their low groups and enter a college prep track in high school[4].

Slower children are far better off, critics of grouping argue, in heterogeneous classrooms and ability groups where they work side by side with swifter children on a common curriculum, only at a slightly slower pace. Instructional techniques such as "cooperative learning" are supposed to make it possible for slower children to succeed in heterogeneous groups and for swifter children to progress just as fast as they would in a homogeneous, accelerated group. Research on these claims is still far from definitive, but it certainly suggests that slower children are harmed by homogeneous grouping and that swifter children are served as well by heterogeneous groups as by homogeneous ones (see e.g., Slavin 1990)[5]. On grounds of equality certainly, and perhaps efficiency, it appears that heterogeneous grouping is a superior instructional strategy.

Research to the contrary notwithstanding, homogeneous grouping is still the preferred practice of American schools. In our analysis we estimated the percentage of eighth graders grouped homogeneously for mathematics. In public as well as private schools, the overwhelming majority of students are taught in homogeneous ability groups. Most schools separate students into groups of slow, medium, and swift learners and risk the inequality in learning that may result thereby. There is a sharp difference, however, between public and private schools in the use of grouping. Private schools place 15 percentage points more students in heterogeneous groups than public schools do. About 30 percent of all private school students take math in heterogeneous groups; only 15 percent of all public school students do.

This difference also appears to operate independent of objective reasons for grouping. Our analysis divided students into schools that are above and below the population median in socioeconomic status, a strong predictor of academic success. Presumably, schools will separate children into homogeneous groups less often as their student bodies decline in diversity: The more similar students are, the less the need to separate them. To a slight extent, this turns out to be true. But basically, public and private schools exhibit different propensities to group heterogeneously, regardless of student diversity. Public schools almost invariably sort and separate children into homogeneous ability groups; private schools are more willing to keep different children together.

Is this really a basic difference between public and private schools? Or does the difference have other explanations? To try to find out, we considered a number of alternative explanations simultaneously. We estimated a probit model designed to predict a student's math group — that is, whether the group is homogeneous or heterogeneous. What we found is that the kind of group an eighth grader is in has many "causes", or at least is influenced in many ways. For example, a student is somewhat more likely to be placed in a homogeneous group (presumably a high one) if he or she comes from a family higher in socioeconomic status, has strong prior math grades, and is Asian. There is very slight evidence that black students are less likely to be grouped homogeneously.

But ultimately there are three strong influences on a student's group. First is the socioeconomic status of the school. Schools serving more affluent children are more likely to group homogeneously, presumably to provide high level groups. Second, schools that are internally diverse, with wide variations in math achievement, are more likely to group homogeneously — though this may be an effect as well as a cause. Finally, all things being equal, a child is much less likely to be grouped homogeneously — to be separated from children of different ability — in a private school than in a public school. For the average child the probability of being grouped homogeneously for math is 0.829 in public school but only 0.556 in private school.

Of course, private schools may be less able to group homogeneously because they are smaller and do not have enough students to offer more specialized classes. In other words, private schools might group heterogeneously if they were larger. Let us ignore for now the question of why private schools are smaller — for perhaps they want to be small so they can avoid internal specialization — and just consider how public and private schools of similar-size group children. We re-estimated the probit model, adding a control for eighth grade enrollment. What we found is that size matters, but that public and private schools still differ greatly. Allowing for school size, an average student has a probability of 0.875 of being grouped homogeneously in public school but only 0.695 of being grouped homogeneously in private school.

Why do private schools show a greater willingness to keep children of different backgrounds together? We think there are two reasons, both rooted in basic differences between political and market institutions. The first has to do with the very different structure of top-down control in the two institutions. Direct democratic control, as it is now structured, promotes the growth of bureaucracy around public schools. As we have argued, this bureaucracy generally tends to rob public schools of the autonomy that they need to exercise professional discretion and to organize themselves as effective educational teams. This bureaucracy also tends to do something very particular that affects how public schools deal with diversity. It prescribes how public schools deal with diversity. Public schools are required by law — federal, state, and local — to run various gifted and talented programs, enrichment programs, remedial programs, vocational programs, and numerous other programs. They are also required to utilize objective criteria — usually test scores — in assigning children to programs. A system of externally imposed rules determines how children will be organized and served.

There is more to this system, however, than the bureaucracy that implements it. The bureaucracy is not the source of the rules that limit how schools respond to diversity. The source is democratic authority, exercised by boards of education, state legislatures, and the US Congress. As we have observed (Chubb & Moe 1990) this authority is exercised in a distinctive way, groups that vie for control of public authority insist on nailing down their political victories in detailed legislation that makes it difficult for subsequent victors to influence how legislation is implemented. Groups want to limit discretion in schools not only because they distrust the professionals that work there; they also want to guard against the uncertainty of what future political decisions may bring.

What this distinctive political process means for equality in public schools is that the public schools are likely to be required to provide different services to different students — even if everyone agrees this practice mainly accentuates differences. For example, there is broad agreement among educators that disadvantaged children are hurt more than helped by programs that pull them out of regular classrooms for remedial work. Yet the politics of "compensatory education" leads to demands for special treatment that can be easily monitored and documented — hence, veritable mandates for "pull-out" programs (Doyle & Cooper 1988). If schools were free to help disadvantaged children however they saw fit, what would guarantee that those children would actually get something extra? The politics of compensatory education is a 25-years saga of ever-tightening constraints. Although it is the classic case, it is not unrepresentative of the limitations that public schools face as they try to manage diversity. Public schools are basically required to treat different children differently.

Private schools are generally not subject to such requirements. They are managed by administrative structures that do not grow out of public authority. There is no politics of comparable consequence and there is no political uncertainty. The administrative structure exists to facilitate the voluntary relationships between individual schools and families — which is to say it scarcely exists at all. On most matters, including the management of student diversity, private schools are far freer than public schools to organize as they see fit. Private schools may treat different students differently or they may try to treat all students the same.

How private schools manage diversity will depend more on demands from families than on the requirements of bureaucracy. Which brings us to a second important difference between educational markets and political institutions, as they are currently structured. In an educational market, parents shop for the school that they believe is right for their child. Placing the child in a particular school is the parents' way of getting "something special" for the child. In today's public system, parents do not have the opportunity to shop or choose to nearly the degree that parents do in the private market. Children are assigned to public schools, and at least within legal jurisdictions, there is not supposed to be anything special about any schools. The schools are supposed to be the same. If public school parents want "something special" for their child they must either work through the political process, which we just described, or they can apply pressure at the school site for special treatment. Public schools are likely to feel more direct pressure from parents than private schools do for the creation of ability groups and special programs, and the placement of particular children in them.

Overall, then, we believe public and private schools show different propensities to separate children of different backgrounds because these schools belong to different institutional systems. Direct democratic control encourages the development of bureaucratic classification systems; market control discourages this. Market control also defuses pressure at the school site for special treatment; political control increases it. Despite these differences, educators in both sectors tend to group children homogeneously. But there is a substantially greater willingness in private schools to keep different children together, to give all the same educational opportunity.

The operation of ability groups

Once schools decide to group children by ability — and most schools do — there are still major steps schools can take to promote equality. They can ensure nondiscrimination in group placement. They can extend the opportunity to participate in accelerated groups to the largest number of students possible. They can work to provide an equally high quality learning experience in groups at every level.

In our analysis we considered how public and private schools differ in all these regards. To begin, we found that private schools place more children — about nine percentage points more children — than public schools do in high math ability groups. Of course, private schools may have a higher percentage of very able students than public schools do. So we estimated a model of ability groups in which the many influences on grouping can be compared. The model is a probit of the placement of a child into one of two categories, the high math group or a lower math group. What the model ultimately revealed is that the chances of being placed in a high group, all else being equal, are noticeably higher in private schools than in public. An average child has a 0.480 chance of being placed in a high group in private school but only a 0.378 chance in public school.

The model also provided some interesting information about how group placement decisions are apparently made. In all schools, public and private, ability grouping depends mostly on individual test scores and prior grades. In both public and private schools, there is evidence of affirmative action on behalf of black students. All things being equal, a black student is somewhat more likely than a white student to be placed in a high math group. Why this occurs is unclear. It may be that schools are more willing to give black children the benefit of the doubt in close cases because their scores are generally low and they are therefore under-represented in high groups. Whatever the case, if we accept test scores and grades as valid placement criteria, there is no evidence of racial discrimination against black students in ability grouping.

Black and disadvantaged students enjoy another edge in high grouping as well. All things being equal, it is harder to get into a high group in a school where there is a lot of competition for spaces, in a school that is high on average in test scores or socioeconomic status. Minority and disadvantaged children tend to attend schools with less competition for high groups and thus, if qualified, are more likely than those in more competitive schools to get a high placement. This, and the lack of blatant discrimination, does not mean that disadvantaged children are grouped ambitiously to any great degree; they still dominate the low groups. Our observation is simply that matters could be significantly worse.

In eighth grade — or actually just prior to it — schools make an additional grouping decision, a decision that can prove far more important than the assignment of students to ability groups. At this juncture, schools decide which students will take algebra during eighth grade and which students will wait until ninth grade or longer. The importance of this decision is that eighth grade algebra tends to be a gatekeeper for the academic or college preparatory track in high school. Students who take algebra in eighth grade usually go on to take an accelerated sequence of mathematics classes through the 12th grade and to follow a curriculum aimed at attending a four-year college. In comparison, students who do not take algebra in eighth grade very often take less mathematics and less advanced mathematics down the road and follow a general or vocational curriculum. Of course, the mathematics courses and high school curriculum that students take may be largely a reflection of student preparation or ability. But schools do make course assignments and schools do have the option of ambitiously assigning students to challenging programs — or of letting students just slide by.

Public and private schools differ significantly in their willingness to place eighth graders in algebra. We found that 60 percent of private school students are placed in algebra while only 42 percent of public students are. We also found that this difference is a result of the schools and not of the students in them. We estimated a model of algebra enrollment that asked: Is the student taking algebra in eighth grade or not? The estimates indicate that an average child has a 0.586 chance of being placed in algebra by a private school but only a 0.401 probability in a public school. In all schools, placement is most powerfully influenced by math scores and socioeconomic status. As with ability groups, it is tougher, all things being equal, to get into algebra in an academically competitive school. And again, there is no evidence of racial discrimination. If anything, there is affirmative action (in both sectors) on behalf of blacks, Asians, and to a very slight extent, Hispanics. All of these influences notwithstanding, there is strong evidence that private schools provide a sizable boost to algebra-taking — and to enrollments in high math groups more generally.

Why might this be? One possibility is that in ways we simply have not measured, private schools students are better prepared for algebra and accelerated math than public school students are. We cannot rule this explanation out completely. However, we controlled for the most direct evidence that schools have of math preparedness — prior math grades and math scores on standardized tests. We also controlled for a host of student background characteristics that might influence course assignments. And we allowed for competition within schools. In light of this, it seems unlikely that the 10–20 point placement probability differences for algebra and high math groups are all due to student differences.

We think the differences more likely derive from the organization of public and private schools and the institutional contexts in which these schools operate. Public schools are organized to be all things to all people, to treat different children differently, and to meet the needs of many and varied constituents. This has come to mean schools that are internally differentiated with lots of programs, levels, and tracks. In such schools, it makes sense to offer different students different mathematics options in eighth grade — none better and none worse, all simply appropriate for different kinds of students. Private schools tend to be more specialized or focused. They have a vision of what their graduates should achieve and they try to do the same for all of their students. Private schools therefore tend to offer few electives and to keep all students on the same path, traditionally an academic one.

Public schools have differentiated organizations because they are subject to multiple layers of political control that impose bureaucratic complexity. Private schools tend to have simpler organizations because they lack political control and must instead meet the demands of clients. In education, the market test is easiest to meet with a coherent teamlike organization, not a loosely connected comprehensive one. Perhaps a private school could meet such a test by focusing on goals other than academic excellence — religion, for example — and placing nobody in algebra in eighth grade. But parents who are choosing schools generally seem to care about academic results. And a pretty good recipe for academic results is a rigorous academic curriculum. Academic opportunity therefore is more widely — and equally — extended in private schools than in public.

Now, it might still be that the differentiated approach to student organization is a sound and equitable one. It may be better for all students not to push slower learners into tough classes or to burden bright students with slower classmates. It may be better for all students if students are grouped by ability and are offered first rate programs appropriate to their different levels of preparation. Since both public and private schools engage in a good deal of ability grouping, it is worth looking closely at the quality of the learning experience in different groups.

In our analysis we looked at a number of indicators of educational quality contained in the student questionnaire. What these indicators showed is that, at least in the eyes of students, the educational environment is better in private schools than in public. On every indicator, a higher percentage of private students than of public students gave highly favorable responses. And on several questions bearing directly on the learning experience — for example, "the teaching is good" and "teachers are interested in students" — the differences seem substantial. Our point in considering these indicators is not, however, to say that the learning environment in private schools is generally better than in public schools. Such a conclusion would require a more sensitive analysis of these questions with special attention to the kinds of students providing the differing responses. Our only point at this juncture is to provide a context for considering how learning environments differ within schools in the public and private sector.

Using these indicators, we took a close look at differences within schools. We estimated regression models of each measure of the learning environment in public and in private schools. The models predicted an individual student's response to each measure as a function of the average response of other students in the student's school, the student's own socioeconomic status, and the student's ability group in mathematics. The models enable us to estimate whether children in high and low groups experience different learning conditions from children in middle groups, controlling for the average learning condition in each school and a proxy for the orientation of each child to learning — socioeconomic status. Estimating separate models for public and private schools enabled us to compare equality across sectors.

The comparative results, though not perfectly consistent, are plain. All else being equal, there are somewhat more disparities in educational conditions within public schools than within private schools. On most indictors, high group children experience significantly better conditions and low group children significantly worse conditions than middle group children in public schools. In private schools, such polarization occurs less frequently. When schools group by ability, private schools appear to do a better job than public schools of giving all children the same experience — and a comparatively good one.

Overall, it seems that private schools practice ability grouping more equitably than public schools do. All things being equal, private schools place more students in high ability groups and in eighth grade algebra. In our terms, private schools are more ambitious in promoting equal opportunity. Private schools also maintain more equal educational environments than public schools do for children in different ability groups. And private schools are especially superior in ensuring that low groups are not educated under inferior conditions. Private schools do a better job, in other words, of providing a threshold of quality below which no student falls and of guaranteeing that separate classes are also equal classes.

Why should this be? Again, we think that the institutions of public and private education provide important clues. As we proposed earlier, public schools are bound by administrative constraints to operate different programs and tracks.

They are not free to push all students into ambitious groups or to ignore objective indicators that would suggest a different or lower placement. Private schools are free to do whatever is likely to satisfy parents and to use their best professional judgment in placing students. Private schools, as we have seen, still judge that many students are not up to the challenge of accelerated programs. But private schools also know that the most ambitious placement feasible is the one that is likely to raise students to their full potential. Of course, getting children to realize their potential is an attractive strategy for satisfying parents. Public schools want to satisfy parents, too. But they must also satisfy higher authorities that mandate how different students must be served. The result, we think, is a higher rate of ambitious placements in private schools.

As for differences in educational experiences within schools, we think institutions also hold part of the answer. Private schools have a powerful incentive to provide the children of all of their families the same educational quality: All parents are paying the same tuition and expecting the same quality. If it is within their power to provide equal quality, private schools will strive to do so. Public schools undoubtedly would like to do this also; parents voting and paying taxes are equally entitled to quality. The problem, we think, is that public schools are not as free as private schools to provide equal quality. In particular, public schools are not free to assign teachers to any class or group they please; collective bargaining agreements often give teachers the right to decline assignments — to teach difficult children, for example — on the basis of seniority or to choose assignments, such as working with the gifted or talented. In general, public schools are not as free as private schools to shape their organizations or to determine the content of their educational programs. Another consequence of these limitations is that public schools cannot as easily as private schools guarantee equality to students, once students are separated.

High school tracking

Ability groups and class assignments in middle school are precursors of broader and more significant divisions in high school. After eighth grade, students are often organized into general programs of study or tracks that determine the amount and level of academic coursework that students take until graduation. These tracks — most often labeled academic, general, and vocational — prepare children for very different postgraduation experiences. Academic track students are prepared for four-year colleges or universities; students in other tracks are schooled for work or less academic forms of higher education, such as community colleges.

Tracking has been the target of enormous criticism because it is perceived to be discriminatory and detrimental to students in nonacademic tracks (Oakes 1985; Ball 1981; Persell 1977). Tracking has also been the subject of extensive research, especially by educational sociologists, who see a potentially strong link between tracking in schools and economic stratification in society (see, e.g., Gamoran & DeMare 1989; Alexander, Cook & McDill 1987; Heyns 1974; Alexander & Cook 1982). That research has plainly shown that racial minorities and poor children are over-represented in nonacademic tracks; however, it has not shown that that over-representation is the result of discrimination. Research has also documented wide gaps in achievement between students in different tracks, but it has not demonstrated that those gaps are caused or exacerbated by differences in educational quality between tracks. More troubling for tracking is the fact that little evidence supports its supposed benefits: Average achievement seems to be no higher in tracked schools than in untracked ones. Evidence on all of these points, however, is quite debatable. The argument over the merits of tracking therefore rages on.

Our interest in the tracking debate derives from the

possibility that schools in different institutional systems may have inherent tendencies to track differently. This is a vitally important consideration if we are concerned with improving schools — raising levels of educational achievement and reducing levels of inequality. If tracking, as it is practised in public schools today, turns out to be so inefficient, inequitable, or both, it is crucial to know why it is so practised and how it can be changed. If high school tracking practices, like middle school grouping practices, have deep institutional roots, it may prove very difficult to change tracking without changing institutions too.

It is well known that private school students are more likely to enroll in an academic track than public school students are. Among 10th graders in the **NELS: 88 First Follow-up Survey**, well over half of all private school students are taking a college prep program while only a third of public school students are. In our analysis we estimated a model that attempts to determine why. The model attempts to predict whether a student is in an academic track or not.

What the model shows, first, is that track placements work very similarly to ability group decisions and class assignments in middle school. The strongest influence on track placement, by far, is prior academic achievement — measured in this analysis by eighth grade tests in math, reading, science, and history. A student's socioeconomic status does have some impact, with higher status children favored. But there is no evidence of racial discrimination against minorities. In fact, there is strong evidence (in both sectors) of affirmative action on behalf of black youngsters. All else being equal, a black child is more likely than a white child to be placed in an academic program. As with middle schools, there is also evidence that competitive high schools, those high in average test scores and socioeconomic status, make it tougher for children to be placed in the top programs. This, too, may provide a boost for disadvantaged children who are usually not in competitive schools.

Once all of this is taken into consideration, however, a strong private school effect still seems to remain. A child who is average on every variable specified in this model would have only a 0.302 probability of being placed in an academic track in a public school but a 0.425 probability of being so placed in a private school. It is true that private school students may be more academically inclined than public school students are. But that greater inclination would have to manifest itself to private schools in subtle yet powerful ways, for the model allows for all of the indicators that schools are said to consider in placing students in tracks. It is therefore unlikely that the public–private difference in tracking probabilities is not due, in some large part, to differences between public and private schools. As we explained before, there are good reasons to expect private schools to be more ambitious in their grouping practices.

Of course, this difference does not mean that the private form of organization is superior on grounds of efficiency or equality. Private schools do extend the college prep opportunity to more students. In so doing, they could compromise the quality of their advanced courses. Public schools might help children just as well or better than private schools do by offering quality instruction, tailored to different levels of preparation, to students in different tracks. In the analysis we tried to explore these possibilities.

First, we compared educational conditions in academic and nonacademic track classes. The models that we employed are essentially the same as those that we used to compare conditions in eighth grade ability groups. With track models we found much the same pattern as in the grouping models but not quite as strong. In public schools, the learning environment is better in the academic track than in the nonacademic track. In private schools, the tracks are not as different. Public high schools appear not only to extend educational opportunity less widely than private high schools do; they appear to offer less equal quality to students who they separate.

The upshot of all this is that private schools appear to do a better job than public schools of providing equal educational outcomes. Private schools accomplish this, we think, by helping students at the bottom of the achievement distribution without hurting students at the top of the distribution. In our analysis we estimated an elaborate regression model of tenth grade combined math and reading scores[5]. The determinants of achievement include prior (eighth grade) math and reading achievement, and prior history and science achievement, which we included as a proxy for academic ability more generally. The model also includes a generous set of controls for individual background — socioeconomic status, race, sex, and the number of parents in the home. In addition, the model allows for regional variations and for differences between urban, suburban, and rural settings. Finally, the model controls for differences in eighth grade schooling — for students who switch between public and private schools at the start of high school.

All of the coefficients in the model behaved as expected. Achievement in tenth grade is best predicted by achievement in eighth grade. The coefficients of greatest interest to us are those distinguishing private schools from public, and academic track students from nonacademic. These coefficients reveal three important things. First, private schools provide an independent boost to student achievement relative to public schools, after all else is considered. Second, academic track coursework increases achievement beyond levels that would be expected from identical students doing nonacademic coursework. And third, students in the academic track end up closer to their nonacademic track schoolmates in the private sector than in the public sector.

Generally, we also find that academic track students do equally well, all else being the same, in public and private schools. In other words, by operating a more inclusive academic track, private schools do not compromise the education of their brightest students. Students in the more exclusive academic track in public schools perform no better. Since private schools are providing the highest academic opportunity to more students than public schools are, private schools are also boosting the achievement of more of their better students than public schools are.

In addition, students who take the nonacademic program fare much better in private schools than in public. As a result, nonacademic students progress at more nearly the same pace as academic students in private schools than in public schools. Over time, the achievement gap within private schools is likely to grow more slowly than it will in public schools. Given the differences in educational conditions that we observed between academic and nonacademic classes in public schools, these outcomes should come as no surprise. The educational experience is more uniform across students of different ability levels in private schools, and the outcomes of this experience are more uniform, too.

Are these differences large enough to be of interest or concern? That is difficult to say because we have only considered the effect of schools and tracks on student achievement over a very brief period of time, eighth grade to tenth grade. And standardized achievement tests, such as these, provide rather limited measures of what students actually learn (Educational Testing Service 1991). Generally, standardized test scores are difficult to raise, and they register substantial improvement only over a long period of time. We do not know for certain whether the effects on achievement that we observe in early high school also precede and follow these two years. But the school and track effects we observe in the nineth and 10th grades are similar to effects observed elsewhere for the 11th and 12th grades (see, e.g., Coleman & Hoffer 1987). In addition, the organizational differences between public and private high schools — in the ambitiousness of tracking and the educational conditions of different tracks — are virtually identical to the differences we observed in middle school. Further, it is well established that groups operating in middle school and high school originate and start to rigidify from the beginning of elementary school.

We therefore extrapolated the predicted differences that we observed in early high school to a 12-years school career. We acknowledge that such extrapolations are fraught with uncertainty, and we offer them merely for illustration. They only tell us how public and private sector experiences differ **if** the differences observed in early high school are repeated every two years from first grade through 12th. What they show is that, all else being equal, a private school student will ultimately achieve about one-third of a standard deviation higher than a public school student on standardized tests — a difference that would separate students by 10–15 percentiles in a standardized distribution. This is no small amount, but it is not our major interest.

What interests us most is how this overall difference is achieved. Basically, the public–private difference is due entirely to the superiority of the nonacademic track in private schools. While top students perform similarly in both sectors, students in the lower half of the ability distribution can be expected to differ by more than one-third of a standard deviation. Within schools, the gap between academic and nonacademic students projects to 0.894 and 0.516 standard deviations in public and private schools respectively. In both sectors, students in different tracks end up some distance apart, due to the inferiority of the lower tracks. In private schools, though, the gap may be more than a third of a standard deviation smaller than it is in public schools.

All things considered, then, the private sector seems to do a better job of promoting equality within schools than the public sector does. Both sectors seem to be nondiscriminating in their internal organizing decisions. But private schools maintain more equality when students are separated, and they succeed more in keeping diverse students together and in extending opportunity widely. Private schools also do a much better job of maintaining a decent threshold of quality for students in the lower parts of the ability distribution. The end result is that students in private schools are more likely to exhibit equal educational outcomes than students in public schools are.

Equality Between Schools

In **Politics, Markets and America's Schools** we argued that the quality of a school depends foremost on the quality of its organization — its focus, expectations, leadership, professionalism, and teamwork. Organization was the one overarching characteristic of schools that most influenced student achievement once student background characteristics were taken into account. School organization, we also argued, is strongly influenced itself most strongly by what we called autonomy. The best predictor of the quality of school organization is the freedom of a school from bureaucratic authority. The freer schools are to staff their own organizations and to define their own educational programs, the more likely they are to be organized effectively. The ultimate problem for public schools, we argued, is that they are part of institutional systems that routinely function to deny schools autonomy and in turn to undermine their organizations and performance. Private schools enjoy an enormous advantage in this regard because the market system of which they are a part promotes autonomy and rewards effective organization and performance.

This does not mean, however, that private schools should consistently outperform public schools, for there are other factors involved. Public schools, we observed, tend to be granted ample autonomy if they are part of suburban school systems, which are usually small, homogeneous, and relatively free of political conflict and uncertainty. Public schools can also be granted substantial discretion if the students they are educating do not have serious academic or behavior problems. By the same token, private schools can find themselves constrained by authorities if their students are unusually difficult. On average, then, we expect private schools to have greater autonomy than public schools and a greater chance for success. But we also expect the performance of public and private schools to vary a good deal and to overlap.

This hypothetical variation has important implications for the distribution of schools that might exist in a system of education based on school choice. As critics have observed, a choice system might concentrate difficult students in a limited number of schools and exacerbate inequality across the educational system. As they point out, our analysis shows that autonomy, organization, and eventually, performance are influenced by the caliber of the student body. A choice system might then breed differences between schools and make inequality worse.

We do not believe this would happen. In part, equality is a matter of system design. A system of school choice can be designed to protect or even promote equality between schools. As we have said and recommended, parents can be prohibited from adding to tuition payments made by the government. Schools can be required to accept certain numbers of educationally disadvantaged children. Parents can be assisted by the government in mailing informed school choices. Equality, however, is more than a matter of design. We also think it is a systemic characteristic, better protected by markets than by political institutions as we now know them.

In our view, the quality of a school is likely to depend foremost on its autonomy to solve its own problems and on the severity of the problems it must solve. In a market system, autonomy will generally be high. In a system of schools subject to direct democratic control, autonomy will vary with the structure of democratic institutions. Autonomy will generally be high in small, politically homogeneous systems and low in large, politically heterogeneous systems. All else being equal, autonomy will always be higher in market systems than in systems subject to direct political control. Overall, autonomy will tend to vary much more in political systems than in market systems, because political institutions vary much more in basic structure than markets do. This is especially true in the United States where 15,000 school districts and 50 state governments create a multitude of institutional settings for schools.

If autonomy varies much more in political systems than in systems based on markets, then so, too, should school quality. Markets should exert a homogenizing influence on schools, providing relatively equal opportunities to schools to organize effectively and weeding out schools that do not organize effectively. Political institutions should present schools a wider variety of organizing situations — from high constrained to relatively free — and should basically tolerate the operation of schools consistent with these constraints, regardless of their performance. Private schools should therefore vary less in quality than public schools do.

There is the exception to this. If school choice worked to sort children into an array of schools that vary dramatically in student composition, then the homogenizing influence that markets exert on school organizations might be offset by the diversifying influence markets can exert on school enrollments. In a market, every school would still enjoy more autonomy than it would have in today's political system, all else being equal. But the schools in a market system might serve a more disparate set of student bodies than the schools in a political system serve, and consequently develop school organizations that vary more, not less, than politically controlled schools vary. For example, if a market were to sort children into schools that range from selective academies for child prodigies to holding pens for juvenile delinquents — a specter often painted by market critics — then the benefits of autonomy might indeed be offset. A politically controlled system, with variable bureaucratic constraints but less variable student bodies, might then provide more equal educational results, from one school to the next, than a market system would provide.

There is no way to know for certain how a public system of school choice would sort children into schools. As we said, sorting is substantially a matter of design. And public authorities can take measures to ensure that student bodies do not vary widely in average ability or behavior. The most important of these measures is guaranteeing that every child has available to him or her the same price of admission. Today, the composition of schools in both the public and the private sectors is strongly influenced by the ability of families to pay — tuition in private schools and housing costs in public schools. Many public schools therefore serve student bodies in which every child lives in poverty while other public schools serve children who are uniformly rich. The same goes for private schools. Although the families in private schools must all come up with some tuition money (unless they are offered complete scholarships), certain private schools charge $10,000 and attract children of economic privilege while others charge $750 and serve children who are decidedly poor.

A look at private schools today therefore cannot tell us everything we might need to know about the composition and organization of schools that a public system of choice might produce. On the one hand, today's private schools serve only a fraction of the student population and provide only a limited view of how choice might work. On the other hand, today's private schools may provide a good view of how choice might operate at its "worst" — where parents are charged widely varying tuitions, where ability to pay influences enrollments, and where the admissions process is completely unregulated. In other words, today's private schools may help tell us how schools will differ in quality when their student bodies are sharply stratified along class lines but when all schools nevertheless enjoy the organizational benefits of markets and autonomy.

Socioeconomic status and school performance

We begin our analysis with an overview of the schools in today's public and private school systems. First, we considered the full distribution of schools in each sector. Some of what we found is completely unsurprising. The average private school has students who are higher in socioeconomic status than the average public school has. This is a sizable difference — 0.865 standard deviations — and surely reflects the fact that parents who make tuition payments, however small, have a higher income than parents who do not. We also found that private schools report average math and reading scores that exceed public school scores by 0.753 standard deviations. Given the public/private difference in average school SES, this test score differences is likewise unsurprising.

What is not unsurprising, however, is what we found when we shifted our attention from the means of the distributions to the variances. We found, first, that the socioeconomic status of private schools varies considerably more than the socioeconomic status of public schools. The variance of the average SES of student bodies in the private sector is two-thirds greater than the comparable variance in the public sector. Even though private school families represent a narrower and more affluent part of the total distribution of socioeconomic status, the schools that they attend vary more than the schools attended by public school families in average socioeconomic status. As we suggested, the fact that private schools operate at widely varying costs and charge widely varying tuitions produces a distribution of schools that varies much more than public schools vary in the class composition of each school. Private schools may then provide a good test of how school quality might vary when schools are socioeconomically stratified but free to operate as they judge best.

What then is the surprise? Despite the greater variation in socioeconomic status across private schools, there is less variation in test scores across private schools than across public schools. From school to school, the private sector is more diverse than the public sector in student background but less diverse in tests scores. This is rather remarkable because these scores are well predicted (and influenced) by family socioeconomic status. All things being equal, private schools ought to have a wider distribution of average test scores than public schools do because private schools have wider distribution of average class backgrounds. It appears, however, that all things are not equal. Test scores and class background do not seem to have the same relationship to each other in the two sectors.

Why might this be? One explanation — the one we proposed — is that private schools are more equal in quality than public schools are. The differences between the best and the worst private schools are not as great as the differences between the best and the worst public schools. Thus, the marketplace can organize children into schools that differ on average more than public schools differ in socioeconomic status. But because private schools are more similar in organization and effective than public schools are, children in socioeconomically disparate private schools end up reaching more similar levels of academic achievement than children in less disparate public schools. If this interpretation of these data is correct, the equalizing effect of a market system may indeed be great. Private schools are two-thirds more diverse than public schools in average class background but slightly less diverse in average student performance.

There are, however, other possible explanations for these distributions. One is that the socioeconomic variance of private schools is deceptively large, inflated mostly by extraordinarily affluent independent schools. The test score variance of private schools only looks narrow by comparison because these affluent schools cannot report the extraordinary test scores that would be required to match their socioeconomic status. Another possible explanation is that the socioeconomic variance of private schools is deceptive in a different way. Private schools may indeed serve groups of children that vary from unusually poor to unusually rich, but the families at the lower end of the distribution may be unrepresentative of the poor. The poor families in private schools may be unusually interested in or supportive of education, and more to the point, the children of these families may bring relatively strong achievement levels to private school with them. In other words, economically poor students in private schools may be something of a disadvantaged elite that private schools are fortunate to enroll but that public schools do not especially help to learn.

To evaluate these explanations, we divided schools into two categories — schools with student bodies above the median SES of all schools, which we label "rich", and schools with student bodies below the median of all schools, which we label "poor". This division enables us to look separately at schools that serve mostly advantaged children and schools that serve mostly disadvantaged children.

Looking first at "rich" schools, we found that private ones in this category are more affluent on average than public ones and that the private distribution of schools is wider than the public distribution. When we look at test scores, moreover, we found precisely the same thing. The average test score of private schools is higher than the average test score of public schools, and the distribution of average test scores is wider in the private sector than in the public sector. In "rich" schools it appears that the socioeconomic make-up of the student body and the achievement of the schools are in much the same relationship in the public and the private sectors. It does not appear that extraordinary affluence in certain private schools is the source of the equalizing relationship found in the sample of all schools.

What about "poor" schools? Several things are of interest. First, the private schools in this category are indeed poor. Their children average nearly one standard deviation below the median socioeconomic status of all schools, and their children are lower in socioeconomic status than the children of "poor" public schools. The "poor" private schools are, however, more socioeconomically diverse than the "poor" public schools, their respective variances differing nearly two to one. If average test scores were perfectly predicted by socioeconomic status, as they nearly are in "rich" schools, the variances of test scores in "poor" private and public schools would differ similarly. They do not. The variance of "poor" private school test scores is less than the variance of "poor" public schools scores — despite the greater socioeconomic diversity of the private sector — and private scores are one-third of a standard deviation higher.

Unless students in "poor" private schools are academically rich in some still undisclosed way, it appears that the public and private systems differ in the distribution of quality schools, with the key difference manifest in schools serving relatively disadvantaged children. In the private sector, "poor" schools are more disadvantaged and more diverse than "poor" schools in the public sector, yet test scores in "poor" private schools are higher and less diverse than scores in their public counterparts. Private schools serving poorer children seem to be doing a better and more equal job of promoting academic achievement than public schools serving comparable children. If this is so, the greater equalizing effect that private schools as a whole seem to have would be primarily attributable to the threshold of quality that private schools appear to meet for the education of even the most disadvantaged.

A closer look at students in "poor" schools

So far we have considered only the average socioeconomic status of families as we have tried to understand potential causes of performance in "poor" schools. But what about other student characteristics? Do public and private school children differ in other ways that might explain their different test scores? We began by considering one leading possibility, family composition. We found that students in "poor" public and private schools are not very different in this regard. Many of these children, more than in the general population, come from homes where the mother and father are not both present. Private school children have a slight edge in living with their mother and father together, but private school children are also more likely to live with their mother alone. If living without both parents in the home is a handicap — and research suggests that it is — children in "poor" public and private schools are equally disadvantaged (United States Congressional Research Service 1987). There are no statistically significant differences.

Much the same is true of family size. Research suggests that children from larger families fare slightly worse in school than children from smaller families, all else being equal (US Congressional Research Service 1987). The reason is not entirely clear, but undoubtedly has something to do with sibling rivalry for parent attention. At any rate, children in "poor" public and private schools come from families that are not significantly different in total size. The story is similar for race. There is not a statistically significant difference in the percentage of white students in "poor" public and private schools. There is a difference, however, in the percentage of black students — though not in the expected direction. Private schools, not public schools enroll a higher percentage of black students — 26.5 percent in the former, only 18.0 percent in the latter.

Finally, we considered parent education, a component of socioeconomic status, but probably the best predictor of student achievement among individual measures of family background. Here we found some difference, in the expected director, between public and private school families. Most noticeably, "poor" private schools enroll only half as many students with parents lacking a high school degree as "poor" public schools do. This difference is then mirrored in small private school advantages at some but not all higher levels of education. Overall, however, the differences in parent educational are not large enough to be confidently labeled statistically significant. The differences are close enough to accepted significance levels not to be ignored, but they may also be meaningless.

In total, the student bodies of "poor" public and private schools look pretty similar. There are no significant differences between them on the measures of family background that are most strongly associated with student achievement. To be sure, the student bodies still differ in one potentially crucial respect: The families of the private school children decided to sacrifice and pay for private school while the families of the public school children decided not to make the same sacrifice. In our analysis we cannot measure or control directly for this difference. But before we concede too much importance to this analytical hurdle, it is important to bear in mind that we are interested not primarily in whether or why private schools are better on average than public schools. We are concerned with whether and why private schools produced less variable results. If we assume that every private school child gets an unmeasurable boost from unusually supportive parents, we have reason to expect higher achievement from every child in private school than in public school, all else being equal. We have no reason, however, to expect the distribution of achievement among those equally advantaged private school children to narrow relative to the distribution of achievement among children in public schools. If we want to understand how and why achievement differs across public and private schools, we will need to look beyond the students in them; we will need to look at the schools themselves.

The organization of public and private schools

Schools perform differently, we believe, because they are not all equally well-organized for the task. This is true of differences between successful and unsuccessful public schools, private schools, or both. The immediate cause of school performance is not the institutional system; the immediate cause is school organization. Institutions are important primarily because of their influence on school organization. Market institutions, we think, produce school organizations that are more effective and more equal than those that current political institutions produce. We have now seen indirect evidence that this may be true. Private schools have higher and less variable average test scores than public schools do, especially among the less advantaged schools, and these differences are not well explained by the kinds of students in the schools.

Might these differences be explained by organization? We do not have space for a compete exploration of this possibility, but the evidence strongly suggests that organization is important. We saw this first when we considered several basic measures of school organization, and the scores of "rich" and "poor" public and private schools on them. Two important relationships emerged from these indicators. The first is that organizational conditions (in the view of school principals) are much better in "poor" private schools than in "poor" public schools. For example, in most "poor" private schools, principals strongly agree that "teachers encourage students to do their best", and that "teachers do not have negative attitudes toward students". The same is true in only half of the "poor" public schools. The differences are equally stark on all indicators. Perhaps these differences explain why "poor" private schools, despite being lower in socioeconomic status than "poor" public schools, report noticeably higher test scores.

The other relationship that interests us bears more directly on our interest in systemwide equality. That is, the organizational conditions in "poor" private schools are very similar to the conditions in "rich" private schools, but the organizational conditions in "poor" public schools are inferior to those in their "rich" counterparts. In other words, school quality is more equal in the private sector than in the public sector — and as we have seen, apparently not because student quality is more equal across private schools.

These findings are reinforced by two additional measures of real organizational importance that we considered. First, we examined indicators of teacher dedication — extra hours spent working with students and teacher absenteeism. Here we found stark public and private differences. But the important differences are nuanced as before. The greatest public/private differences occur in "poor" schools. There we found 15.7 percent of public school teachers giving less than five hours a week in extra time to students, but only 1.6 percent of private school teachers being so stingy. The modal number of extra hours spent by private school teachers is 10.5 to 14 per week; the mode for public school teachers is 5.5 to 10. Among teachers in "poor" private schools, only 6.3 percent are absent more than two days in a semester; among teachers in "poor" public schools, the figure is 37.3 percent. Teachers in "poor" private schools exhibit greater devotion to their work than teachers in "poor' public schools do.

Private school teachers are also far more similar in their behavior from "poor" schools to "rich" than public school teachers are. Indeed, there is some evidence that teachers in the "poor" private schools are even slightly more devoted than their privileged counterparts. This again suggests that private schools are more equal in quality than public schools are. The market provides a setting in which all schools, "rich" and "poor", are granted substantial autonomy and are disciplined by parental choice. Subject to these relatively constant influences, private schools are relatively constant in organization. Public schools function in a variety of institutional contexts, depending on the scope and complexity of political jurisdictions, and experience a range of bureaucratic constraints. Their organizations, highly disparate in apparent effectiveness, reflect this institutional stamp. The public sector consequently produces schools that are rather unequal in basic quality. All things considered, markets appear better suited to producing schools of equal quality.

Conclusion

The analysis summarised in this paper is the beginning, and not the ending, of an investigation that we hope will eventually clarify the sources of inequality in American education. Our suspicion is that the institutions responsible for running American education bears major responsibility for this inequality — not because of ill-intent or poor design, but as a result of their normal functioning. Driven by political conflict, these institutions have developed a bureaucratic control structure that seems to have institutionalized inequality. Within schools, the structure reinforces a system of student classification and educational specialization that exacerbates differences in achievement among students — differences that often run along class and racial lines. Between schools, the structure imposes the most burdensome constraints on the schools that can least tolerate them — schools serving the disadvantaged — and pushes those schools even further behind schools serving the advantaged. Public education consequently produces results that are not only lacklustre, they are unnecessarily unequal.

We say unnecessarily because, as we saw in our look at private education, it is plainly possible to do better. Public education is not overtly discriminatory in its dealings with disadvantaged students; indeed, there is some evidence of affirmative action on their behalf. But public education could do much more. It could place children of different abilities together in the same groups and classes more often than it does now, exposing slower students to the curriculum and expectations provided to swifter students. It could provide accelerated programs to a higher percentage of students. It could ensure that children organized into different groups, classes, and tracks are provided programs and teachers equally high in quality. And it could work to provide schools now stymied by urban bureaucracies the kind of autonomy that promotes effective organization in urban private schools. If public education took these steps, it would surely raise the achievement of its most disadvantaged students and, without jeopardizing the achievement of its best students, raise the average achievement of the system as a whole.

The problem is, these steps cannot easily be taken. If we are correct, the qualities of organization that breed inequality within and between public schools are rooted in the institutions that govern public schools. Unless these institutions are changed, the organization of public schools is unlikely to change — at least in any substantial way. School reformers, after all, have been criticizing tracking and bureaucracy for years, yet efforts to solve these organizational problems have been largely to no avail. We believe institutional reform is necessary. And we believe the market provides the most promising model for carrying it out.

In **Politics, Markets, and America's Schools**, we made the case for institutional reform, based on markets, primarily on grounds of educational efficiency and effectiveness. Now we would consider taking the case one step further. Markets appear better suited than institutions of direct democratic control to promoting educational equality. Markets seem to encourage the more equitable treatment of students within schools and the more equal distribution of quality across different schools. With public regulation aimed at further promoting their fair operation, markets would appear to provide a more promising foundation on which to rebuild public education than the institutions that now govern them. In any case, if the inequality and ineffectiveness that now plague American education are to be substantially reduced, research and reform must work harder to get to their roots.

Notes

1. For a critical introduction to the literature on ability grouping and tracking, see Oakes (1985).
2. For a brief review of the research literature on grouping and tracking, see Gamoran and DeMare (1989). Among the very few studies comparing grouping and tracking in public and private schools are Lee and Bryk (1988), and Camarena (1990).
3. A data appendix describing the variables used in this paper is available from the authors, on request.
4. For critical views, see Oakes (1985), and Persell (1977).
5. We did not model the history and science tests because scores in those subjects can be strongly and legitimately influenced by school-to-school variations in curriculum sequences.

References

Alexander, K. and Cook, M. (1982). Curricula and coursework: A surprise ending to a familiar story. *American Sociological Review*, **47**, 626–640.

Alexander, K., Cook, M., and McDill, E. (1978). Curriculum tracking and educational stratification: Some further evidence. *American Sociological Review*, **43**, 47–66.

Ball, S. J. (1981). *Beachside comprehensive: A case study of secondary schooling*. Cambridge: Cambridge University Press.

Camarena, M. (1990). Following the right track: A comparison of tracking practices in public and Catholic schools. In R. Page and L. Valli (Eds), *Curriculum differentiation* (Chapter 8). Albany, New York : SUNY Press.

Chubb, J. E. and Moe, T. M. (1990). *Politics, markets and America's schools*. Washington, DC: The Brookings Institution.

Coleman, J. S. and Hoffer, T. (1987). *Public and private high schools: The impact of community*. New York: Basic Books.

Doyle, D. and Cooper, B. S. (Eds) (1988). *Federal aid to the disadvantaged*. London: The Falmer Press.

Educational Testing Service (1991). *Psychometric report: NELS: 88.* Princeton, ETS.

Gamoran, A. and DeMare, R. D. (1989). Secondary school tracking and educational inequality: Compensation, reinforcement, or neutrality. *American Journal of Sociology*, **94**(5), 1146–1183.

Gutmann, A. (1987). *Democratic education.* Princeton: Princeton University Press.

Heyns, B. (1974). Social selection and stratification within schools. *American Journal of Sociology*, **79**(6), 1434–1451.

Lee, V. and Bryk, A. (1988). Tracking and achievement in public and Catholic schools. *Sociology of Education*, **61**(1), 78–94.

Oakes, J. (1985). *Keeping track. How schools structure inequality.* New Haven: Yale University Press.

Perself, C. H. (1977). *Education and inequality.* New York: Free Press.

Slavin, R. E. (1990). *Cooperative learning: Theory, research, and practice.* Boston: Allyn and Bacon.

United States Congressional Research Service (1987). *Student achievement: An explanation of trends.* Washington, DC: Government Printing Office.

Chapter Ten

THE DESIGN OF SCHOOLS AS OUTPUT-DRIVEN ORGANIZATIONS

JAMES S. COLEMAN
Department of Sociology, University of Chicago

In this chapter, I want to discuss the design of organizations. The aim of the chapter is first to analyze the functioning of a particular type of constructed organization, elementary and secondary schools in the United States — and then to take some steps toward the design of schools as organizations. To begin, I will focus briefly on the theory and practice of formal organizations generally.

Innovations in Organizations

Max Weber was one of the first to attempt to develop theory for non primordial, constructed organization. He described such organizations a "rational authority system", and specified its characteristics, with the pure type being the bureaucratic form.

This form, with variations, has been widespread as the paradigmatic form in which the rationalization of society proceeded. However, in recent years, organizational practice has initiated new forms, some of which have little in common with the theory and practice of bureaucratic authority as described by Weber.

One of the first of these was the development of the multidivisional firm, initiated by General Motors in the 1930s. This structure gave both more autonomy and accountability of divisions of the firm, by introducing relative autonomy of divisions, along with internal pricing of goods and services produced and consumed by divisions, making each division in effect a profit center.

Another innovation, one which goes beyond the multidivisional firm, is franchising. Although franchising, which has recently become widespread, takes various forms, they have in common a contractual relation between the central franchising organization and the franchisee (usually a retailer), with a portion of the income going to franchiser and a portion to franchisee.

Some organizations have developed a modus operandi which involves funding of spinoffs by employees who develop a possibly profitable product idea, with the spinoff a joint venture between the parent firm and the persons who developed the idea. One of the earliest firms to engage in the activity was 3M of Minneapolis, seeing itself in effect as a central bank and research and development center to finance ideas generated by employees, particularly those engaged in research and development.

Other organizational innovations can be found in the computer hardware and software industries, as well as in genetic engineering firms. The latter involve creative arrangements between university faculty members and biotechnical firms, either independent startups or spinoffs from larger firms.

There is one innovation that has arisen in a quite mature industry, the automobile industry. This innovation, which began with Japanese firms and is being imitated in some American firms, can be describes as a reallocation of rights.

The rights that are involved are those having to do with rejecting parts and stopping the production line. In the classical manufacturing organization, the supervisor in an assembly line may have the right to reject input parts that are out of specification, or may only have the right to request from a higher supervisor in the plant that the parts be rejected. The decision may depend on whether it is possible to work around the defect without shutting down the assembly line. If it is a defect that can be repaired, a tag is put on the car, and at the end of the line, all cars with tags on them are shunted to a repair facility. Whatever the decision, the right to make that decision is never in the hands of the assembly line worker. It is in the hands of either the line supervisor, or a superior of the line supervisor.

In the Japanese auto assembly line, the rights are allocated somewhat differently. First, many of the rights and responsibilities that would ordinarily be held by the line supervisor are held instead by a collective consisting of the members of the line (called a quality circle, or QC). The supervisor is in most cases altogether missing, the positions being absent from the firm's organization. This specific right, the right to reject out-of-specification inputs, is held by the individual worker on the line who must use these parts in accomplishing the task. The right includes the right to shut down the line. Execution of the right, that is, rejecting out-of-specification parts, is made more likely by the fact that the next worker who handles that component of the automobile has the right to reject the component for any out-of-specification parts. Thus the right to reject input parts is accompanied by the responsibility for output being **in** specification. The worker thus has an incentive not to accept input parts that are out of specification as well as to insure that labor input does not result in an out-of-specification output.

The general character of the incentives that are introduced by this structure of rights is fairly obvious. It is an incentive on the part of the worker carefully to inspect input parts and to be vigilant about possible errors introduced during processing. The set of rights that are involved in a production process like that involved in building automobiles is a string of rights which mirrors the production process itself. The workers who receive worker A's output have the right to reject worker A's production, and thus to affect A's pay, which is partly based on bonuses (and, depending on other aspects of the organization of work, possibly also to affect the pay of others in the same workergroup as A). The holder of the ultimate right is outside the organization altogether, that is, the shipper, the dealer, and finally, the customer, each of whom can reject the finished product.

An additional right that accompanies these rights of workers is the right to extra pay (paid as a bonus), depending on quality and quantity of output by the workgroup. As much as a third of a worker's pay may be in the form of a bonus.

The allocation of rights that I have just described produces what I have elsewhere called "backward policing", that is, a policing of the quality of the product that reverberates backwards step by step through the production process of the organization. The feedback process from the production of a defect to the discovery of the defect and then back to the producer of the defect is an extremely short loop. This contrasts with the long loop of the feedback process that is characteristic of the classical hierarchical structure of rights that could be described as downward policing.

There is one important aspect of the allocation of rights that I have just described. The quality circle, constituting the members of an assembly or sub-assembly line, has rights as a group which give it collective authority over its members, replacing the authority of the line supervisor. This, together with the fact that members' bonuses are dependent on the group's quality-quantity productivity, means that norms develop in the group that are consistent with and reinforce the organization's goals. For this to occur, it is important that the group not be too large, and the quality circles typically have 8–12 members[1].

All innovations that I have described can be seen as attempts to overcome the incentive problems that arise in a centrally directed, hierarchically structured organization. By making benefits (income, autonomy, responsibility) contingent on the quality and quantity of what is produced, the organization creates a direct link between effort and reward. In the multidivisional firm, the link is at the divisional level, in franchising it is at the franchise owner level, in spinoff joint ventures, it is at the entrepreneur level. In backward policing, it is at the workgroup (QC) level.

At least in this last case, and very likely in the others as well, the organizational innovation does more that overcome the incentive problem of hierarchical organizations (as, for example, payment according to piecework pay rates would do). The innovation creates a **social group** in whose interest it is to encourage high effort and careful work on the part of the members. Social norms come into being to reinforce organizational goals. These norms go beyond overcoming incentive problems of hierarchical organization. They provide extra incentives of their own, because each member's work benefits all. And this comes about because rights which, in the hierarchical organization, are held centrally and delegated vertically are instead given to subgroups — divisions, franchisees, quality circles.

Incentive problems arise in any centrally managed and directed system. The collapse of state socialism in Eastern Europe and the Soviet Union was principally due to failure to solve these incentive problems. The continuation of state socialism in China results from China's capacity to solve, at least for the present, the incentive problems through the introduction of a market.

The different allocations of rights which exist in these organizational innovations involve in effect a **reconstitution** of the firm. The Weberian bureaucracy can be described as "authority-driven"; these innovations create, to differing degrees, organizations that are "output-driven". The degree to which the organization as a whole can be said to be output-driven depends on the degree to which the external criterion on which the rewards to a part of the organization are contingent is transmitted to subparts within. This occurs by explicit design in backward policing, and may or may not occur in the other forms.

School Design

From the organizational innovations that I have discussed, I want to take two elements that can be useful for the redesign of schools. The first of these elements is the replacement of authority-driven organization by output-driven organization. I will make use of this in its most extensive form, in which the output driven principle is applied not merely for the organization as a whole but internally for subparts of the organization — as shown in the chart.

The second element I want to take from these innovations is the explicit creation of social capital in the organizational design. This is most evident in the quality circles seen in Japanese automobile manufacturing firms, although it is present to varying degrees in the other innovations I described. It is brought about primarily by the allocation of rights and obligations, not merely to individual positions in the organization, but to subparts of the organization — which themselves are not hierarchically organized from without (though they may create such a structure from within). These subparts must be small enough and with dense enough interaction that they can be norm-generating bodies. The norms, of course, are not imposed by the central authority. Rather, if the rights are allocated appropriately, norms will arise that support and reinforce the goals of the organization. If the rights are allocated inappropriately, the norms that arise will act against the organizational goals.

Diagnostics of American Education

With these organizational innovations in mind, I want now to turn to the question of what schools might be modified to improve their functioning. I will first give a brief diagnosis of American education.

First it is important to recognize that there are a number of dimensions along which children develop during their formative years — and thus a number of dimensions to the goals which schools have. There is, however, a justification for laying out a design for schools which focuses on one goal alone. When an educational system (including the full social environment of the child's family, school, community, popular culture) is seriously deficient with respect to an important goal, then it is reasonable to carry out designs which attend to that goal alone. Inertia and countervailing forces will serve as counterweights when any such design is implemented.

Educational systems in the United States are seriously deficient in academic achievement. Relative to students in other countries of the developed world, American students score at or near the bottom on standardized tests.

Measures of performance both before and after the test score decline of the 1970s show the extraordinary deficiencies with which American children leave the system of elementary and secondary education. The deficiencies are found throughout the distribution of performance. At the lower levels, the ineffectiveness of inner-city schools is well known, and the weak performance of the children who attend them is extensively documented. These low levels of performance of American students at the high end of the achievement distribution are ordinarily forgotten. But the well-known SAT test score decline of the 1970s and 1980s shows the greatest loss at the top end of the distribution. In Minneapolis, Minnesota (a state which scores near the top of American states on standardized tests), performance in mathematics in 20 fifth grade classrooms was compared with that in 20 classrooms in Taipei, Taiwan and 20 in Sendai, Japan. The average for the **highest**-performing classroom in Minneapolis was below that for the **lowest**-performing classroom in Sendai, and below that for 19 of the 20 in Taipei. The performance of Americans in foreign languages is miserable compared to that in almost any other developed country. In mathematics, over half of the new assistant professor appointments in American college and university mathematics departments are persons born and educated at least through high school in some country other than the United States. Our scientific and technological achievements depend heavily on a "brain drain" from other countries to the United States. If another country became a more attractive destination for these immigrants, so that we had to depend on our own system of bringing the young into adulthood, we could not maintain current technological levels, much less innovate technologically.

This low level of performance in American schools is reinforced by another aspect of most schools: in middle schools and high schools, across the socioeconomic spectrum and among all racial and ethnic groups, the informal norms that develop among students are **not** norms that extol achievement, but are norms that scorn effort, and reward scholastic achievement only when it appears to be done without effort.

This need not be true, of course, and in fact there are some high schools in which it is not true. There are even more schools in other countries where it is not true. It is a mark of incorrect organizational design that such norms exist in schools.

The sources of the low effort put into high achievement are several, but one is important: the lack of well-publicized external criteria of performance from which meaningful consequences flow. In the absence of such criteria, there can come to be an implicit compact to reduce the strain imposed by high standards (including extensive homework), and parents unable to enforce extensive homework. The implicit compact can result in quite low levels of performance. An example can be seen in an incident reported by Arthur Powell in a high school he visited:

> Students were given time (in class) to read **The Scarlet Letter**, **The Red Badge of Courage**, **Huckleberry Finn**, and **The Great Gatsby** because they would not read the books if they were assigned as homework. Parents had complained that such homework was excessive. Pressure from them might even bring the teaching of the books to a halt . . . (as one teacher put it). If you can't get them to read at home, you do the next best thing. It has to be done . . . I'm trying to be optimistic and say we're building up their expectations in school.

Theodore Sizer reports the implicit compact more directly in describing one teacher's class:

> He signaled to the students what the minima, the few questions for a test, were; all tenth and eleventh-graders could master these with absurdly little difficulty. The youngsters picked up the signal and kept their part of the bargain by being friendly and orderly. They did not push Brady, and he did not push them . . . Brady's room was quiet, and his students liked him. No wonder he had the esteem of the principal who valued orderliness and good rapport between students and staff. Brady and his class had agreement all right, agreement that reduced the efforts of both student and teacher to an irreducible and pathetic minimum.

How can a teacher gain the friendly and orderly classroom climate that Brady has if at the same time the kind of effort and involvement that generals high standards and sacrifices the good climate and risks a rebellion — reduce the standards and sacrifice achievement? One of the few ways out, in the absence of strong parental demands that hold a rebellion in check, is for the teacher to be freed from the task of setting the standards. Then the teacher and the class can be engaged in a common task, that of batling the externally established standards. The standards may be those of an external test, or those imposed by a contest with students in another school, or perhaps in a different way. But so long as the standards are out of the teacher's control, the teacher and students have a common interest which generates both the effort needed for achievement and a positive climate.

There is an interesting "natural experiment" involving the absence of any external demands whatsoever toward school learning. A Scottish educationist, A. S. Neill, established a boarding school, Summerhill, in which **no** external demands are made upon the children: no classes, no assignments, no school. According to accounts of Summerhill by Neill and by others, Summerhill is a very pleasant place, one in which the children manage their own affairs individually and collectively, but as Neill himself put it, one in which "the children of Summerhill are not much interested in book learning".

There is one exception to the lack of external demands at Summerhill. When the children reach their upper teens, they see that in order to manage in the outside world, they will need a school leaving certificate, one which requires passing a set of externally imposed examinations. Thus in their last year at Summerhill, they spend a considerable amount of time studying, in order to pass these examinations. This last year studying is the exception that proves the rule: When an external criterion is imposed, efforts toward learning begin.

It is interesting to point out also that a much cruder and more partial natural experiment can be found in the comparison between American and Japanese schools. In Japanese schools, where there are strong externally imposed standards, the achievement is much higher that in American schools, but there is lower satisfaction with the schools, and the children and parents are less satisfied with the child's performance than is true for American schools. This is sometimes seen as paradoxical: American children learn much less but they and their parents are more content with their educational performance. But as in Brady's classroom described by Sizer, American children are content because they are satisfying the demands made on them, demands which are depressed by the majority's unwillingness to meet stronger demands.

Young people have a time-and-attention budget, jut as adults do. In that budget, the time spent on homework has declined to about three and a half hours per week, half of what it was in the 1960s before the SAT test score decline. Time spent in sports, in part-time jobs, in TV viewing, in consumption of popular culture with friends all compete with homework for the teenager's time. While the time on homework has declined, the time spent in part-time jobs (about 10 hours on average for high school seniors) and in television watching has increased. Television, at 20 hours per week, is twice to three times that in other OECD countries. It is clear that since the 1960s, other activities, such as watching television and earning money at part-time jobs, have come to occupy a more important place in the average teenager's life, and schoolwork has come to occupy a less important place. Parents, meanwhile, appear unable to prevent this.

Schools as Authority-Driven and Schools as Output-Driven

The organizational innovations that were described earlier constitute a change from an authority-driven organization to one that is output-driven. Schools as they exist are not pure forms of either of these. Most American schools, as any teacher knows, are hierarchical authority systems, with the principal the immediate authority. Principals in turn are under the school district superintendent as the central authority, often with little discretionary authority over resources, over hiring of teachers, and other decisions. The structure of schools follows that of the bureaucratic authority driven organization quite well.

At the same time, there are various examples, some the result of intentional design but others not, which show the power of output demands in education. One of these is the importance of college entrance requirements for high school courses taken. College entrance requirements in the United States have traditionally been a requirement for 32 "Carnegie units" of high school courses, including specific requirements, such as two years of a foreign language. These entrance requirements have dictated the "college prep" curriculum in high schools. In the 1960s and 1970s, with the onslaught of the youth revolt and the civil rights movement, colleges liberalized their entrance requirements. Most elite colleges dropped the foreign language entrance requirement. A consequence of this was that foreign language course-taking in high school dropped precipitously. Only in the late 1980s, as colleges reinstated entrance requirements, has course-taking in foreign languages revived. Note an additional point: This is an external demand, which imposes requirements **not** upon output but upon actions. By "passing courses", the student meets the demand, and the teacher is free to determine the criteria for passing. Even though the Carnegie-unit demand has this crucial flaw, the example of foreign languages illustrates how relation of a demand has its effects.

There are other examples. An interesting one is handwriting. High schools, as the recipients of the products of middle schools, set the external requirements for middle schools, and middle schools set the external requirements for elementary schools. When, in postwar America, high school teachers no longer required good handwriting, middle school teachers did not either; and when this occurred, elementary school teachers stopped teaching children to have good handwriting.

As these examples indicate, educational systems are in

part output-driven, despite their hierarchical authority structure. In fact, containing these two conflicting modes of organization, schools create an untenable situation for teachers. They are under the constraint to meet certain output demands, yet are subject to authority which eliminates their autonomy in meeting these demands. However, as I will suggest, the character of the output demands is defective, which means that the system reverts to an authority-driven system, and the teachers set flexible internal standards — as in Mr Brady's class — to make the situation tolerable.

The absence of external standards for actual performance has several negative consequences. It puts teachers in the position of establishing the requirements, by deciding what levels of performance will be necessary in order to receive a given grade. By placing this task upon the teacher, the school puts the teacher in two roles with conflicting interests: as the person who sets the standards, and as the person who tries to get students to meet the standards. This creates an ambivalent relation between teacher and students, quite different from that between coach and performers. The coach can devote undivided efforts to improving the performance of the team members; the teacher's efforts must be on two fronts: improving students' performance, and to struggling with students over the level of performance required for a given grade. This can poison the student–teacher relationship, voiding the effort on the first front. It is also an important element in generating the norms among students to restrict output. Students would have no reason to develop such anti-achievement norms if the standards used to measure performance were not under the teacher's control.

Another negative consequence of the absence of external standards is that it creates an incentive for a student to take easy courses. A student's grade point average (GPA) creates a rank in class, which is well known by students and widely used for college admission. But a GPA can be augmented by taking easy courses, with less effort and greater certainty than by performing well in difficult courses. Various devices have been introduced to signal to colleges that grades are based on hard or easy courses, but none of this would be necessary if the rank in class were based not on teacher-established standards in specific courses, but on externally established standards in specific areas of performance.

A third defect of internally established standards is that they introduce noise into the information the receiving institution (further education or employer) gets from the school about the student's performance. Employers seldom use the information at all, and colleges augment rank in class by scores on the SAT or ACT test, and sometimes by a ranking of the high school itself. None of this would be necessary if standards were externally set rather than set internally.

A major reason for schools' use of internal standards of achievement rather than externally established ones is the great disparities between performance of students from different backgrounds and with different preparation. The use of internal standards is clearly a defective way of addressing this problem. I shall show later another method of doing so that has none of these defects.

Definitions

In approaching the design of schools as organizations, a number of concepts will be useful. While I have already used some of these concepts, it is useful to list them and give brief definitions.

(1) **Output-driven**: An organizational form in which the rewards and punishments for performance in productive activity come from the recipient of the product. Applied to intermediate products within the organization, this

means that the recipient of the intermediate product has the right to monitor the quality of that product, and thus to determine the rewards and punishments for the part of the organization from which it receives intermediate products — and in turn the obligation to satisfy the requirements that its own products must meet — as monitored by the recipients of **its** products.

(2) **Authority-driven**: An organizational form in which the rewards and punishments for performance in productive activity come from the central authority in a hierarchical structure, through an immediate supervisor.

(3) **External standards**: Standards that are set outside the organization, ordinarily by the recipient of the product of the organization.

(4) **Internal standards**: Standards for performance of organization members set within the organization.

(5) **Stages of schooling**: Elementary school (grades 1–4; ages 6–9), middle school (grades 5–8; ages 10–13) high school (grades 9–12; ages 14–17).

(6) **Value added of performance gain**: These two terms will be used interchangeably. They mean the difference in the performance level at the end of the prior stage. The terms are also used to refer to yearly differences in performance levels.

(7) **Rights allocation**: The allocation of rights in the organization by the central authority to positions in or subdivisions of the organization.

(8) **Incentive**: Any source of motivation to apply effort toward a goal. Here the term is applied to teachers, students, and parents.

(9) **Informal norms**: This term is used to apply to

norms that arise within an organization, held by all or some of its members. The norms may be facilitated through structures established by design, or they may be unanticipated and unintended by the organization's designer. Informal norms held by the teachers in a school and informal norms held by students in the school are discussed separately.

(10) **Social capital**: This term refers to the informal social relations, basic institutions (families, religious groups), and other informal institutions that exist in the environments of the children in a school), insofar as they can serve to augment the education of the children. It is also used to refer to social relations and norms that arise spontaneously within a constructed organization, when these augment the organization's goals.

(11) **School**: A set of 4–16 nonspecialist teachers with primary responsibility for a set of children at elementary, middle, or high school levels. In addition to the teachers, the school includes a building and a building custodian.

(12) **Specialist**: This term refers to a teacher, who, under contract with nonspecialist teachers in schools, teaches or coaches students from more than one school in noncore special areas. These include music, drama, other performing arts, advanced topics in mathematics, sciences, sports, technical skills, and other areas.

Working Backwards

In the design of an output-driven system, it is necessary to begin at the end and work backwards. The necessity arises from the fact that it is the standards at the final output which, properly employed, provide the motive force that reverberates back through each of the earlier stages to energize and focus actions. What is necessary, then, is a reward for performance at the end of the process of elementary and secondary education. It cannot, however, be a reward which comes only to teachers, or only to students, or only to parents. It is clear that there are three parties whose actions are directly relevant to achievement: the child, the parent, and the teacher. High achievement requires that each of these three parties be motivated to bring it about.

There are several requirements that must be met by the evaluation system on the basis of which teachers, students, and parents are rewarded.

> **The standards must be externally designed by those parties that will be the Institutions receiving the "graduates" of the school.**

For the high school, there is such a test for college admission, the SAT test and the ACT test. There are no such tests for high school that are also collegiate sports there are good evaluation systems. These result from the extensive interscholastic competition which allows high school students to exhibit their competence. In areas where interscholastic debate is widespread, the same is true to a lesser extent for debaters. In those states which have statewide interscholastic competition in music, drama, and other areas, there is the basis for such an evaluation system in these areas as well. Advanced placement tests provide a starting point for such evaluation in particular academic areas.

There are international "olympics" at high school level in some specialized technical areas, in which students from some countries (not the United States) participate. There is also an academic olympics at least in the field of mathematics, in which some countries (Hungary and Poland, for example) participate. Thus in many specialized areas, there already exists the basis for externally designed performance tests.

There exists little basis for externally designed standards for the end of middle school and the end of elementary school. Following the logic of output-driven organizations, and more specifically backward policing, the design should come from, or at least be extensively informed by, teachers at the next level: high school teachers for the tests at the end of middle school, and middle school teachers for the tests at the end of elementary school. These tests, however, can hardly be designed without knowledge of what the criteria at the end of high school will be.

The system must evaluate not "basic abilities", but achievement.

The SAT and ACT tests, currently used in the United States for college admission, fail this criterion. From the outset, the SAT test was designed to be "curriculum-free", and students were told that it is not possible to study for the SAT. It is, in fact, possible to increase one's performance on the SAT test; there are commercial organizations which provide courses which do just that. But the fiction was strongly maintained by the SAT test-markers at Education Testing Service (which constructs the SAT), until it lost a court case in which one such organization proved its claim to be able to improve performance substantially.

One original aim of the SAT was to give students from weaker high schools the same chance to score a good grade as students from stronger high schools. The effect, however (in contrast to English O- and A-levels, French Baccalaureate, the German Matura) was to deprive the SAT of any motivating power toward achievement. Note that the SAT and ACT tests, in transmitting the notion that what is tested is immutable, are less egalitarian than a test which is explicitly tied to curriculum, because it creates the belief that the score measures one's self, fixed and immutable, relative to others.

The system must evaluate the performance *level*, but also the performance *gains* over the period between test administrations (1-year or a 4-year school).

If any evaluation provides rewards only for level of achievement, there are at least two negative consequences. First, it leads only to interpersonal comparison (rather than to comparison with the students' own prior performance), which tends to differentially motivate persons who are at different points in the distribution of achievement. Only those at the high end of the distribution are highly motivated. Second, it can be seen as unfair to teachers, parents, and students. If we accept that different children have different abilities, then to measure all against a single standard of performance is not to have a level playing field: different abilities give, in effect, differentially efficient equipment to students, parents, and teachers, to reach the same goal.

But if an evaluation provides rewards only for gains in achievement, then there are at least two negative consequences. One is that this creates a moral hazard: gains in achievement can be brought about not only by high final levels of achievement, but also by low initial levels of achievement. Thus this creates a motivation to have low initial scores as well as a motivation to have high final scores. Second, a system of evaluation with rewards only for gains in achievement lacks the authenticity that one having rewards for level of achievement automatically carries. Thus it would lack the legitimacy necessary for a stable evaluation system.

Note that not only do the evaluations used in the United States — SAT and ACT — fail this criterion, but the Baccalaureate, Matura, and A-levels fail it as well. Each measures level of performance, and thus serves its motivating function only for high performers; there is no measure of gain in performance.

The evaluation must provide rewards that motivate teachers, parents, and students.

Certain evaluation systems, such as those in the Japanese and Hungarian systems, and to a lesser extent those in various European countries, meet this criterion, despite failing others. The Hungarian system does this through contests, local, regional, national, and international. Teachers gain prestige and the opportunity to move to a more prestigious school by having high-performing students, just as coaches do in the United States. Students and parents are motivated to win the prize. The defect, of course, is that the effectiveness of this reward differs greatly at different points in the distribution of performance. It is most effective for high achievers, and for teachers and parents of high achievers.

Teachers can be motivated by pay, by career opportunities, and by autonomy in their jobs. Students and parents can be motivated by various benefits to be discussed, and also by recognition for doing well; but recognition for doing well academically is not automatic.

Yet recognition depends on conditions under the control of educational policy. For example, schools now have informal rankings in parents' and teachers' minds in terms of the **level** of performance of students. This is reinforced by the reporting of standardized test score in local newspapers. But if the year-to-year **gains** in performance (for example, gains in percentile position or standard score) were reported, there would be a shift in the rankings, as a consequence of publication of the gain scores.

Possible Designs

There are six elements of a new school design that are central, each of which is not present in the design of current schools for American children: Externally imposed standards as the basis for all evaluations of student performance, evaluations based on two measures: **level** of performance, and **performance gain** or **value added**, yearly rewards to teachers, students, and parents for level of performance and performance gain, using the final output criteria (the externally imposed standards) as the starting point for designing evaluations at each stage of the education of a child, creating a system with short feedback loop, allocation of rights and responsibilities not only to individuals, but also to groups of teachers, groups of students, and groups of parents, to encourage the development of social capital, that is, informal norms that support educational goals, the use of a core of academic achievement plus an area of specialized performance (which may be academic, but need not be) as the performance criteria.

I can anticipate that not everyone will embrace eagerly the three school design principles I have laid out. In part this is because I have not been able here to do more than sketch the bare outlines of what I have in mind. But the disagreements will certainly not vanish by greater design detail. I recognize this.

What I do hope to have achieved, however, is to have convinced you that careful application of well established sociological principles can be the basis for design of a school — or of another type of organization. I hope also to have convinced you that the task of engaging in such design is an important task for sociologists. I hope that I may have challenged some of you who do disagree with my designs to draw upon sociological knowledge in developing competing designs.

Both social research and social theory are important inputs to this task, but the design, and its realization, constitute the crucible which tests social theory. It is a crucible that sociologists should not avoid, and one which if we do not avoid it, constitutes a major contribution of sociology to the future of society.

Notes

1 A note by editors — As mentioned earlier, James Coleman was not able to attend the workshop because of illness. He had prepared the following paper and we have included it in

this volume. In his chapter, Coleman argues that the key to transforming schools is to reorganize them so that they are not paralyzed by authority-driven forms of organization, which he describes as dated and counterproductive. He suggests, instead, that schools become "output-driven". By output-driven he means that students are evaluated by external standards, and that schools should be reorganized so that rights are reallocated from central authority figures to teachers and learners.

The relation of Coleman's perspective to school choice is the linkage between school improvement and creating schools of sufficient quality so that choice becomes meaningful. Professor Coleman had written a great deal on school choice and had he had the opportunity, there is little doubt that at the workshop he would have made the linkage between his theory of school improvement and school choice clear.

We are fortunate to be able to include his chapter in this volume.

2 One worker's cooperative, planting trees under contract to the US Forest Service, organized work in teams of 12 members who monitored one anothers' work (Mackie 1991). They concluded that for purposes of workgroup monitoring and maintenance of productivity-oriented norms, 12 members was an optimum size for the workgroup.

Chapter Eleven

SCHOOL CHOICE AND THE CREATION OF COMMUNITY[1]

PETER W. COOKSON, JR.
Teachers College, Columbia University

Introduction

Few school reform movements have aroused public passions as deeply as school choice. If educational reforms can ever be said to be hot, then school choice must be scorching. Sides are being taken and lines are being drawn. Much of this debate is emotional and ideological, although there is an evolving body of research (Chubb and Moe 1990; Coleman, Schiller and Schneider 1991; Crain 1992; Driscoll 1992; Lee 1991; Plank *et al.* 1992; Wells 1991; Willms and Echols 1992; Witte 1992; Menahem *et al.* 1993; Cookson 1994). Generally, this research addresses four major issues: the relationship of choice to student achievement, to school improvement, to educational equity, and to community building. As to the first issue, there is little convincing evidence that choice leads to higher achievement scores; the simple act of choosing a school bears little relation to cognitive outcomes. There is some evidence, however, that choice may lead to school improvement — the evidence concerning the relationship between choice and equity is mixed. In the end, it may be that choice's most significant contribution to educational improvement is the degree to which family choice is a building block for developing stronger communities. Choice allows families to feel a sense of "ownership" of schools and choice enables families of similar demographic, religious, and ethnic backgrounds to come together in a common effort: school choice rests on social trust, and if designed properly school choice plans can build social trust. There is no more critical issue facing American education today than reconnecting families and students to schools. For at least 30 years, schools have become increasingly isolated from their communities, partly due to changes in demographics and partly due to the types of enrollment policies that have evolved over time.

This chapter is an examination of how school choice can create more vibrant educational communities. The effects of school choice can be direct and/or indirect. An example of a direct effect would be if school choice had a measurable effect on student achievement. Indirect effects are more global in nature, less measurable perhaps, but significant nonetheless. "Ownership" is not a concept that easily lends itself to measurement. Ironically, research seems to indicate that the impact of school choice is greatest outside the classroom rather than within; in other words, choice policies have greater impact on the culture of schools, and the relationship between schools and communities, than classroom practices (Cookson 1994). While the data for these conclusions are not as strong as one would like, there is sufficient evidence to date to lend credibility to this chapter's central position: properly conceptualized choice plans can renew schools and empower families in the education of their children. I should say at the outset, however, that in my estimation school choice is a tactic of educational reform and not an overall strategic plan. To truly improve schools, reformers will have to address such issues as purpose, personnel, curriculum, and resources. This chapter is divided into three parts. Part one examines the educational and social context within which the school choice movement developed and through example seeks to demonstrate the limits and possibilities of school choice. Part two discusses some ways in which choice can build communities, and part three is a proposal about how choice can be used in the overall effort to reform and reinvigorate American education.

Rustbowl Schools/Rainbow Dreams

Since the 1980s Americans have experienced what amounts to a national panic attack about the condition of their children and their schools. Squeezed by completely inadequate resources and waves of poor immigrant children, public school systems in virtually every large American city are collapsing: dropout rates hover well above 50 percent, truancy is the norm rather than the exception, violence is common, students struggle for basic literacy often without success, a great deal of teaching is uninspired, and the physical condition of the schools is a disgrace. Most inner-city schools have the look of crumbling fortresses; windows are left unrepaired, and the paint is peeling in so many places that the schools would appear abandoned if it were not for the children. The playgrounds of these schools are asphalt patches that are cracked and littered. These are the "savage inequalities" Jonathan Kozol (1991) writes about so eloquently and so poignantly.

The condition of American children is also so profoundly serious that at a very basic level it could be said that all American children are at risk. To cite but a few examples: one-fourth of all preschool children in the United States live in poverty; each year roughly 350,000 children are born to mothers who were addicted to cocaine during pregnancy; 15 million children are being reared by single mothers whose family incomes average a little over $11,000 a year; only 6 percent of American children are raised in the traditional household with a working father, a housewife mother, and two children of school age; on any given night approximately

200,000 American children have no home; each year child protection agencies receive over two million reports of child abuse and neglect (Hodgkinson 1991: 10). In a **Time** magazine article, Nancy Gibbs (1990 p. 42) observed: "Every eight seconds of the school day, a child drops out. Every 26 seconds, a child runs away from home. Every 47 seconds, a child is abused or neglected. Every 67 seconds, a teenager has a baby. Every seven minutes, a child is arrested for a drug offense. Every 36 minutes, a child is killed or injured by a gun. Every day 135,000 children bring their guns to school". Clearly, education is operating in a cultural milieu that is hostile and even violent toward children.

There is little wonder, then, that critics across the political spectrum have called for the reform of American education. Some critics argue that without a complete restructuring of the governance of public education there can be little hope of genuine reform. The school choice movement arose out of this frustration with the status quo. While choice plans vary significantly, they share a common assumption: freedom of educational choice is essential if we are to lift the bureaucratic weight that currently threatens to crush the spirit of public education. Some of the claims made by choice advocates border on the utopian because they remove schools from society and thereby overlook the complex relationship between social structure and school processes. School choice cannot overcome social inequalities even in states that are willing to experiment educationally and have an ethos of equality.

The state of Minnesota, for example, has experimented widely with a variety of school choice plans. Many of these experiments have been ground-breaking, but as the case studies below indicate we should be hesitant to claim that school choice can break the cycle of educational deprivation. The Miltona Science Magnet School and the Urban League Street Academy in St Paul represent the two worlds of choice. Miltona is in the heart of the Minnesota farm country, and the

Street Academy is in the heart of the Minneapolis inner-city ghetto. Miltona is a tiny agricultural community two and a half hours northwest of Minneapolis. Agriculture in Minnesota is somewhat depressed, and in the past several years the State Board of Education has been encouraging small communities to consolidate their schools. This consolidation movement, however, has been resisted because small towns in Minnesota use their schools as community centers. Schools are often the heart of the towns, and the citizens of the towns and the nearby farms take pride in the accomplishments of their schools and the students who attend them. According to Tom Andert, principal of the Miltona Magnet School, "School is the culture of the community".

During the period of consolidation it looked as though the Miltona School might be closed, so Andert and a group of teachers wrote a proposal that turned this small elementary school into a science magnet school. Essentially, Miltona is a school of choice in a rather large geographic area and has attracted families because of its innovative curriculum, the centerpiece of which is its emphasis on science. The whole school feels rather like a science lab and includes a weather station, a greenhouse, and innumerable child-made scientific instruments. According to Andert, most of the students come from Norwegian farm families, with an orientation that he refers to as "Be nice to your neighbors". Interestingly, these families have progressive ideas about education and are supportive of what amounts to a progressive school approach to teaching and learning. There is an air of informality and innovation at Miltona that would have made John Dewey feel at home. Andert and his colleagues have devised an elaborate strategy of getting parents to come to the school, which includes newsletters, radio announcements, open houses, and visits with teachers. The school is a community in itself and part of a larger community.

The Urban League Street Academy is a privately organized alternative program which seeks to personalize the education of students at risk so that they may complete high school. Located in a poor neighborhood of St Paul, the Street Academy is in the basement of an old building. Unlike Miltona, the Street Academy has not had a fresh coat of paint in years and has no funds for capital improvements. The principal, Perry Price, is an African-American educator who is struggling to create a learning environment for at-risk African-American students who have either dropped out of or been dismissed from public school. There are 75 students enrolled in the school; approximately 45 attend each day. The curriculum is African-American centered, with an emphasis on black literature. There are five teachers and a librarian and no substitute teachers, which means that if a teacher is absent, students have no teacher for the duration of the absence. The school relies primarily on donations for equipment, including paper and books, and there is a constant struggle to keep enough toilet paper in the restrooms, to keep the classrooms clean, and to provide the students with the basic necessities. Price needs psychological support for his students, and he needs lockers. At the Urban League Street Academy there is no place for students to hang up their coats. The Academy's library consists of inexpensive paperback books that have been lovingly catalogued.

To me, however, the visual experience that best symbolizes the stratification of educational opportunity is the science lab at the Academy. Tucked away in a corner of the building, the lab consists of a counter punctuated by sinks and faucets. Underneath the counter are drainpipes that hang in mid-air; they lead nowhere because they are unconnected to a water main. The Academy cannot afford to connect these pipes, so students cannot actually do science experiments at the Academy; they can only pretend. According to Price, he never turns down any student, but students do often get lost in the system. What these examples demonstrate is that school choice can build communities if two essential ingredients are present: trust and resources. Without these qualities there cannot be meaningful choice; without good schools the ability to choose is empty.

Changing Hearts, if not Minds

What makes a school good? Schools are not only sites for learning, they are also sites for living. When a hundred or more children or young people and adults unite to create a learning environment, a dynamic of its own takes place. If this dynamic is positive, the school can be an oasis; if this dynamic is negative, the school can feel like a jail. The fact that schools create strong cultures is overlooked by too many policy-makers. These cultures affect the living experiences of children. Educator Mary Driscoll (1992), in her study of parent and student beliefs about public schools of choice, found that, on average, parents in choice schools were more satisfied with their children's education, and that the culture of the choice schools were more positive than that of "zone" schools where attendance is mandated. According to Driscoll, choice parents

> ... believe more strongly that their child enjoyed school and was challenged by it; that the homework assigned was worthwhile and that the school was a safe place. They spent more time talking to their child about high school plans and were contacted less frequently about their child's behavior. Surprisingly, they did not exhibit higher expectations for their child's education in terms of level of completion (BA, Ph.D., etc.). In other words, although they had significantly positive beliefs about their child's school experience, they did not translate these beliefs into higher expectations for completion of college or professional degrees.
>
> (Driscoll 1992: 11)

Driscoll also found that while students who attended schools of choice did not have higher test scores than other students, they "clearly believed that their experiences with teachers were better than did the students in the matched schools". As Driscoll points out, there is a paradox here: whether or not they change minds, choice schools may change the "hearts" of students and parents. Driscoll offers several explanations of why this may be the case, but perhaps the most compelling is that choice parents believe that the school their child is attending is better because it is selective. That is, the very act of choosing creates an aura of specialness.

My research confirms Driscoll's findings. Schools of choice may not be objectively better than other schools, but there is a sense of being special. I have already mentioned the rural magnet school in Miltona, Minnesota. The very fact that families chose this school provides them with a sense of ownership. The sociologist John Meyer discussed the "chartering effect" in education more than 20 years ago. According to Meyer (1972, p. 111), "A charter is an institutionalized agreement that a given program, college, or system of colleges is to produce and does produce a given kind of person". In effect, a choice school can create its own charter; that is, choice schools can appear to be selective and thus adopt some of the cultural characteristics of academically and socially "superior" schools. Driscoll continues:

> It is possible, then, that selective or choice schools establish a charter that seems to promise greater access to superior academic opportunities. Meyer's theory would suggest that this special sense of mission in the school is transmitted to students and parents (perhaps through the selection process) and is reflected in their relatively greater "investment" in academic values (i.e., agreement with the statement "the homework assigned is worthwhile").
>
> (Driscoll 1992: 15)

Driscoll concludes, "There is little evidence in this data that making schools more selective makes them better, if we use academic achievement as our criteria for evaluation. Merely improving choice in education will not serve as a panacea for improving school performance" (Driscoll, 1992, p. 22). Perhaps the most important outcome of school choice is the sense of ownership that students and families have when they are able to choose schools creatively. Choice is a form of empowerment, an overused but nonetheless important term. There is something dignified about being allowed to choose a school rather than being compelled to attend a particular school.

Choice may well lead to new forms of educational organization; certainly, the development of parent information centers has played a significant role in binding families to schools and schools to families. Parent information centers not only provide information about schools to parents, they also provide information about students to schools. The act of going to a parent information center is symbolically rich; families are integrated into their communities by participating in choice. In this sense, school choice, especially managed school choice, can be a breath of fresh air. Schools are required to open their doors to families and in doing so break down the barriers that have isolated schools from their communities and isolated parents from schools. Almost all research indicates that parent involvement in schooling is beneficial to students' cognitive and affective growth.

The first parent information center began in Cambridge, Massachusetts. To avoid court-ordered bussing in the 1980s, the city of Cambridge organized a controlled choice plan which allowed families to choose their children's elementary school. It is critical that the racial balance of the schools reflect the racial balance of the district. The heart of the Cambridge experiment is the Parent Information Center. Essentially, the Center operates as a community service organization which provides information to families about the elementary schools in the community. Parent information centers have multiple functions. They compel parents to think about their children's education, they provide information about the schools in the community, they encourage schools to improve and sharpen their missions, and they provide feedback to all members of the community concerning the schools in which the children are enrolled. The Cambridge Parent Information Center provides much printed literature about the process of school choice and how parents can best choose a school for their child. Parents are encouraged to visit schools and principals and faculties are encouraged to examine their curricula and educational philosophies. When I visited the Parent Information Center in Cambridge, I was impressed by the dignity surrounding the process and the care given to each family, no matter what its racial, ethnic, or socioeconomic status. The Cambridge Parent Information Center has been duplicated in other cities including Fall River, Massachusetts and White Plains, New York.

Fall River is a small industrial and maritime city of approximately 90,000 people, located in southeastern Massachusetts. During the nineteenth and early twentieth centuries, Fall River was an important center for the textile industry. By the 1960s the city had sunk into an economic depression from which it has only recently begun to recover. For many years the city has attracted immigrant Portuguese families, coming predominantly from the Azores, Portugal, and Brazil. Nearly 250 Portuguese families arrive in the city every year. There is also a recent wave of emigrants from Cambodia and South America. In the fall of 1987, the Fall River public schools embarked on an educational venture aimed at accomplishing two major goals: to eliminate all vestiges of minority isolation in the schools, and to make dramatic improvements in the education of all children. They instituted a controlled choice plan very similar to that of Cambridge. Parents played a significant role in planning and implementing the choice program in Fall River. As in Cambridge, every school in Fall River is a magnet school and a school of choice. A city-wide Parent Advisory Council was established, and parents throughout the city were surveyed about what kinds of educational options they would like to see for their children. A parent information and outreach system was established and an evaluation system was designed to assess and track the plan's progress. Thirty-eight percent of those Fall River parents who chose schools since the inception of this program have chosen non-neighborhood schools. Eighty-two percent of the students were admitted to their first-choice school, and 90 percent to one of their first two schools of choice. There is more racial balance in the schools than there was in the past, and there has been a general educational renaissance in Fall River that has been remarkable. Some of these breakthroughs have included literacy/media centers, adopt-a-school programs, mentor programs, computer-assisted education, a transitional bilingual education program, special arts programs, a whole-language approach to reading, and an environmental/marine science program. Part of the push toward school improvement in Fall River is the evolution of school-based management and the continuing pursuit of genuine diversity. Recently, there has been some discussion of expanding choice to other nearby suburban school districts.

Most parents visit the Parent Information Center in the month of February. There are several documents that they must bring with them, including the child's birth certificate and evidence of a physical examination of the child by a physician. School assignments are based on space availability, sibling preference, linguistic minority balance, proximity of school to home, and a random lottery. As if to emphasize the importance of education, every child receives a free book and an application for a library card as he or she leaves the Center. The Center itself includes learning materials, desks for interviews, and a general air of enthusiasm and a love for children.

The outreach program is extensive. It includes literature

about each school participating in the choice program, contact with preschools, telephone surveys, direct mail, and a monthly newsletter called **The Magnet Memo**. When choice was being introduced in Fall River the superintendent, other administrators, and teachers were out almost every night discussing choice with teachers, community groups, civic organizations, and community activists. Another key element in making choice work is to let the power flow to those who must own the choice program — the parents. Without parental involvement, choice cannot succeed.

A parent information center was also established in White Plains, New York. White Plains, located approximately 30 minutes from New York City, has a highly diverse student population of approximately 5,000 students. Most of the white population lives in single-family housing in middle-class or upper-middle-class neighborhoods, while most of the minority population lives in subsidized housing projects. An increasing number of students come from homes where English is not the primary language, and the District educates scores of children from homeless families. In 1988, the Board of Education approved an ambitious plan for restructuring the entire District. The key features of this plan are racial and ethnic balance, a new student assignment plan, school-based management, and decentralization. The Board's current policy on racial/ethnic balance is as stringent as in any district in the country. The goal is to achieve, at each elementary school, a mix among African-American, Hispanic, and "other" students that is "within 5 percentage points of the District average for each of these groups in each of the grade levels". Because all White Plains children attend the same middle school and high school, the issue of racial/ethnic balance is related only to the five elementary schools in the District.

In 1988 White Plains opened its Parent Information Center to help all parents in the District understand the procedures and the rules of the choice program. Parents are asked to choose three schools and rank them in order of preference; incoming kindergartners are guaranteed their school of first choice if they have siblings attending that school. Incoming students are also given preference for their first choice school if they live in the school's former catchment area. After the guaranteed placements are assigned, a public lottery is held and students are placed on waiting lists which are publicly posted at the Parent Information Center. An average of 89 percent of families are assigned their first-choice school, and all families who complete the process are assigned one of their three choices. In 1991, 36 percent of parents with children entering kindergarten chose to send their children to schools outside their former attendance zone. The outreach program includes putting up posters throughout the community in laundromats, bodegas, libraries, health and day-care centers, and churches. These posters include tear-off reminder cards encouraging parents to visit schools. During the first year of the program, 400 visits were scheduled; in 1991, 675 visits were scheduled. Like the Parent Information Center in Fall River, the Center in White Plains has become a community center for families and children.

This discussion of parent information centers has been extensive because I believe that without school choice there would be no parent information centers and without these centers the rift between families and schools would remain a chasm of misunderstanding. As Driscoll's work shows, school choice can alter the way people look at their educational opportunities and as a consequence these policy options have ripple effects that extend beyond issues related strictly to school governance. Thus, it seems sensible to integrate the concept of school choice into a strategy for the revitalization of the public education. Below is a brief schematic outline of how I believe school choice can help create and improve school communities and the communities in which schools are located.

A New Educational Covenant

A justice-driven educational system requires that all children have an equal share of the community's educational resources; it also requires that we provide incentives to educational innovators to revitalize their schools while at the same time holding them accountable for their students' intellectual growth, physical safety, and moral development. I am suggesting a plan for the revitalization of public education that includes an educational trust fund for every American child, managed public school choice, and the creation of model schools. In the proposal detailed below, the rationale for these policy suggestions and a plan for their implementation are outlined. The underlining principle of the plan is that we must enter into a new educational covenant that gives high priority to the welfare of children, to the revitalization of the democratic vision, and to the celebration of learning.

A fundamental principle of this alternative educational design is that every child has the same right to educational opportunity as every other child. This does not mean that all children have equal talent, or that we should expect all children to reach the same level of academic proficiency. The right of access refers to the social obligation of the community not to discriminate against a child because of family characteristics or personal limitations. Every child is entitled to an educational share. This share can be used at a public school of choice. In a society that is highly stratified and where the financing of education is also highly stratified, it is necessary in a redesign plan that the monetary worth of an educational share be in inverse relation to the family's income. The poorer the child, the greater his or her educational share. In effect, every child receives at birth an educational trust fund that guarantees equal access to the public schools of his or her choice. How these trust funds will be financed will be discussed in the proposal itself. The fundamental concept of the educational trust fund is that we, as a community, make a commitment to children that their life opportunities will not be foreshortened for reasons of their birth.

Experience with controlled choice has shown that managed public school choice can create racial integration, involve parents in the education of their children, and lay the groundwork for the rejuvenation of public schools. Inter-district public school choice makes every public school a magnet school. Public schools in a managed choice system would continue to receive revenues raised through taxation and state-operated gambling. However, these revenues would cover only 70 percent of the school's operating budget. The remaining 30 percent would have to be raised by attracting poor students who have educational shares to spend. A public school that receives such educational shares from students must ensure that 20 percent of its student body is composed of those whose family's incomes are at or below the poverty line. It is important to remember that these students receive larger shares than others and are thus economically attractive. The key incentive for public schools to recruit share students, especially share students from disadvantaged homes, is that the schools retain any surplus accrued during the school year in order to improve their programs, reimburse their faculty, or maintain the school itself. This method of financing public education balances the need for a strong, publicly supported school system with the need to stimulate competition among public schools. Managed school choice is a method of encouraging schools to innovate by rewarding those schools with educational vision and social commitment.

Equity is a necessary but not a sufficient condition for a true educational transformation. Just as a community has the obligation to protect its children and to guarantee equal educational opportunity, so it also has the obligation to promote individual and family liberty. Thus if teachers and parents are able to found schools that do not violate the conditions of the educational covenant discussed above, these may be designated "model schools" and made eligible to receive shares from children's educational trust funds. Model schools must have 50 or more students and must enroll at least 20 percent of their students from families whose incomes are at or below the poverty line. If the founders of model schools do not violate the rule of equal access, then the state should have only a small role to play in monitoring and regulating the schools.

The Proposal

The central purposes of this educational share plan are to promote educational experimentation, to provide equal access to educational opportunity for all American children, and to create a superior system of public schools in the United States.

The educational trust fund

Who is entitled to shares in the educational trust fund?

(1) Every native-born American child has a right to shares in the educational trust fund. Each child's fund will be established one year after his or her birth.

(2) Non native-born children aged one year or more are entitled to a trust fund from the date of their first entrance into the American educational system.

How will the shares be distributed?

(1) Every child must receive a social security number by his or her first birthday. An educational trust fund cannot be established until a child has acquired a social security number.

(2) A computerized data base will be established for each one-year-old child and will be updated

as needed. This data base will include name, date of birth, home address, and notation of any handicapping conditions.

(3) Each child will receive an educational share from the Department of Education in the state in which he or she resides. State Departments of Education may lawfully appoint certain financial institutions to act as their agent for the distribution of shares.

(4) The monetary worth of a share shall be determined by a formula that reflects the following principle: the lower the family income of a child, the greater the value of his or her share. Family income will be determined according to federal tax returns. If parents do not file an income tax return their yearly income may be determined by alternate means.

(5) At no time shall the monetary worth of a share exceed the annual average per-pupil cost of the state in which the child resides.

(6) Shares will be sent to students' homes. If a student has no known address or is without a home his or her share will be sent to the central office of the last school district in which the student was a resident.

(7) Unclaimed shares will be sent to the central office of the last school district in which the student was known to reside. These shares will remain for three months in the school district and if unclaimed will be returned to the State Department of Education.

(8) Shares cannot be accumulated from year to year, nor are they transferable between recipients.

(9) All interest accruing from educational shares will revert to the state and not to the child or his or her family.

How will the educational shares be financed?

(1) On the premise that the present method of financing public education is unconstitutional because it denies students equal protection under the law, states would revise their current method of collecting property taxes as it relates to school finance. The purpose of this revision would be to equalize the amount of money available for spending on education between and within school districts. (Very recently the state of Kentucky passed a law which provided for equal financial support of the school districts in the state.)

(2) Part of the revenues raised through property taxes would finance 40 percent of the educational trust funds. An additional 25 percent would be financed through other state forms of taxation, including the sales tax. Another 20 percent would be raised through public lotteries. Fifteen percent would be contributed by the federal government either through a special bond issue or by budgetary allocations. Other revenues needed to support public-sector schools would be raised through property taxes, user taxes, and state-supported gambling.

(3) All revenue will be kept in trust by the State Departments of Education or their fiduciary designees.

Managed choice and model schools

What types of schools may receive educational shares?

(1) All currently operating public schools may receive educational shares. Public schools will receive 70 percent of their revenues directly

from the State Departments of Education; 30 percent of their revenues must be raised through educational shares. All public schools that receive educational shares must make a commitment to accept at least 20 percent of their students from disadvantaged backgrounds. Any monetary surplus that a public school accumulates may be retained by that school for purposes of school improvement.

(2) Newly founded model schools may also receive educational shares. These model schools may be founded by teachers and parents in collaboration with other educational providers. A model school must have a population of at least 50 students and provisions for fiduciary responsibility. Model schools may not espouse hatred or hostility toward others on the basis of race, ethnicity, social class, religion, handicapping condition, or gender. Model schools must also meet the requirement of equal educational access. All model schools must make a commitment to accept at least 20 percent of their students from disadvantaged backgrounds.

(3) Model schools may set whatever admissions criteria they wish as long as these criteria do not discriminate against students in terms of race, ethnicity, class, or gender. Schools that do not admit students with handicaps may have limits set on the amount of educational shares they receive.

(4) All model schools will be subject to state requirements concerning health, safety, and student welfare. The state reserves the right to close a school if that school violates standards of health, safety, and student welfare.

(5) All schools that are eligible to receive educational shares must fully disclose publicly their missions, their tables of organization, their curricula, and such pertinent financial information as may be required by the state. This information will be made available to parents and children through published reports distributed through parent information centers.

(6) Every model school must be insured and bonded to protect consumers from fraud, incompetence, and malfeasance.

Issues of Implementation

This proposal is meant to provide a conceptual framework upon which individual state plans could be developed, although the essential components could not be substantially changed without weakening the entire structure. To my mind there are three major areas of concern in terms of implementing this proposal: (1) the distribution of the educational shares, (2) consumer information, and (3) consumer protection.

The distribution of the educational shares

Clearly, one of the primary requirements for the transformation of American education is ending bureaucratic control over individual schools. Part of the reason that state and city educational bureaucracies have grown in the past 45 years is that they have taken on a wide variety of functions. Some of these functions are directly related to teaching and learning, but many are related to the maintenance of the bureaucracies themselves. This has proved to be an obstacle to genuine educational reform. It would make little sense to dismantle the bureaucratic regulation of schools only to substitute another bureaucracy to distribute educational shares. The role of the state in the regulation of education and in the distribution of educational shares should be minimal.

The responsibilities of the state unit that would administer educational trust funds are to be extremely circumscribed. These responsibilities would include: (1) to collect and "bank" the trust funds from the four sources of income mentioned earlier; (2) to collect the names, addresses, levels of family income, and information about possible handicapping conditions of all students who have received a social security number and are state residents; (3) to determine the value of each child's educational share according to the formula determined by the state legislature, and (4) to create the shares and distribute them. In states with populations in excess of 10 million many of these tasks could be divided into geographical subsections in order to reduce the error and delay often associated with organizations servicing a large number of clients. Before the advent of the computer, the distribution of educational shares might have been cumbersome; sophisticated computer systems, however, have made large-scale data processing feasible. Undoubtedly, for children who come from highly transient families or from families without homes, there would be difficulties related to administering the share system. Yet, using computerized data bases, children would not be lost in the system as long as each child had a social security number. If a child sought enrollment in a school district other than that in which he or she had originally enrolled, a computerized data search would reveal the size of the child's share. When families relocate across state lines they may apply for a new share. Students who transfer during the school year will have the value of their share determined on a pro rata basis. With the advent of computers, there are many possible efficiencies that could be built into the system. In any case, because the authority of the units administering the educational trust funds is focused and limited, it would be relatively easy for outside auditors to evaluate their operations.

Parent information centers

School choice and social trust are contingent upon accessible and accurate information. One of the main benefits of encouraging student and parental choice is that it compels schools to identify their missions, examine their strengths and weaknesses, and make this information available to families. Research in parent information centers indicates that a critical element in ensuring that parents and children use the centers fully is an intense effort by those who manage the centers to reach every family in the community, through mailings, radio and television announcements, posters, and visits to churches, nursery schools, laundromats, shopping centers, and other locales where families are likely to gather. Information should be made available in the languages of the families living in the community. Moreover, parent information centers should treat each family as unique and dignify the process by which families choose schools. Information collected at the centers can be used to supplement other information the state may need in order to determine the residences of children. In essence, parent information centers are community resources that bring schools and families together and act as benign brokers of educational choice. Without investments in these centers, the process of school choice becomes chaotic, uninformed, and potentially destructive to children.

Consumer protection

It is the responsibility of the state to ensure that taxpayer dollars are not wasted through fraud or incompetence. It is extremely important that educational trust funds be used to support legitimate enterprises that fulfill the educational promises they make. It is essential that every public and model school that accepts educational shares report annually to the state on its educational mission, organizational framework, the characteristics of its faculty and students, and its financial condition. It is also important that the state reserve the right to investigate complaints by visiting schools and subpoenaing relevant records. Serious complaints of fraudulent practices could be channeled through consumer protection officers working in parent information centers. These officers could turn their findings over to the local district attorney's office if it was deemed that a case warranted further investigation. Parents who wish to educate their children at home may not receive educational shares because there is no reasonable way to monitor whether or not these shares are used for children's education.

Conclusion

Scholars have differed as to the reasons why school choice has become the dominant reform idea of the current era. Sociologist Kevin Dougherty and educator Lizabeth Sostre have attempted to explain the rise of the school choice movement through the "state relative autonomy" theory of political power and the "garbage-can theory of organizational decision-making" (Dougherty and Sostre 1992, p. 160). The former theory hypothesizes that the state is a neutral arbiter, weighing and directing interest group pressure; the latter theory, first hypothesized by James March and his associates over 30 years ago, makes the commonsensical point that people choose policies for conflicting reasons and because policies are available. Organizational leaders have short-term objectives and do not actually know what the consequences of their decisions will be. My experience in studying school choice, however, is that this reform symbolizes something very important in American society. The ideals of individualism, autonomy, and competition run deep in the American character. I believe that we are in the midst of a profound cultural transition; the core consensus that united public opinion since the Great Depression has all but evaporated. The debate over school choice is in some ways a metaphor of what constitutes not only the good school but the good society. Much of this debate has been unproductive because it has ignored the real world of schools, families, and children. I believe that choice can be a mechanism for achieving a school system that is just, innovative, and academically productive. It is a method by which we can reinvent public education, reintegrate schools into their communities, and redefine community. Schools can become oases of authenticity in a troubled and often alienating world. If school choice policies can aid in creating a greater sense of community within schools and between schools and families then these policies will have contributed to a safer and more intellectually challenging world for children.

Note

1 The data for this paper have been collected for four years. As part of the background for this study I conducted an extensive review of the literature, and every state in the union was surveyed in 1992. Field visits were made to a number of states including California, Minnesota, Massachusetts, and New York. Interviews with leading opponents and proponents of school choice were conducted over a two-year period.

References

Chubb, J. E. and Moe, T. M. (1990). *Politics, markets, and America's schools.* Washington, DC: Brookings Institute.

Coleman, J. S., Schiller, K., and Schneider, B. (1991). Parental involvement and school choice. In *Resources and actions: Parents, their children and schools.* Chicago: NORC/University of Chicago Press.

Cookson, P. W. (1994). *Contested ground: School choice and the struggle for the soul of American education.* New Haven, Connecticut: Yale University Press.

Crain, R. (1992). *What can New York City's career magnet high schools tell us about the choice debate?* Paper delivered at "Choice: What role in American education"? Sponsored by the Economic Policy Institute, Washington, DC.

Dougherty, K. and Sostre, L. (1992). Minerva and the market: The sources of the movement for school choice. *Educational Policy*, **6**(2), 160–179.

Driscoll, M. E. (1992). *Changing minds and changing hearts: Choice, achievement and school community.* Paper delivered at "Choice: What role in American education"? Sponsored by the Economic Policy Institute, Washington, DC,

Gibbs, N. (1990). Shameful bequests to the next generation. *Time, October* **8**, 42–46.

Hodgkinson, H. (1991). Reform versus reality. *Phi Delta Kappan, September*: 9–16.

Kozol, J. (1991). *Savage inequalities: Children in America's schools.* New York: Crown.

Lee, V. (1991). Educational choice: The stratifying effects of selecting schools and courses. *Educational Policy*, **7**, 125–148.

Menahem, G., Spiro, S. E., Goldring, E., and Shapira, R. (1993). Parental choice and residential segregation. *Urban Education*, **28**(1), 30–48.

Meyer, J. (1972). The effects of the institutionalization of college in society. In K. A. Feldman (Ed.), *Selected Readings in the Social Psychology of Higher Education*, pp. 109–126. New York: Pergamon.

Plank, S., Schiller, K., Schneider, B., and Coleman, J. S. (1992). *Choice in Education: Some Effects.* Paper delivered at "Choice: What Role in American education"? Sponsored by the Economic Policy Institute, Washington, DC.

Wells, A. S. (1991). The sociology of school choice: A study of Black students' participation in a voluntary transfer plan. Unpublished doctoral dissertation, Teachers College, Columbia University, New York.

Willms, J. D. and Echols, F. H. (1992). *The Scottish Experience of parental choice of schools*. Paper presented at "Choice: What Role in American Education"? Sponsored by the Economic Policy Institute, Washington, DC.

Witte, J. F. (1992). *The Milwaukee Private-School Parental Choice Program*. Paper delivered at "Choice: What Role in American Education"? Sponsored by the Economic Policy Institute, Washington, DC.

Chapter Twelve

CREATIVE NON-COMPLIANCE

SEYMOUR FLIEGEL
The Manhattan Institute, USA

This chapter describes and analyzes the widely publicized reforms in District 4 of the New York City school system beginning in 1974. In demographic terms, District 4 is one of the poorest school districts in the nation. The chapter describes the historical development of alternative conception schools, beginning with three schools in 1974 and expanding to include 23 schools and nine bilingual schools today. The reforms as described were premised on the assumption that by encouraging professional staff members to develop innovative programs based on interests, talents, or philosophical beliefs, and by giving them the opportunity to realize their dreams and participate in the decision-making process, a district can succeed in maximizing the potential of both staff and students. The results provide students and parents with a choice in selecting a school geared to their interests and abilities.

Creative Noncompliance in East Harlem Schools

The current furor over corruption and incompetence in local school districts in New York City has provoked another outcry for dismantling the decentralized system instituted in the 1960s and "recentralizing" power within the school system so that the Central Board of Education exerts increased authority over all of the city's schools. Before addressing the issue of relocated authority, I would like to describe one program which has evolved in East Harlem, made possible by the latitude afforded local school districts under the current arrangement.

Over the past two decades, District No. 4 in Manhattan has developed a system of "choice" at the junior high school level which permits students and parents to select an intermediate school from a list of differentiated alternatives available to them. Our thinking followed these lines: youngsters in private and parochial schools have long enjoyed systems that allow for individual preferences; why should children in public schools fail to benefit from the opportunity to select a quality school that meets their interests and abilities? In District No. 4, we proceeded on the belief that what is good for the children of the rich is good for the children of East Harlem.

Our motive was to better our schools. Parental choice can provide the catalyst for educational reform by introducing a market mechanism into the public school system — if parents are permitted to choose which school their children will attend, each school has reason to question how its own educational program might be improved. If this questioning becomes widespread, and a reasonable degree of support is forthcoming, a school district can become a marketplace for ideas, innovations, and investments. In addition to the bracing effect of "competition" for informed students, parental choice in public education can increase the sense of ownership in a school for parents, students, teachers, and administrators. While choice in itself is not a panacea, it can stimulate a framework for improvement efforts in districts and individual schools.

In this chapter I will describe our experience establishing a system of parental choice, draw some lessons we learned in the process, and consider current prospects for similar developments given the current attitude toward decentralized reform in the City of New York.

The alternative school approach in District No. 4

When I was Principal of P.S. 146 in Manhattan, I found one year-end ritual particularly disheartening. Each June, a number of parents of graduating sixth-graders would come to my office to express concern about the local junior high schools to which their children would be assigned according to the "zoning" of the community. These were invariably the best-informed parents, whose information about the local junior high schools provided some grounds for genuine concern. I knew the reputations of the schools to which their children would be sent: they had reason to worry. And yet, the best suggestion I could make — the best the system could offer — was, "Why don't you use someone else's address"?

Teachers and principals in other school districts have certainly given similar advice to worried parents of elementary school graduates, but the situation in East Harlem at the time was especially troubling. In 1973 the schools in District No. 4 ranked 32nd out of 32 districts in the City of New York by reading and mathematics scores. Less than 16 percent of the students in the district were reading at or above grade level.

These numbers describe a school district that is not working to educate its children and an unlikely site for ground-breaking educational reform. Expectations of improvement were not high at the central Board of Education, and funds for that purpose were not forthcoming. Nor was assistance likely to come from parents of children enrolled in the schools: families in the surrounding community lack the financial resources which are usually needed for any sort of enrichment at the individual school level or innovative experimental program.

District No. 4 is located in New York's East Harlem neighborhood, sometimes called Spanish Harlem or "El Barrio". Approximately 65 percent of the population is Hispanic, 34 percent Black, and 1 percent nonminorities. Its geographical boundaries extend from 96th Street to 125th Street and from Fifth Avenue to the East River. The poverty and ethnic isolation of the community is in stark visible contrast to the well-heeled Manhattan neighborhoods to the south.

Statistics make dull reading unless they are understood as referring to people — in this case, to suggest the obstacles confronting the students of District No. 4, and the environment surrounding the schools. East Harlem has the eighth highest rate of welfare recipients among the 26 poverty area of New York City. Thirty-one percent of the population receives assistance from the Department of Social Services; this number is double that of New York County and the City as a whole, and nearly three times the median for New York State. Twenty-seven percent of the households in East Harlem are headed by single parents; this compares with 9 percent in New York County and a State median of 7 percent. Almost a third of the people in East Harlem are adolescents, twice the percentage in Manhattan as a whole.

Conventional wisdom and demography suggests District No. 4 would be a difficult place to produce any significant educational change. And yet the desperation of the educational and economic prospects of the community provided something of a spur to reform in East Harlem schools. I believe in the judo principle — using the force of your opponent or opposition to make headway, turning a negative force into a positive one. A failing system allows for innovation and risk taking that a mildly (or inconsistently) successful system might not allow. The leadership in District No. 4 chose to take risks and be innovative, in part because it had nowhere to go but up.

Meaningful change can take place in the education of youngsters in what appears to be the most difficult situations.

In East Harlem we had the opportunity to implement changes, and we began slowly. We did not have a long-range plan that has brought the district step-by-step to where it is today. If I had to characterize the process. I would say ours developed organically, growing out of the needs, the aspirations, the energies, and the visions of a number of dedicated educators, parents, and community leaders.

The first phase: Creating alternatives

In 1974 we started three small alternative concept schools. One of those schools was the BETA School — Better Education Through Alternatives. BETA is the kind of alternative school that is always supported by everyone, including those people who do not support alternative education, because the school offers "Give us your troubled youngsters and we will take them off your hands". I have often thought that you could raise support for this kind of alternative school if it consisted of no more than the facade of a school building sitting on the edge of a cliff. Principals in other schools would be happy to transfer youngsters who are acting out in class, who take too much of their time, and who are not succeeding in their academic programs.

The BETA School was a success from the first day it opened. We were able to provide those "acting out" youngsters with a learning environment that accepted them as individuals who had difficulties adjusting to the regular school program. Three years after opening, it was cited as a model program. Enrollment in the BETA school is one measure of the improvement that has taken place in all the schools in District No. 4: at one time there were 240 students enrolled, while today there are less than 60 "acting out" youngsters in the district.

The BETA School was designed for a specific student population — troublesome students — and its usefulness for people teaching and learning in the regular schools was clear from the outset. Not all new alternative concept schools are so warmly welcomed. At the same time as the BETA School was started, we opened the East Harlem Performing Arts School and Central Park East — neither with a mandate to take on disruptive youngsters. In **Quality Education in the Inner City: The Story of the Central Park East Schools**, David Bensman of Rutgers University describes the early days of Central Park East:

> In the summer of 1974, staff members of what would soon be Central Park East School (CPE) began posting fliers in East Harlem Day Care and Head Start centers to announce the opening of their new elementary school that fall. Their school-to-be had no constituency, no student body, no base of support in the poor, largely Hispanic community where it would be located. Its physical facilities — two floors of a seventy-five year old, run-down elementary school on 103rd Street between Fifth Avenue and Madison — could hardly have been less hospitable to the open classrooms the staff aimed to create.
>
> (Bensman 1987)

The building was already occupied by a regular school, whose principal outranked the fledgling Director of Central Park East; no grant money was available, nor were special provisions made for exemptions from the many rules of the Central Board of Education and the United Federation of Teachers. Education reformers in the district, then engaged in developing bilingual programs, regarded the new venture as a rival for resources; conservatives were suspicious of its "permissive" educational philosophy. Its reputation had to be built on the experience of children in its classes — word of which spread rapidly through the district and beyond. Bensman (1987) reports:

> By 1984, as the school celebrated its tenth anniversary, CPE was widely hailed as a success. Well-known

> educators from across America extolled its innovative practices and its startling successes. At the local level, the school's reputation ensured it had more applicants than it could accommodate, while parents from distant neighborhoods desperately sought admittance for their children.
>
> (Bensman 1987)

The third of our original alternative concept schools, the East Harlem Performing Arts School, has been renamed the Jose Feliciano School, and has been adopted by the performer as a cause for benefit performances. It is one of the largest alternative schools in the district and is recognized as one of the outstanding performing arts schools in the City. Students are accepted on the basis of their interest in the arts rather than their proficiency, an admissions policy decision with many benevolent consequences for the academic curriculum.

By 1976 there were six alternative schools in District No. 4 and I was appointed the first Director of the program as a whole. My job, as a I understood it, was to nourish, support, and protect the existing alternative schools and to develop new schools which would offer students in East Harlem real alternatives to the local junior high schools their geographic location would otherwise select for them. Having broken the mindset of one school per geographic location, we were free to design schools as we felt they should be designed, reflecting our own biases and imaginations.

I believe planning is essential. Central bureaucracies, however, tend to plan too much and do too little. In New York City, they take three years to plan a failing high school. Much of the energy and excitement of a new project can be exhausted before the school opens; too often, the people responsible for running the school have not played major roles in the development of the plan. We wanted to make certain that our own bureaucrats in the district office did not get in the way of new ideas and enthusiasms. The role we played at the district office was to create an environment that encouraged dreamers and planners to come forward with their visions, and then to help them develop their dreams into realities. We made judgements with regard to the credibility of a plan, and the role of the proposed school in our overall vision of the district. We were aware of the kinds of schools we needed to broaden our appeal and meet the needs of students we were not reaching. But the idea for each school had to arise from the local level as an expression of the personnel whose personal commitment would eventually determine the success or failure of the enterprise.

The process of change in District No. 4 was a process in which teachers or principals came forward with proposals for new alternative schools, each with a philosophy or organizing theme for the new alternative schools, and were met with approval and encouragement from the leadership of the district. Revealingly, despite the diversity of personalities and philosophies of the teachers who presented ideas and plans for the new programs, their reactions, when given the final okay, were always the same. They were shocked and surprised and usually asked, "Do you mean it"? Their expectation of the educational system was that even when people express admiration and support for a new idea, the establishment would always find a reason for not being able to do it at that time. This expected attitude of complacency about the status quo is far easier to maintain in districts where enrollments are guaranteed by geographical zoning; when students and parents are given a choice among schools, enthusiasm and imagination on the part of teaching staff are quickly more highly esteemed.

Over the next 10 years, 23 alternative schools were approved and established. We build them one by one, expanding earlier schools or adding new ones where teachers in the district were willing to work together to develop a more exciting place to teach. Each prospective director was carefully screened and had demonstrated success in a previous position.

The further development of the Central Park East schools provides a good example of the process by which parental choice was extended throughout the district. The original Central Park East school was so successful that CPE II was opened, and River East became the third elementary school following the philosophy and practices of CPE I. Debbie Meier, Director of the original Central Park East and a gifted educator, was awarded a MacArthur "genius" grant for her work in creating the schools. In an article in the January 1986 issue of **New York Magazine** entitled, "The middle-class parents' guide to the public schools", Katherine Davis-Fishman ascribed the growth of the Central Park East Schools to the strong word-of-mouth earned by the first school among parents in the community:

> Three magnet schools, each with two or three grades and about 90 students, were launched in District No. 4 in the mid-seventies. Each was given a few classrooms in a larger school, and the teachers who ran them agreed to take on directors' responsibilities with no salary increase . . .
>
> One of the District's original magnet schools, Central Park East at 19 East 103rd Street, has earned considerable recognition. As an elementary school, it received such a strong park-bench press that it had to reproduce itself twice in order to meet the demand without losing intimacy.
>
> (Davis-Fishman 1986)

In 1985 we opened the Central Park East Secondary School, starting with grade seven. Eventually the school will include grades seven through 12. The school is guided by the principles advanced by the Coalition of Essential Schools, a national organization of public and private high schools directed by Theodore Sizer, Dean of Education at Brown University. Briefly stated, these principles include the idea that less is more: It is better to know some things well than to know many things superficially. High standards are set for all students, with students required to be able to demonstrate mastery of a subject. Teaching and learning are personalized; students are perceived as workers, and teachers as coaches. Students discover answers and solutions, and learn by doing.

CPESS offers a common core curriculum for all students in grades seven through 10, organized around two major fields: mathematics/science for half the school day, and humanities (including art, history, social studies, and literature) for the other half. Every effort is made to integrate academic disciplines, so that students recognize and understand the relationships between different subjects of study. Communication skills (both writing and public speaking) are taught in all subjects by all staff members, regardless of disciplinary focus.

Different educational philosophies, of course, attract different students, whose parents may feel strongly about one or another approach to their childrens' learning. Programs which allow choice among public schools provoke staff to define, for themselves as well as for parents of potential students, the ideas about education which inform their teaching practices. What is essential is not whether a given school subscribes to a more progressive or more traditional philosophy, but that the schooling reflect the enthusiasm and commitment of its teachers in an academic program that is sound, integrated and rigorous.

The second phase: Opting for choice

By 1982, the Alternative Schools Program had enough of a reputation to invite all of the students in the district to select a junior high school from a list of offerings whose diversity and quality amounted to a real choice. The move to a district-wide system of parental choice was noted in an editorial in the **New York Times** on 25 March 1982:

> New York's District No. 4 has done it again. For nearly a decade, the East Harlem Community District has been a font of innovation, offering "alternative" junior high schools that give students unusual choices. Now it will take the further step of allowing seventh-graders to attend whatever district school they like.
>
> (**New York Times** 1982)

The program does not in fact allow students to choose to attend any junior high school they like — each child of age for junior high school is sent an application form with three alternatives listed, among which the student and his or her family may choose. All of the schools are described in an accompanying pamphlet, which outlines the educational theme or philosophy of each, its special rules and requirements. Prospective students must apply for entrance into the alternative schools, who decide whether a given student is granted admission. But choices are fairly distributed, admission policies are reasonably drawn, and the quality of the alternative schools is high enough for an estimated 90 percent of students to gain entrance into at least one school of choice.

Results of the move to choice were dramatic: today, more than 64 percent of students in District No. 4 read at or above grade level, and since 1981, the district has ranked 16th–18th among 32 districts citywide in reading, and 21st in mathematics. While these tests are viewed as problematic by some of our directors, who question their accuracy and usefulness, most accept them as necessary and realize that improved scores can buy freedom and autonomy for an alternative school.

Before the development of alternative schools, only one child in six in District No. 4 was reading at or above grade level; after the introduction of parental choice, more than

Table 12.1. Community School District No. 4, Manhattan percentage of students reading at or above grade level

1973	15.9
1974	15.3
1975	28.3
1976	27.9
1977	29.7
1978	25.9
1979	25.9
1980	35.6
1981	44.3
1982	48.5
1983	52.3
1984	48.1
1986	62.2
1987	62.6
1988	64.9

Note: The citywide testing instrument for the years 1982–85 was the California Achievement Test in Reading. The population reflected includes grades two through nine. The citywide testing instrument was changed in Spring 1986 to the Degrees of Reading Power (DRP), testing grades three through nine.

three out of five are doing so. Over the last few years, as news of our success spread beyond the borders of the district, students from other areas have been drawn to our schools. These have not, however, unduly influenced our measures of improvement. Of those who came from outside the district, 49.4 percent were reading below grade level; our highest ranked elementary school had fewer than 2 percent of its students coming from outside the district (Table 12.1).

Though less striking, mathematics scores have also improved, nearly 10 points over four years in the district (Table 12.2).

A measure which directly reflects junior high school achievement is the rate of admission into New York City's specialized high schools: the Bronx High School of Science,

Table 12.2. Community School District No. 4, Manhattan percentage of students performing at or above grade level in mathematics

1982	41.8
1983	49.0
1984	50.6
1985	51.0

Note: The citywide testing instrument for the years 1982–85 was the California Achievement Test in Reading. The population reflected includes grades two through nine.

Stuyvesant High School, Brooklyn Tech, and LaGuardia High School (the former Music and Art and Performing Arts High School), each of which requires an entrance examination. The test is basically one of reading and mathematics ability. While in 1973 only 10 youngsters from District No. 4 were accepted into specialized or private high schools, in 1987 over 250 youngsters were accepted into these schools. In the same year, 75 percent of the district's eighth graders were found to be competent writers on statewide examinations.

Today, District No. 4 serves approximately 15,000 students, enrolled in prekindergarten to grade nine. The district is noted for its range of quality educational program choices. Twenty-one school sites house 16 regular elementary school programs, nine bilingual schools, 23 alternative schools, and two high schools.

Each alternative school program has a thematic core, from science to sports, around which curriculum is structured. These quality educational option programs have attracted students from all over the city. The bilingual schools incorporate programs that support the retention of Spanish language and culture while promoting English language skills and content area learning. As a result of these option programs, District No. 4 has experienced what counts as success in New York school districts: dramatic improvements in reading and mathematics scores and in student placement in the city's academically enriched high schools.

What Makes a Good School?

Extrapolating general "principles" from our experience in District No. 4 inevitably leads to conclusions which make social scientists uneasy. Good experimental conclusions require optimally the neutral context of a laboratory, or minimally some effort to control for variables, but we had the advantages of neither as we set about improving our schools. Most classroom reformers, however, find themselves in similar circumstances, since schools are never neutral, and variables are impossible to control. For an excellent summary of the research on choice, see Raywid (1989); given the constraints on experimental design of our work in East Harlem, I prefer to label my experiential conclusions not "findings" but "biases" — a term which connotes the personal edge these prescriptions include:

- **Bias 1**: I believe strongly in public education; when I speak about choice, it is in regard to public schools. Youngsters attending private schools are already part of a choice system. Public education cannot take its constituency for granted; our responsibility is to provide a variety of quality educational opportunities for children.
- **Bias 2**: Choice is only meaningful when quality and diversity are involved. Diversity is developed through increased autonomy — school site management and shared decision making.
- **Bias 3**: In order to get somewhere, you have to know where you want to go. Without a vision, dream, or mission for a school, the school

will go nowhere. A school is an idea, not a building. But it is important to keep in mind that the theme of a school is only a vehicle for motivating students to higher achievement. A school dedicated to the performing arts can raise reading and math test scores by providing an environment that encourages and respects students' interests.

- **Bias 4**: Teaching and learning are so intertwined that it is impossible to separate the two. In lecturing about teaching and learning, Neil Postman produced the sentence, "I taught that but the students didn't learn it", and asked his class to define the word "taught" as it was used in that sentence. The word meant something like "announced" or "presented" — verbs which refer only to half of the rhetorical transaction.

I do not think teaching takes place until learning occurs. For example, I would never say that I taught my son to swim but he cannot stay afloat. I may have attempted to teach him, but if he cannot keep his head above water, I have not succeeded in my teaching. In good schools, teaching is understood as inextricably linked to the learning that accompanies it.

- **Bias 5**: It is how you treat people that determines how they behave. George Bernard Shaw illustrates the idea elegantly in **Pygmalion** when the difference between a lady and a flower girl is defined not by her inheritance but her treatment.

When children, parents, and professionals are treated with respect and high expectations, they will respond accordingly.

On the other hand, when youngsters are treated without respect and with low expectations, you can be sure they will not disappoint you.

- **Bias 6**: Smaller schools are better for children than larger schools. Youngsters do not get lost easily in small schools. Teachers and students get to know each other in a personal way; students do not feel alienated from their own education.
- **Bias 7**: Extending ownership of the schools to students, parents, and professional staff enhances academic performance. People treat what they own better than what they do not own. Ownership is intimately connected to the experience of the environment of a school.

"Ownership" is the feeling among teachers and students that a given school is really theirs. The Program of Alternative Schools in District No. 4 was designed to foster this sense of ownership in every possible case. Rather than drawing up a program of alternative schools in the district office and then finding people to staff the relevant roles, we invited those teachers and principals who held particular educational points of view to propose their own alternative schools, which would be provided building space and the resources necessary to put their ideas to work.

Just as ownership develops when students and parents are given opportunities to select a school geared to their own interests and abilities, so too does it develop when teachers can create and work in a school in accord with their own beliefs about effective education. By encouraging professional staff members to develop innovative programs based on interest talents, or philosophical tenets, and by giving them chances to realize their dreams and participate in the decision-making process, a district can succeed in maximizing the potential of the teachers and students working together in its schools.

Implementing Choice: Suggestions and Cautions

I have already referred to some of the resistance faced by the new alternative schools, and have cited Bensman's (1987) account of the obstacles faced by Central Park East as its staff began to build a different kind of school in the midst of an existing public system. Each school faced its own challenges in that process; those with sufficient commitment on the part of its staff, and interest on the part of its students, prospered. I generally hold to the principle: If a major change or innovation is not threatening to stakeholders, it is probably not worth doing. And the corollary: Growth without some friction will result in an easy backslide. The changes we made in the district were threatening to some teachers and principals, who had to be involved in the process, if possible, or reassured, if not. But throughout the history of the program, only one alternative school failed in its attempt.

The program itself, of course, encountered resistance, but most of it has been overcome by the improvements in mathematics and reading scores and by the enthusiastic support of students and parents in our district. We did not allow anxiety over unforeseen consequences of the change or the attitudes of educational authorities to dictate the form and adequacy of the schooling we provided.

I do want to share some cautions and suggestions regarding free parental choice systems. As choice (in one form or another) becomes popular, it becomes attractive as a way to solve larger educational problems. A system of parental choice such as the one I have been outlining can help stimulate creative programming on the part of district administrators, stir teachers to take a more active part in helping students learn, and motivate children to engage in a more personal way — but it cannot solve all problems confronting the schools. Overly ambitious expectations can kill any new idea; we are always looking for panaceas but seldom find them.

Choice has to be approached carefully and slowly. In District No. 4, we started with three alternative schools, and concerned ourselves with quality and diversity within those three. Our development was organic; we did not begin with a grand plan that envisioned 49 different schools in 28 buildings. When I hear about a school district deciding to become a complete choice system overnight, I worry. Quality schools — with a guiding vision, a staff committed to that vision, a curriculum reflecting that commitment, and a student body intellectually engaged with the issues of that curriculum — take time to develop. If you already have a wide variety of quality schools, choice can be offered more broadly. In New York, we needed to institute choice as a mechanism for school improvement. None of the proponents of choice that I know consider it a panacea — it is the opponents of choice who use that word in denouncing the idea.

Without quality and diversity of educational offerings, choice is a meaningless policy. Choice does not encourage diversity when it provokes increased school site management and planning. But if choice is to be the catalyst for diversity and quality, it should be instituted with sufficient care and time. It is much better to have, at first, a few schools that are oversubscribed because of the quality and diversity they offer than to have a larger number of poor choices for everyone. Once those few are established, demonstrating that a quality educational environment can be provided, the development of others becomes easier, until a network of schools emerges that can eventually open choice to all students.

Start in a small way, but make each start a success. Determine what kinds of schools are attractive to the parents and students in the community to be served by the school. Our experiences in East Harlem (and reports I have heard from people around the country) suggest it is often easier to start an open-education or progressive-education elementary school than some other kinds because there are always parents interested in that philosophical approach who are often dissatisfied with the existing traditional schools. On the junior high level, performing arts schools or schools for science and math have demonstrated strong popular appeal.

In District No. 4 our experience has underscored the benefits of starting small, with only an opening grade. On the junior high level, two classes of seventh graders make a good beginning. Over the course of the next two years, take in another two classes of seventh graders. At the end of three years, allow six classes in seventh, eighth, and ninth grades. Avoid the trap of too-rapid growth — for some reason, almost everyone likes to expand when they have a well-organized, productive school that is able to maintain a family atmosphere and allows staff and students to get to know and care for each other. If you must, add two more classes and stop.

The question of exceptional leaders is raised whenever the possibility of replication is discussed. I feel that every school district in this country has the potential leaders necessary for effective and stimulating educational opportunities. In District No. 4, we established a climate that encouraged teachers to come forward with their ideas about how to educate youngsters. Most of our successful schools grew from a concept by a director and/or principal who had some personal vision and the chance to develop that vision; they were able to communicate their ideas to teachers, students, and parents, who then took ownership of the dream. Our requirement was that a new director had to be successful at what he or she was already doing and willing to put in extra time, energy, and enthusiasm to make the dream a reality.

Once a school director is identified, a sympathetic administrator must be responsible to provide real support: adequate space, funds, and personnel. Any new school needs time, encouragement, and protection for its philosophical base. The district leadership must have confidence in the new school's leadership even when they do some foolish things — mistakes will be made whenever risk-taking occurs. Wherever possible, visible symbols of support and recognition should be awarded the new school. And sometimes this support must be more than symbolic: if, for example, the alternative school is sharing space in an existing school building, at some point the staff of the new school will infringe upon the prerogatives or otherwise slight the principal of the original school in the building. It becomes the supervisor's role somehow to soothe the injured party, and nothing is as soothing as some additional funds or equipment.

All of the schools in District No. 4 were required to recognize that the mission of the district is to improve learning. Principals and directors must provide leadership by ensuring an instructional environment that is conducive to intellectual, emotional, and social development. Schools have to be good places for youngsters and adults to share experiences and engage one another for their mutual growth — a requirement which is true, of course, for any school, progressive or traditional.

I have thus far discussed the steps that must be considered in setting up a new alternative school intended to participate in a system of educational choice. Similar steps can be taken when preparing an existing traditional school to join a system of choice. In many ways it is better and easier if an established principal and staff opt to become a school of choice — it can help avoid a good deal of maneuvering for power and territory. Of course, the best way to support a new school is to demonstrate to the community and to critics of all persuasions that the school provides quality education for its students; high scores on standardized tests never seem to hurt a developing school.

Parental involvement at the school's inception can help protect a new school and build a constituency rapidly. Some critics insist that choice cannot work in inner-city school systems because inner-city parents lack the education needed to make informed choices ... but those critics are wrong. Inner-city minority parents are no less concerned than their middle-class counterparts to see their children educated in stimulating, orderly, vigorous schools, and no less capable of choosing those schools when information is made available to them. Their lack of resources rarely presents them with educational choices but their aspirations for their children are no more limited than those of more affluent parents. And schooling often represents to an inner-city student the only avenues of escape from a situation much more desperate than that confronting a middle-class peer.

It has been a long time since I was disheartened by the inadequate response we would give to parents who were dissatisfied with their zoned schools. Today, parents in East Harlem do not have to present phony addresses to enroll their children in superior educational programs. To help them choose among alternatives at the intermediate level, they receive a simple application after an orientation process that includes classroom lessons in decision making, a booklet that describes all the schools, meetings where school representatives and parents discuss each of the choices available, and school visits by parents and prospective students. This process allows them to make the best choices for and with their children.

District No. 4 has received recognition from many sources outside of our community. The media and educational observers tend to focus on reading and mathematics scores. The greater satisfaction to me comes from the pride in and support of the district from the parents, students, and staff members who meet in its schools. Their imagination, hard work, and persistence gave rise to the strong academic alternatives that made choice at the intermediate level an important opportunity for children in the community. How this potential was enlivened, liberated, and directed must be counted among the important lessons of the East Harlem experiment.

Current Prospects: Choice and Recentralization

In some respects, the prospects for choice in public education have never been better than they are at this moment: at the White House Conference on Choice in January 1989, President Bush declared his support of choice in public education when he said, "expanding parents' rights to choose public schools is a national imperative", an opinion which has been noted and seconded by politicians across the country. It is ironic, then, that the prospects for choice in District No. 4, and in New York City more generally, have been placed at grave risk by the call for recentralization, for returning greater power over local schools to the Central Board of Education.

The success of District No. 4 in improving reading and mathematics scores, in preparing our students for special high schools and beyond, has stemmed from our ability to elicit from our staff the vision and energy to create better educational programs for the children of East Harlem. Our ability to do so derives from a willingness to allow people to express their imagination and commitment as they saw fit, by allowing them some measure of autonomy over the schools in which they teach. This autonomy, a sharing of decisions that affect teachers and students, was possible in East Harlem because authority had already been shared with the district under the system of decentralization. We were able to allow individual teachers to participate in the policy and philosophy of their schools because the district had won the authority to decide where that decision-making power could be trusted. And where we did not have the authority, we took it.

John Chubb, Senior Fellow of the Brookings Institute, argues that administrative bureaucracy is itself a significant obstacle to effective schooling. "The more centralized a school system, the worse the achievement of its students", he writes.

> Schools are more likely to pursue ambitious missions and develop strong leadership and professionalism if they are relatively free from constraints on hiring, firing, curriculum, instruction, and discipline ... much like other professionals, educators seem to function more effectively if they are given the freedom to organize themselves and serve their clients according to their best judgments.
>
> To be sure, central administrators are reluctant to grant Principals and teachers the autonomy to shape their own schools, because more autonomy seems to mean less accountability. But autonomy disciplined by parental choice provides substantial accountability.
>
> (Chubb and Moe 1990)

Chubb wants to do away with bureaucracy in order to improve the schools. Another alternative available to school district administrators in my experience has been to ignore the central administration. In some respects, a large bureaucracy can be helpful because of its size: no one is willing to take responsibility for much that happens. A vacuum is created that allows an innovator room to begin to implement a change. Once something is started, if it makes some sense, it is very difficult for centralized administrators to stop it, especially if it crosses lines of bureaucratic responsibility. In New York City it is not difficult to make change in a Community School District if you have the support of the Community Superintendent — the Central Board is not usually interested in the day-to-day operation of a district. If the Superintendent is a forceful or likeable person who had established good relations with key Central Board administrators, a good deal of innovation can take place. If the good relationship happens to be with the Chancellor, a great deal of innovation can occur. It is always helpful if the Superintendent knows what he or she is intending to do;

if the changes raise reading and math scores, you can walk away with the store.

The paralysis of the New York City school bureaucracy has allowed individual educators the freedom to do what they think is best. What has been necessary is the recognition that it is our own choice, undertaken at your own initiative, to take that autonomy in hand. It might have meant giving up personal ambitions to climb the bureaucratic ladder, but if educators were willing, and had some support for their efforts, it was possible to take real risks in the system.

The proposed recentralization would reduce those possibilities, limiting opportunities for individuals to engage themselves in new educational ideas. Where power is all in the same hands, the pressure for schools to conform within the district, and for districts to conform within the system, is very great. The risk goes up on risk-taking; resistance to innovation increases. Recentralization closes the door on ownership, too, so the motive to improve a school dries up with the opportunity.

I think it is fair to say that the success of District No. 4 owes something to the "creative noncompliance" of teachers and administrators with the strictures of the Central Board of Education. In slipping between the cracks of bureaucratic hierarchy, we were able to improve the ability of children in East Harlem to read and write and handle mathematics at a rate that has drawn the approbation of scholars, political leaders, and individuals of every calling with some concern for public education in this country. The move to recentralization threatens the program of choice in District No. 4, and the emergence of other programs that might enrich the education of our children.

The advantages of recentralization, in addressing the corruption and incompetence of some local school districts, are altogether unclear. Corruption does not discriminate; it takes place at all levels. In a centralized system, we do not know the corrupters until they are exposed by the media; in a decentralized system, the likely corrupters are known by all and have to be more careful. The case has yet to be made that large bureaucracies in the City of New York are guarantees against individual vice and mismanagement.

References

Bensman, D. (1987). The story of the Central Park East School. In D. Bensman (Ed.), *Quality education in the inner city*. New Brunswick, NJ: Rutgers University Press.

Chubb, J. and Moe, T. (1990). *Politics, markets, and American schools*. Washington, DC: the Brookings Institution.

Davis-Fishman, K. (1986). Middle-class parents' guide to public education. *New York Magazine,* January.

New York Times (1982). Editorial, 25 March.

Raywid, M. A. (1989). The mounting case for schools of choice. In J. Nathan (Ed.), *Public schools by choice: Expanding opportunities for parents, students, and teachers.* St. Paul, Minn.: The Institute for Learning and Teaching.

Chapter Thirteen

SCHOOLS OF CHOICE AND THE REVIVAL OF URBAN COMMUNITY

CHARLES L. GLENN
Boston University

The central contention of this chapter is that a new form of relationship between working-class families and schools could play an important role in restoring healthy community life in inner-city residential areas, and that this could in turn help to provide a context of mutual responsibility within which more adults in these areas could begin to live productive lives and to function effectively as parents, workers, and citizens.

The reader would be justified in doubting that such a bold claim can be made good in a few pages, and might grow even more dubious upon learning that I hope to show that the contentious issue of school choice is a key element in reaching this happy result. It is true that I cannot prove my contention beyond a reasonable doubt, but I will show by a few examples that the possibility deserves further exploration, particularly in view of the acknowledged failure of many programs of top–down intervention.

Parent Choice of Schools

There has been a great deal of discussion — perhaps too much — in recent years over whether allowing parents to choose the schools their children will attend is a good idea. Those who grow very heated on either side seem often willfully to ignore the fact that most parents **do** choose or at least exert considerable influence over where their children will go to school.

This is true in the many nations — almost all of the industrialized democracies — that have formal policies supporting parental choice through funding of any approved school (whether government-operated or not) that parents select. It is true even in the United States, where an ideological commitment to the government-operated "common school" has prevented the adoption of these policies.

It would indeed be possible to conclude that the **worst** way to organize school choice is the way we have done it in the United States. In principle, parents have no right of choice: pupils are assigned to a public school on the basis of where they live. In fact, parents with resources are active choosers of schools for their children, whether by deciding where to live, or by paying tuition to nonpublic schools, or by manipulating the assignment process through their sophisticated access to information and influence. Those who have money have choice, while those who do not, do not. Because of the high degree of residential segregation based upon income that is characteristic of a highly suburbanized society, poor children (disproportionately African-American and Latino children) are concentrated in certain schools, and these schools are in turn characterized by low expectations and low achievement.

It is approaching the point, at least in our cities, at which "public" schooling has come to have the same negative connotation as "public" housing. Sociologist William Julius Wilson's comment that "... the Great Society programs established what amounted to separate legal and medical systems — one public and predominantly black, the other private and predominantly white" (Wilson 1987, p. 119) applies with equal force to public schooling, and the exercise of choice by parents within a policy context that seeks to pretend that choice is not occurring is the major factor in this unhappy result.

The exceptions to the general rule that choice is merely tolerated but neither promoted nor regulated by public policy have mostly come into existence as a way to balance legal requirements for race desegregation, with political resistance to mandatory integration. Thousands of "magnet schools" attract a racially diverse enrollment through the special character of their programs, while in a few metropolitan areas there are voluntary desegregation programs under which urban children may attend suburban schools. In 16 cities in my own state of Massachusetts, **every** pupil attends school through a parental choice process that is linked with a strategy to strengthen those schools that are least successful in attracting applicants (Glenn 1991a).

In addition, there are various small-scale efforts to extend the range of publicly-sponsored school choice beyond the context of desegregation, including interdistrict programs in several states, a very limited private school tuition voucher program in Milwaukee, and a growing number of privately funded education voucher programs for poor children[1].

In sum, there are many opportunities for school choice in the United States, but they are not guided by consistent policy to assure that their effects will be benign, and there is reason to believe that, except when there is a deliberate effort to the contrary (as in some Massachusetts cities), the effect of parent choices is increased social class segregation and unequal educational opportunity.

As a result of the widespread exercise of school choice by parents in the United States and in other nations, however, the question as to whether this is a good idea is neither very relevant nor very interesting. There is no way to stop it, nor should a free society attempt to do so,

A better question would be, should government seek to create a policy framework that will enhance the benign effects of the parent choice that will in any case occur and limit its negative effects to the greatest extent possible? It is not self-evident that government should seek to manipulate the private decisions of its citizens at the very heart of their family lives, nor is it clear that policymakers and legislators always have a light enough touch to know how to go far enough but no farther in their interventions in the civil society.

The answer to the question, in fact, may have to wait upon that to a further question, one that is in many respects even more interesting. Perhaps it would not be inappropriate, in Israel, to pose this question, as Samson did, in a riddle:

> Out of the eater, something to eat;
> Out of the strong, something sweet.
>
> (Judges 14: 14)

Is it possible that something sweet, something to heal a broken society, to "strengthen the feeble hands and steady the knees that give way" (Isaiah 35, p. 3), can come out of a process in which schools are competing for pupils and, to do so successfully, are becoming more distinctive rather than more similar one to another? Does this require abandoning what we have come to think of as one of the principal functions of state-supported schooling, to reduce the differences among future adult citizens in the name of nation-building and of social reconciliation?

Put in more prosaic terms, is it possible to devise school choice policies that are benign in their effect, not only for the pupils participating but also for the society as a whole? If not, surely government should leave well enough alone, and seek to discourage, though it cannot prevent, the exercise of choice by those parents with sufficient resources, dissatisfaction, or sophistication to make use of the possibilities already available.

What I propose, therefore, is to discuss **how** school choice may best be organized, and what effects can reasonably be anticipated from such measures. If the case is convincing, it should not be difficult to agree that it would be legitimate for government to exercise its authority to put these measures into effect.

When we ask what effects we are seeking from the organization of school choice, we must confront a further question: what needs are we trying to meet, what problems are we trying to solve? Most discussion of school reform places the focus upon improved effectiveness, as demonstrated by pupil achievement on various measures. Proponents of school choice usually argue that all schools will become more effective through improved clarity of focus and through the chastening disciplines of the market (e.g., Chubb and Moe 1990; Hill, Foster, and Gendler 1990, 1990). While I am convinced that this **can** happen, if the system of choice is appropriately organized, I do not intend to make that case here (see, e.g., Glenn 1990/91, 1991b, in press a).

The rights of parents are the basis for much of the case for policies supporting choice of schools, and this right in particular has been recognized by the United States Supreme Court and by a number of international covenants[2]. For example, the **Universal Declaration of Human Rights** states that "parents have a prior right to choose the kind of education that shall be given to their children" (article 26, 3). According to the **International Covenant on Economic, Social and Cultural Rights**:

> the States Parties to the present Covenant undertake to have respect for the liberty of parents ... to choose for their children schools, other than those established by public authorities, which conform to such minimum educational standards as may be laid down or approved

> by the State and to ensure the religious and moral education of their children in conformity with their own convictions (Article 13, 3).

Similarly, the First Protocol to the **European Convention for the Protection of Human Rights and Fundamental Freedoms** provides that "in the exercise of any functions which it assumes in relation to education and teaching, the State shall respect the right of parents to ensure such education and teaching in conformity with their own religious and philosophical convictions" (Article 2).

The same principle is enshrined in the recently adopted constitutions of a number of post-communist societies[3]. For example, the new Bulgarian Constitution (1991) stipulates that "the raising and the education of children until they come of legal age is a right and an obligation of their parents; the state provides assistance" (Article 47, 1). That of Estonia (1992) provides that "parents shall have the final decision in choosing education for their children. The provision of education shall be supervised by the state" (Article 37). Croatia (1990) provides that "parents shall have the duty to bring up, support and school their children, and shall have the right and freedom independently to decide on the upbringing of children" (Article 63). Hungary states in its new Constitution (1989) that "parents shall have the right to choose the type of education they wish to ensure for their children" (Article 67, 2)[4].

The Challenge of Persistent Poverty

Important as increased school effectiveness is to a modern economy, significant as respect for the rights of parents should be to a democracy, my focus in this chapter is upon another question, the relationship of educational policy to a challenge faced by all advanced industrial societies: the persistence of a population that does not connect successfully with civic or economic life, a group that is not only poor but seemingly incapable of doing anything about its own poverty. Lawrence Mead estimates this "underclass" in the United States as consisting of about nine million among three times that number of Americans who are poor in the sense of lacking sufficient income. The smaller group is characterized by "functioning problems such as difficulties in getting through school, obeying the law, working, and keeping their families together ... traceable to an unstable family life, marked by absent fathers, erratic parenting, and low self-esteem and aspiration" (Mead 1986, p. 22).

A hundred years ago, pioneering social reformer Charles Booth made the same distinction, in his study **Life and Labour of the People in London**, between the working poor and those below them (and threatening always to pull them down) "who from mental, moral, and physical reasons are incapable of better work", a group who were the "leisure class" of the poor, whose leisure was "habitual to the extent of second nature" (quoted in Himmelfarb 1992, pp. 109, 123).

There is a growing recognition, among liberals as well as conservatives, that an answer to this persistent poverty and social disorganization will have to be found in a change in the attitudes and behavior of the poor themselves, as well as in the circumstances which surround and affect them. Thus Marian Wright Edelman of the Children's Defense Fund told an interviewer that "What scares me is that today people don't have the sense that they can struggle and change things" (quoted in Mead 1992, p. 164). New government programs to bring the "underclass" or "truly disadvantaged" into the social mainstream, absent that conviction among the intended beneficiaries, seem certain to produce results as meager as those of the programs that went before.

How can the persistently poor develop the **competence** — "not only intelligence, but foresight, energy, discipline, and the ability to sacrifice for the future"? (Mead 1992) — that would enable them to escape from poverty? Do schools have a part to play in helping in the development of this competence, of what could equally well be called, in Aristotle's sense, the habit of virtuous action? I believe that they do, in several ways.

Schools are complex phenomena, producing effects that go far beyond the manifest content of classroom instruction. The two effects that will concern us here are those upon pupils through the internal life of the school, and those upon parents and communities through how the school functions in relation to them.

Schools as a Focus of Community Action

Inner-city schools have had their mission defined for them by opinion-makers steeped in the liberal belief that the poor are simply victims of circumstances, without ultimate responsibility for their own actions, and must not be "blamed" or expected to find their own solutions. As a result, staff of these schools have tended to operate from low expectations and to concentrate, with pupils, upon various forms of remedial education and, with their parents, upon an attempt to minimize "interference". The operative assumption is that both parents and pupils can be defined adequately by what they lack of the qualities that would make them worthy of real respect. Kenneth Clark's recollection of the schools he attended in Harlem, in the 1930s, presents a very different picture:

> The teachers were concerned with holding **all** of us to high standards, because they were convinced **all** of us were educable. We had not yet come to that particular breakdown in the public education of minorities which is due to ... the sloppy, sentimental good intentions of certain educators, who **reduce** learning standards for low-status youngsters, because they believe ... it is impossible for them to learn as much as suburban children... But when I was going to public school we had teachers who did not consider themselves ... rationalizers of educational

inequities. They were asked to teach reading, arithmetic, grammar, and they **did**[5].

The prevalent attitude today, by contrast, has developed as part of a larger unwillingness to base social policy upon solid convictions about responsibility. It is not that educators became more judgmental toward the poor; to the contrary, they have become systematically **un**judgmental, in the sense of adopting an attitude of not expecting anything in particular from those whom they regard as society's victims. Wilson notes that "the liberal perspective ... on the ghetto underclass has become less persuasive and convincing in public discourse principally because many ... have been reluctant to discuss openly ... the sharp increase in social pathologies in ghetto communities" (Wilson 1987, p. 6).

Out of this reluctance to fall into "blaming the victim", the contributing role of cultural factors to accomplishments or failures in life became a topic that could not be discussed without charges of cultural imperialism and even racism. "Liberal scholars in the 1960s argued that cultural values do not ultimately determine behavior or success. Rather [they insisted], cultural values emerge from specific social circumstances and life chances and reflect one's class and racial position" (Wilson 1987, p. 14).

The beliefs that individuals hold about the need for and probable results of sustained effort and responsible behavior **do** matter, however, as does the way they act upon these beliefs, and there is an urgent need to ask what role schools should play in the development of such values. This will require that educators abandon their squeamishness about the moral dimension of education, a dimension that is as unavoidable as it is desired by most parents. Schools invariably communicate **something** about right and wrong, and individual effort and the behavior of groups, even if what they communicate is that these are unimportant questions.

Too many leading educators, in recent decades, have been

influenced by a loss of confidence in values other than tolerance. With other well-meaning liberals (in the American sense), they:

> no longer had enough self-confidence to know which, if any, values they wanted the state to encourage. Liberals increasingly forgot that there is a distinction between using the state to **enforce** a rigid moral code – an effort as impractical as it is unappealing – and insisting that programs **promote** certain values that are beneficial to both society and the individuals who practiced them. Over time, liberals were no longer certain what kind of family was worth encouraging; they feared welfare programs that required recipients to work; they nearly always saw understandable worries about law and order as covert forms of racism.
>
> (Dionne 1992: 76)

It is clear that there are powerful "school effects" through the interactions that occur between teachers and pupils during the course of classroom and other activities. Pupils may learn a great deal of the rewards of disciplined effort and the merits of mutual respect and cooperation through the daily life of a good school (see, e.g., Wynne and Ryan 1993).

The extensive research on **effective schools**, over the past two decades, has been concerned to identify the internal practices of schools that can contribute to overcoming the effects of social disorganization and other factors that inhibit the educational success of children from poor families. The research of Rutter, Mortimore, and their colleagues in London schools has been particularly fruitful in this respect (Rutter *et al.* 1979; Mortimore *et al.* 1988). Schools **can** make a difference. But too many schools do not make a positive difference in the lives of children and youth for whom the world outside of school is characterized by disorder and lack of purpose.

Families can also make a difference, as Reginald Clark has illustrated in a moving account of how some of the poor black families in Chicago, whose daily interactions with their children he studied were able to provide support and structure that led to success in school. "It is the family members' **beliefs**, **activities**, and overall **cultural style**", he found, "not the family units' composition or social status, that produces the requisite mental structures for effective and desirable behavior during classroom lessons" (Clark 1983, pp. 1–2).

There is reason to believe, however, that it is only when schools and families function in a mutually supportive fashion that children are effectively educated. Unfortunately, the linkages between families and schools have grown weaker in all sectors of society, leading to frequent complaints by teachers that their efforts are no longer supported adequately by the home. More than half of those surveyed in a recent poll, for example, responded that "strengthening parents' role in their children's education" should be the highest priority for public education policy (Richardson 1993).

Sociologist James Coleman has argued that there is a general decline in the ability and willingness of parents, in a postindustrial society, to provide a consistent structure of expectations and attitudes that will lead to success in school. Even parents whose "human capital" — educational level and income — has grown have a diminished "social capital" of legitimacy in the exercise of authority in relation to their children and willingness simply to **pay attention** to their children. The decline of residential communities and of the institutions of face-to-face socialization have placed a burden upon schools that they cannot adequately bear. It is poor children who are most vulnerable to the effects of this loss, and who most urgently need the creation of new institutions beyond the nuclear family within which they can find both emotional safety and encouragement (Coleman 1987).

Such support cannot be provided effectively by bureaucratically structured institutions like human service agencies and public schools as now constituted. The authors of the influential new book **Reinventing Government** have suggested a way of classifying the functions that are now largely monopolized by government according to which of three "sectors" is best able to carry them out. Government itself, they write, is best at "policy management, regulation, ensuring equity, preventing discrimination or exploitation, ensuring continuity and stability of services, and ensuring social cohesion . . ." Business is best at innovating and adapting to rapid change. It is what they call the "third sector" of voluntary, nonprofit organizations that is "best at performing tasks that generate little or no profit, demand compassion and commitment to individuals, require extensive trust on the part of customers or clients, needs hands-on, personal attention . . . ; and involve the enforcement of moral codes and individual responsibility for behavior" (Osborne and Gaebler 1992, pp. 45–46).

The strengths of "third sector" organizations are very much those called for by Coleman, who has similarly argued that schools outside the public sector that are freely chosen for their religious or pedagogical character are better able to set and enforce expectations that lead to achievement, especially for poor and vulnerable children, than are public schools enrolling pupils on the basis of residence.

> Family background makes much less difference for achievement in Catholic schools than in public schools. This greater homogeneity of achievement in the Catholic sector (as well as the lesser racial and ethnic segregation of the Catholic sector) suggests that the ideal of the common school is more nearly met in the Catholic schools than in the public schools.
>
> (Coleman 1990: 247)

This leads to his conclusion that "the strict separation of church and state, as practiced in America, has been harmful to the least advantaged and particularly harmful to children in the black community (Coleman 1987, p. 37). It is important that it becomes more feasible for new kinds of schools to be established that are based upon shared beliefs about education and the formation of character, whether these have a religious basis or not, and whether the schools are operated by government or by "third sector" organizations. Such schools would, as Coleman suggests, provide a far more intensive and extensive environment for children than is possible for public schools, as now constituted and staffed.

This will require an increased commitment of funding so that caring — not custodial — adults will be available to provide a range of learning experiences and support for children and youth 12 hours a day or more. It will require summer programs of equal intensity, it will require the option of residential programs for some, it will require outreach to families in their homes (routine in Europe but almost unheard of in American education). In short, it will require "comprehensive and intensive services" provided by staff "with the time and skill to establish relationships based on mutual respect and trust" (Schorr and Schorr 1989, p. xxii).

When the idea of "urban boarding schools" was first raised, it seemed an insult to the adequacy of inner-city parents. Now many of those parents are themselves crying out for help, for anything to give their children a safe environment within which to grow. Public policy should no longer ignore what parents tell us by putting their children in nonpublic schools which they cannot well afford.

My own children have attended public schools in Boston for more than 20 years, and we are able to provide them with the support — including a mother who has put her career on hold to be actively involved in school and after school — to make that a success for them. For many of our neighbors, though, another sort of sacrifice has seemed necessary, to keep their children in parochial and other nonpublic schools that have some of the qualities Coleman describes. Recently in the mail I received the brochure of Parkside Christian Academy, a

K-6 school located in the inner-city neighborhood where I live and enrolling around 80 percent minority pupils. PCA provides a good example of what many urban parents are convinced (correctly) that their children need.

In addition to offering (for $2.90 an hour) before and after care between 7 am and 6 pm, PCA is explicit about its educational mission:

> In the best interest of your young child, PCA seeks to foster a climate of Christian concern and caring in which moral and spiritual values are integrated with a quality academic education emphasizing the potential of each child. We strive for academic excellence with Christian values. . . . PCA provides opportunities, both in and out of the classroom, for students to acquire a strong positive self-image and to develop appropriate decision-making skills. Each student is taught the value of learning from one another as well as with one another. PCA believes that spiritual, social, and emotional growth is equally important as academic achievement. . . . School discipline is approached from a positive, personal perspective as the staff clearly define and consistently maintain limits. Students are treated with respect, acknowledging their right to form opinions, express ideas, make choices and accept consequences . . . We are committed to excellence in all facets of the curriculum. PCA will uphold the values and standards you expect in a Christian environment . . . Fully integrated culturally, socio-economically and denominationally, PCA seeks to build a strong sense of community among a student body from diverse backgrounds.

It sounds remarkably like the terms used by elite boarding schools to explain what they offer, and why not? As Seymour Fliegel is fond of saying, "What's good enough for rich kids is good enough for the children of East Harlem!"

Talk is cheap, of course, and so are promises in school brochures, but it is clear that many parents in our low-income neighborhood are willing to find $3,465 a year to send their children to Parkside, which is without the subsidies that enable some Catholic schools to charge lower tuition; this is still only half of the expenditure in the public schools. These parents are not avoiding the presence of black children — most of them are black! What they are avoiding is (as some have told me) their perception of disorder and lack of values in the public schools. Within radical changes in the institutional and political framework within which public schools must operate, it is difficult to see how they could provide the sort of morally coherent environment that many inner-city nonpublic schools like Parkside offer.

As Coleman has pointed out, "A principal of a [public] school today in which attendance is based on residency has no set of dominant community values to uphold. Instead, there are a number of contending values, each claiming legitimacy, and at least some of them capable of being backed up by legal suits in court" (Coleman and Hoffer 1987: 11–12).

Policy Imperatives

To enable public and nonpublic schools to provide a truly supportive environment for children with severe physical, social, emotional and other disabilities, and to do so for at least twice as many hours a year as schools now operate, would require a very substantial additional investment. It would not be an investment in perpetuating and extending mediocrity, however, but in making a dramatic difference for children and their families. Such an investment would not produce the desired results, however, unless the dynamic of the relationship between family and school was changed. It is not enough that schools start **doing** certain things differently; they must **become** different kinds of institutions in relation to the families that they serve.

Linkages between the school and the home in the inner city tend to founder on passivity on the one side and an immovable bureaucratic and professional culture on the other. The poor are not alone in keeping at a distance from public schools as they function in large school systems. Herbert Gans notes that most "middle Americans" practice" **organizational avoidance**. They seek mainly intimate and informal relations and groups, and try to avoid all but the essential contacts with formal organizations, not only the national ones of the 'larger society' but also local ones" (Gans 1991, p. 43).

It is very appropriate that some inner-city public schools have begun to provide parent centers, staffed with outreach workers and providing educational and support services for the parents of pupils[6]. It is also appropriate that many schools must now consult with parent representatives as they make certain kinds of decisions. The well-known structural reforms in Chicago offer the most notable example of mandatory involvement of parents in decision making.

Such measures are not enough, however. They do not do enough to change the internal dynamic of the school. Equally seriously, they do not make the school an agent of positive community change. The relationship of schools with families has an impact which goes beyond what children experience in school; it either increases the alienation of inner-city life or helps to build healthy community.

American inner-city residential areas (it will not do to call them "communities") have been experiencing over the past three decades a collapse of the civil society that had earlier provided structure and dignity even in the face of poverty[7]. Wilson attributes this collapse to the exodus of middle-class and stable working-class black families from areas to which they had previously been confined by housing discrimination and other barriers. Their loss:

> made it more difficult to sustain the basic institutions in these neighborhoods (including churches, stores, schools,

recreational facilities, etc.)... As the basic institutions declined, the social organization of inner city neighborhoods (sense of community, positive neighborhood identification, and explicit norms and sanctions against aberrant behavior) likewise declined.

(Wilson 1987: 138)

The "very presence of these families" according to Wilson, "provides mainstream role models that help keep alive the perception that education is meaningful, that steady employment is a viable alternative to welfare, and that family stability is the norm, not the exception" (Wilson 1987, p. 56).

But the causality runs the other way as well; the exodus of middle-class and working-class families from the most blighted areas is in part also a result of the weakening of the neighborhood institutions which had linked them to a stable community. It is not simply that those who might have sustained the institutions abandoned them instead, but that many of the functions of settlement houses, churches, lodges, and neighborhood associations were taken over by government, leaving increasingly hollow shells that had once been the center of neighborhood life.

Tocqueville warned of this possibility more than 150 years ago, asking:

> what political power could ever carry on the vast multitude of lesser undertakings which associations daily enable American citizens to control?... The more government takes the place of associations, the more will individuals lose the idea of forming associations and need the government to come to their help. That is a vicious circle of cause and effect. ...The morals and intelligence of a democratic people would be in as much danger as its commerce and industry if ever a government wholly usurped the place of private associations. Feelings and ideas are renewed, the heart enlarged, and

> the understanding developed only by the reciprocal action of men one upon another
>
> (Tocqueville 1988: 515)

I was in a position to observe the great extension of government action in inner-city neighborhoods in the 1960s, and its effect upon the institutions which had long served a vital social role of maintaining "the morals and intelligence", the ability to work together and to confront and solve problems, of the people who lived in those areas. From 1961 to early 1966 I was a community organizer and minister in Boston's Roxbury section, from 1966 through 1968 I was the national Episcopal Church's staff member responsible for helping local church-related efforts to participate in the War on Poverty, from 1968 through 1970 I was a community representative in Boston's Model Cities program; in 1970 I became the education coordinator for Model Cities programs statewide, and then the state's official responsible for urban education for the next 20 years.

One irony about these efforts is that they were launched with the intention of building upon elements of community action that already existed, and with a mandate to include "the maximum feasible participation of the poor". In 1967 I conducted a study of how "maximum feasible participation" was functioning in 200 poor areas across the country, and found that it had generally had a profoundly distorting effect upon the existing community structures. An army of "poverty hustlers" had come into existence, employing confrontational tactics to obtain large amounts of government and charitable funding for programs which produced very few lasting benefits. In the process, many neighborhood-based institutions which could not or would not adapt themselves to the new order were discredited if not swept away.

Part of the difficulty was that those who set policy for and administered the new government programs had very little respect for the ways in which the poor were already organizing their lives: through storefront churches and other institutions whose focus was on mutual support in the difficulties of life at the bottom, rather than on radical political change. The men and women exercising leadership in these organizations created by the poor themselves were usually morally conservative and cautious toward the structures of power; they were considered "Uncle Toms" and "Aunt Sallies" by the "poverty warriors".

> So the poverty professionals were always on the lookout for the bad-acting dudes who were the "real leaders", the "natural leaders", the "charismatic figures" in the ghetto jungle... From the beginning the poverty program was aimed at helping ghetto people rise up against their oppressors... It was no accident that Huey Newton and Bobby Seale drew up the ten-point program of the Black Panther Party one night in the offices of the North Oakland Poverty Center[8].

An illustration: the Boston Model Cities Program decided that parents in my racially mixed neighborhood should be "organized", and hired two young women with impeccable radical credentials to do the job. Since I represented the area on the citywide antipoverty board, they came to me for suggestions about what issues would most effectively dramatize the need for fundamental change. I suggested that they begin by helping the existing parent association of a local school to organize the up-coming spaghetti supper to raise funds for the school, as a way of becoming known and trusted by the leadership. They rejected the idea: those leaders were precisely the compromising flunkies who must be swept away. Instead, the professional organizers put up posters all over the neighborhood announcing a rally to attack "educational genocide" against black pupils in the same, racially integrated school. No one but me attended their rally, but the parent association's efforts to draw minority parents into school activities was severely hindered by the resulting climate of suspicion, and it soon disbanded. The would-be community organizers (themselves white) denounced the servile spirit of the neighborhood for their failure to "heighten contradictions", and soon moved on to other employment.

Schools and Community

What we have been learning over these past decades of repeated failure to reverse the degradation of our inner cities is the importance and value of the ways in which people seek to manage their lives. Sociologist Amitai Etzioni describes the importance of the institutions that, in the inner city, have largely been lost:

> Social institutions — the family, the school, neighborhoods, voluntary associations — serve to countervail excessive individualism, to sustain mutuality and civility in three ways:
>
> 1 They stand between the individual and the government, protecting each from the other.
>
> 2 They set patterns and make arrangements that discharge at least some of the tasks that otherwise are loaded on the government or overwhelm individuals.
>
> 3 They "educate" individuals. They introduce and reinforce a mentality that sustains individuals' mutual and civil commitments (Etzioni 1984: 94).

The loss of these institutions is a downward spiral, since they depend upon each other for support. As community institutions are no longer capable of performing their socializing and supportive functions, individuals no longer develop the commitments and the social skills that in turn sustain institutions.

As a result of this downward spiral, the dominant social types[9] in some inner-city areas are now single women with children who are "married to the government" through welfare support, young men who are marginalized and doomed, and older people who are disconnected, living in extreme isolation compounded by fear. As a result, it is extremely difficult for children to grow up in a more positive pattern, and adults find it almost impossible to unite to solve community problems or to link into the wider society.

Bureaucratic organizations, as Tocqueville pointed out, cannot replace the social organizations that people create for themselves. This is not to suggest that government has no role, but that it should not be seeking to do what can better be done by non-governmental organizations. This leaves government free to **govern**, to set the general rules that assure fairness and that protect the common interest. As Osborne and Gaebler put it, ... governments ... may no longer produce services, but they are still responsible for making sure needs are met. When governments abdicate this steering responsibility, disaster often follows. The massive deinstitutionalization of mental patients ... was a perfect example (Osborne and Gaebler 1992, p. 73).

Government — that is, the society as a whole — must articulate clear expectations and create procedural safeguards, and beyond that direct its efforts to encouraging a vigorous civil society. Michael Walzer writes that the "words 'civil society' name the space of uncoerced human association and also the set of relational networks — formed for the sake of family, faith, interest, and ideology — that fill this space" (Walzer 1991, p. 293). This is the sphere of life that has largely been squeezed out of inner-city residential areas.

The social integration of the nonworking poor is an important object of public policy. This will depend upon the health of the social institutions with which they interact on a daily basis. Healthy institutions — schools, churches, associations of all kinds — will also help to maintain the social class mix whose loss, as Wilson has pointed out, is having such devastating effects. The problem is that inner-city areas are losing the stable working class and lower middle-class families whose leadership and resources is essential to those institutions. Improvement in the lives of the poor will therefore depend upon making the areas where they live communities that can sustain social institutions. This will require, as Wilson puts it, "a massive effort to restabilize institutions and create a social and economic milieu necessary to sustain such institutions" (Wilson 1987, p. 158).

There are two parts to this, as there are to making a garden. One is pulling out the weeds, reducing social disorder, the other, creating a fertile soil for the growth of the associational life that sustains social order.

Public schools do not generally play a positive role in this garden-making. On the one hand, they are perceived (not always fairly) as places of disorder, unable to bring their weeds under control, while on the other they offer a harsh and sterile soil for the growth of association through cooperation and mutual trust. Schools typically function in a bureaucratic fashion which increases the alienation of poor parents and children. Public schools do not belong to the communities that they serve; families are **members** of inner-city churches, but they are **clients** of inner-city public schools — all the difference in the world.

Why not establish policies that make schools more like the associations that people organize for themselves? Coleman and Hoffer suggested, along these lines, that:

> policies which would bring about expansion of choice should contain provisions that encourage the growth of social structures that can provide the social capital important to a school. Policies ... that would facilitate the creation of schools by institutions in which parents of prospective students are already involved exemplify this (Coleman and Hoffer 1987: 243).

Schools started by inner-city churches, by neighborhood associations, by organizations based upon a common homeland (the Trinidad and Tobago Association or the League of Haitian Families) would provide a reason for cooperation over time and an opportunity to develop trust and the habit of being trustworthy.

There is ample evidence that inner-city parents are deeply concerned about the schooling of their children, and that this concern extends both to the quality of academic standards and also the climate and character of the school. They are more dissatisfied with the schools to which their children are assigned than are other parents, and more supportive of school choice measures, including vouchers (Glenn, McLaughlin, and Salganik 1993). It is in the nature of schools that they are **daily** and thus promote continued involvement; the educational process is face-to-face and lends itself to developing habits of cooperation.

Postcommunist Eastern Europe provides an example of how parents and teachers can come together to build schools when the bureaucratic hindrances are swept aside. The systematic destruction of civil society under Communist rule and the manipulation of schooling for political mobilization left a profound dissatisfaction with government schools, despite their relatively high standards as measured by test scores. One of the first manifestations of the spontaneous re-emergence of civil society, as totalitarian regimes collapsed, was the creation of a host of alternative institutions that were independent of government control. Among the earliest of these were supplemental education programs at all levels, through which habits of independent cooperative action developed. As soon as it was legally possible (1988 in Poland, 1989 elsewhere) full-time schools were started by groups of parents and teachers. These have increasingly served as training grounds for other forms of participation in social reconstruction.

The idea that the same process could occur in the blighted

areas of American cities may seem far-fetched, but hundreds of community-based inner-city schools already survive (it would not do to say that they flourish, given their state of perpetual financial crisis in the absence of any public funding)[10]. John Witte's careful evaluation of the voucher program in Milwaukee makes clear that, while by no means performing miracles, the non-government schools taking part are functioning reasonably well, and involving parents far more effectively than did the public schools that their children previously attended. Similar schools serve poor and immigrant communities in the Netherlands, England, and Ireland, with government funding in some cases.

This is not to suggest that the "disconnected poor", the "underclass", or "truly disadvantaged", are poised and ready to create and operate their own institutions. Such a romantic notion ignores the profound impact of social disorganization, the mistrust and fear that prevail in many inner-city neighborhoods. Despite the process of middle-class exodus described by Wilson (1987), however, there continue to be thousands of functioning social institutions in America's inner cities, particularly churches, and these are continually developing new leadership through their complex organizational life.

My own church may serve as an example: there is a board that makes all major decisions, associations of women, of men, and of youth, several choirs, ushers, altar guild, Sunday School, classes in English as a second language, a tutoring program, a scholarship committee, another to look after the church building, a hot meals program for the elderly, regular visiting of the sick, fund-raising activities, a thrift shop, a food pantry for those in crisis, and so on and so forth. All of this activity depends upon the unpaid efforts of women and men, many of whom are looked upon with scant respect by the wider society. The congregation has recently taken over and renovated an abandoned supermarket, which now houses government-funded programs including early childhood classes for more than a hundred children, and is currently discussing the development of housing for the elderly: We help support a school in Honduras and programs in Haiti and Jamaica, since many of us come from those countries. Such activity is by no means exceptional, though it is rarely noticed by social commentators and reformers.

Those whose efforts keep this and thousands of other churches and associations in the inner city functioning are not the "disconnected poor". By definition, they are well-functioning men and women whose economic circumstances may be difficult but whose personal competencies are considerable. The existence of these social institutions in the midst of social disorganization, however, is a sign of hope and a source of potential community leadership. Their inability to extend their impact more widely, particularly in relation to the children of the city, may be attributed in large part to the lack of resources and of legitimation for such efforts. The situation was quite different two generations ago, when it was taken for granted that such indigenous institutions would be the mainstays of communities organized around their activities.

The most effective way for this church and others like it to have a consistent impact upon the lives of children whose families are too disorganized to be active participants in associational life would be through providing a day school, as Catholic parishes did when their members were mostly poor and confined to the same inner-city neighborhoods. Why doesn't this church, with so much energy available for cooperative effort, operate a school?

The main reasons may be the monopoly of public funding and prestige by established systems of schooling. There is a sense that it would be very difficult to provide a substitute for public schools. Each new initiative to spend more on inner-city public schools, in fact, makes it harder for nonpublic schools (like Parkside, described above) to survive.

There is an additional reason, the widespread belief that all pupils should be required to attend a "common school" in which they will become real Americans (Glenn 1988). The fact that American public education is even more segregated by race and class than is nonpublic education seems not to have shaken this belief. Public education enjoys a presumption of legitimacy that makes it difficult for public-spirited citizens to work to create alternatives, whether within or outside of the public system.

Perhaps the strongest reason to do so is because a school that truly belongs to the parents who send their children can provide a setting of unparalleled intensity for the development of the habits of responsible activity on the part of adults and children alike.

This is not the place to describe the framework of public policies that would provide a balance between school-level autonomy and the common needs of the society, between free initiative and public accountability (see Glenn 1991a; Coons and Sugarman 1992). These are the key elements of such a framework:

(1) Inclusion of all schools and all parents, with effective parent information and counseling about the choices available;

(2) A simple process of starting new public and nonpublic schools, as in the charter school provisions included in recent American education reform legislation, and in the new Russian education statute;

(3) Start-up loans available to new school associations that can demonstrate (say) commitment from the parents of a sufficient number (50?) of prospective pupils;

(4) Protections for teachers that provide benefits now offered by union membership without the associated rigidities of seniority, negotiated terms of work, etc.;

(5) Definition of expected outcomes and a process

for periodic assessment of how much each school has benefitted its pupils (e.g., "value-added"), with consequences for unacceptable performance (below that of comparable schools).

Our experience in Massachusetts has taught us that these elements are the most important in developing a system of school choice benefiting all children and thus society as a whole.

Notes

1 For an overview, see Glenn (1993), to be published in a six-nation OECD study.

2 The passages that follow are taken from the very useful collection entitled **Liberté d'enseignement. Les textes**, Geneva: Organisation internationale pour le développement de la liberté d'enseignement, no date; texts are provided in French, English, and Spanish.

3 The passages that follow are taken from the very useful collections by Siegfried Jenkner of Hannover, entitled **Das Recht auf Bildung und die Freiheit der Erziehung in Europäischen Verfassungen**. Part 1 (May, 1992) provides extracts from the constitutions of members of the European Community, and Part 2 (June, 1993) from those of non-EC members of the Council for Cultural Cooperation. Both include English translations.

4 A full discussion of the development of educational freedom in postcommunist Europe will be found in Glenn (in press b).

5 Interview with Nat Hentoff (1982), quoted by Mead, 1986, pp. 248–249.

6 My colleague Professor Vivian Johnson is conducting a nationwide study of such parent centers in urban schools.

7 Something of the same effect has occurred in some French **banlieux** occupied primarily by North African immigrant families, and in such residential areas elsewhere in Western Europe.

8 From Tom Wolfe's amusing **Radical Chic and Mau-mauing the Flak catchers** (1970, p. 122), which captures the spirit of the times very well. Pressman and Wildavsky (1973) gives a good account of the failure of an attempt at comprehensive renewal of Oakland.

9 Not by any means universal, or perhaps even in the majority, but shaping the cultural milieu and the perceived alternatives available.

10 See **The Journal of Negro Education**, theme issues 61, 2 (Spring) and 4 (Fall), 1992 and studies by Joan Ratteray and others.

References

Chubb, J. E. and Moe, T. M. (1990). *Politics, markets, and America's schools.* Washington, DC: The Brookings Institution.

Clark, R. M. (1983). *Family life and school achievement. Why poor black children succeed or fail.* Chicago: University of Chicago Press.

Coleman, J. S. (1987). Families and schools. *Educational Researchers,* **16**(6), 32–28.

Coleman, J. S. (1990). Quality and equality in American education: Public and Catholic schools. In *Equality and achievement in education.* Boulder, CO: Westview Press.

Coleman, J. S. and Hoffer, T. (1987). *Public and private high schools: The impact of community.* New York: Basic Books.

Coons, J. E. and Sugarman, S. D. (1992). *Scholarships for children.* Berkeley, CA: Institute of Governmental Studies Press.

Dionne, E. J. Jr. (1992). *Why Americans hate politics.* New York: Simon and Schuster.

Etzioni, A. (1984). *An immodest agenda.* New York: McGraw-Hill.

Gans, H. J. (1991). *Middle American individualism.* New York: Oxford University Press.

Glenn, C. L. (1988). *The myth of common school.* Amherst, MA: University of Massachusetts Press.

Glenn, C. L. (1990/91). Boston: The proof of the choice pudding. *Educational Leadership, December/January*, 41–43.

Glenn, C. L. (1991a). Controlled choice in Massachusetts public schools. *The Public Interest*, **103**, 88–105.

Glenn, C. L. (1991b). *Perestroika* for American education. In J. R. Bliss, W. Firestone, and C. Richards. (Eds), *Rethinking effective schools*. Englewood Cliffs, NJ: Prentice-Hall.

Glenn, C. L. (1993). *Parent choice of schools in the United States*. Organization for Economic Cooperation and Development, Paris.

Glenn, C. L. (in press, a). Choice and urban school reform. In B. Jones (Ed.), *The new American schools: Alternative concepts and practices*. Norwood, NJ: Ablex.

Glenn, C. L. (in press, b). *Educational freedom in Eastern Europe*. Washington, DC: US Department of Education.

Glenn, C. L., McLauglin, K., and Salganik, L. (1993). *Parent information for school choice: The case of Massachusetts*. Boston: Boston University and Johns Hopkins Center on families, communities, schools, and children's learning.

Hill, P. T., Foster, G. E., and Gendler, T. (1990). *High schools with character*. Santa Monica, CA: Rand Corporation.

Himmelfarb, G. (1992). *Poverty and compassion*. New York: Random House Vintage Books.

Mead, L. (1986). *Beyond entitlement*. New York: Free Press.

Mead, L. (1992). *The new politics of poverty*. New York: Basic Books.

Mortimore, P. *et al.* (1988). *School matters*. Berkeley: University of California Press.

Osborne, D. and Gaebler, T. (1992). *Reinventing government*. Reading, MA: Addison-Wesley.

Pressman, J. and Wildavsky, A. B. (1973). *Implementation*. Berkeley, CA: University of California Press.

Richardson, J. (1993). Teachers in poll seek greater federal push for parent involvement. *Education Week, May 19*, 10.

Rutter, M. *et al.* (1979). *Fifteen hundred hours: Secondary schools and their effects on children*. Cambridge, MA: Harvard University Press.

Schorr, L. B. and Schorr, D. (1989). *Within our reach: Breaking the cycle of disadvantage*. New York: Doubleday Anchor Books.

Tocqueville, A. de (1988). In: J. P. Mayer (Ed.), trans. G. Lawrence, *Democracy in America*. New York: Harper and Row.

Walzer, M. (1991). The idea of civil society. *Dissent*, **38**, 293.
Wilson, W. J. (1987). *The truly disadvantaged.* Chicago, IL: University of Chicago Press.
Wolfe, T. (1970). *Radical chic and Mau-mauing the flak catchers.* New York: Farrar, Straus and Giroux.
Wynne, E. A. and Ryan, K. (1993). *Reclaiming our schools: A handbook on teaching character, academics, and discipline.* New York: Macmillan.

Chapter Fourteen

PARENTAL CHOICE: CONSEQUENCES FOR STUDENTS, FAMILIES, AND SCHOOLS

ELLEN GOLDRING, WILLIS HAWLEY, ROBERT SAFFOLD AND CLAIRE SMREKAR[1]
Vanderbilt Institute for Public Policy,
Vanderbilt University Nashville, TN 37203

Introduction

The debate over the use of choice to achieve school improvement, and the need to rely increasingly on magnet schools to secure school desegregation, continues. In addition, two significant demographic trends complicate the issues further. First, the nation's school population is becoming increasingly diverse, racially and ethnically. Second, the proportion of the nation's children who live in poverty is increasing. These two developments intensify the need to understand the consequences of parental choice and to ensure that increasing parental choice does not work in ways that further disadvantage children who are most in need of high quality education.

The purpose of this chapter is to review the central conceptual theories and empirical research which dominate current discussions on choice. Additionally, we will present an analytical model and set of research questions which could further our understanding of the impact of increasing the opportunities parents have to choose the schools their children attend.

As our review of research (below) indicates, there is limited evidence that can be used to assess the validity of the claims made by either the advocates or opponents of policies that would increase parental choice in determining what schools their children will attend. This chapter focuses attention on the development of specific research questions aimed at addressing the consequences of extending the options parents have for their children's schooling.

Choice and Magnet Schools

In the ongoing national debate over education reform, proposals to provide parents with greater choice regarding the schools their children attend are receiving increasing consideration. Choice is hailed by its proponents as an essential step toward fundamental reform, and is called by some advocates a "panacea" for all that ails the nation's education system. On the other hand, choice is criticized, even vilified, by its opponents as a strategy that constitutes a possible threat to the very existence of public education. Some critics argue that it is a policy that will likely result in intensified racial and socioeconomic stratification as well as further fragmentation and erosion of shared values that transcend race, class, and location.

The recent surge of interest in choice has been fueled by frustrations over the pace of change initiated by the regulatory reforms of the mid-1980s and by disappointments over the inadequacies or apparent failures of these earlier reform efforts to bring about desired improvements in educational quality. In addition, concern persists over what many regard as the lackluster performance of our educational system at a time when America appears to be losing its competitive edge in the global economy. Although the recent change in presidential administrations has apparently moved choice off the active federal policy agenda, the debate over choice remains hotly contested at the state level. According to the U.S. Department of Education, 13 states have enacted some form of choice legislation to date (vouchers, tax credits, open enrollment, etc.), and more than 36 states are currently involved in some form of activity related to choice, including pending legislation and task force recommendations (U.S. Department of Education 1992). Moreover, according to a recent Gallup poll, choice is favored by nearly a two to one majority of virtually every population segment sampled, including parents of public school children, although a majority of respondents do not favor using public funds to support parental choice of private schools (Elam, Rose, and Gallup 1991).

Advocates of choice argue the merits of this reform strategy on several grounds[2]:

(1) Parents should have the right to choose where their children attend school.

(2) Choice will enhance educational equity by increasing the access of poor and middle class families to learning opportunities currently available only to the more affluent.

(3) Choice will dismantle the power of the "education establishment" by empowering parents.

(4) Choice will force schools to compete for students and resources, breaking what is effectively a government monopoly on schooling. Thus, schools will develop more focused goals and define for themselves "market niches" leading to greater diversity among schools and

enhanced quality, productivity, efficiency, and responsiveness to parent and student needs and/or wants. All of this will lead, in turn, to higher student performance and persistence.

(5) Choice will result in greater parent and student satisfaction, commitment, and involvement in schooling by fostering an enhanced sense of ownership among parents.

(6) Choice will foster greater building-level autonomy, stronger leadership, and enhanced teacher collaboration and professionalism.

(7) Choice will result in an enhanced sense of community within schools by fostering shared values among parents, students, teachers, administrators, and staff.

Opponents of various forms of choice argue that the risks and uncertainties associated with choice as a policy strategy should at the very least direct policymakers to construct choice systems so as to minimize those risks (through careful regulation) or to look to alternative strategies to promote educational change. Their major concerns are[3]:

(1) Choice will undermine instruction in the core values that undergird the American civic community and that minimize cultural and political divisiveness.

(2) Choice will increase racial and social class isolation.

(3) Choice will increase existing inequities among schools and public school districts.

(4) Choice will perpetuate, and perhaps intensify, the educational disadvantage experienced by many low income families who will not have the information, time, and access to

transportation needed to make effective use of the highest quality options available.

(5) Choice will force schools to place undue stress on marketing.

(6) Choice will violate the separation of church and state (if sectarian schools are included in the plan).

"Magnet schools", sometimes called "alternative schools" or "schools of choice", are public schools that provide incentives to parents and students through specialized expectations for students, curricular themes, or instructional methods. Typically, magnet school enrollment is regulated to ensure that attendance is racially balanced. While some magnets have selective admission criteria, pupil assignment may be established through any combination of first-come-first-served application, lottery, and/or percentage set-asides for neighborhood/attendance area residents.

Over the past 20 years, magnet schools have been created largely by urban school districts as a primary tool for achieving desegregation goals and encouraging innovation. In many instances, districts have supported magnet schools with a considerable investment of resources so as to optimize their consequences.

The first magnet schools — at least the first schools to be so labeled — were developed in the early 1970s, and by the 1982/83 school year magnet schools could be found in one-third of the largest urban districts. Today, most urban districts use magnets to promote desegregation through parent and student "choice" (Blank 1989). Increasingly, mandatory reassignment of students has been used as a back-up measure to a magnet strategy in order to encourage choices for desegregation and to assure that desegregation is achieved when magnets, by themselves, fail to accomplish the goal. As a result of the recent decision by the U.S. Supreme Court in the DeKalb County, Georgia, desegregation case (**Freeman v.**

Pitts 1992), it is likely that magnet schools will increasingly become the strategy of "choice" for achieving whatever desegregation is pursued.

Advocates of magnet schools argue that magnets: (1) attract pupils of different racial and socioeconomic backgrounds with similar educational interests, so as to create racially heterogeneous schools; (2) provide unique sets of learning opportunities; and (3) encourage innovation. In other words, magnets are viewed as an effective way to enhance diversity and equity among schools, increase educational quality in a school district (Blank 1989), and stabilize enrollments.

Critics of magnet school programs charge that magnets can exacerbate existing class or socioeconomic cleavages, especially when the magnets are academically selective and few in number. They assert that middle-class parents are more motivated and more informed regarding the availability of educational options, while lower-income parents opt for or otherwise "end up" in conventional attendance area schools with no specialized offerings and fewer resources. Thus it is claimed that magnets tend to "cream off" more academically motivated and able students, as well as more effective and innovative teachers, resulting in diminished educational opportunities (e.g., less rigorous curricula, lower expectations by teachers, and different school climates) for those who do not attend them. Further, critics of magnet schools argue that magnet schools divert resources that should be used for system-wide improvements and that there is little "spill over" from magnets because systems do not use them to foster the dimusion of innovation.

Theoretical Perspectives on Choice

The theoretical case for choice is robust and the claims made for the effects of parental choice on school improvement are both ambitious and controversial. However, the empirical

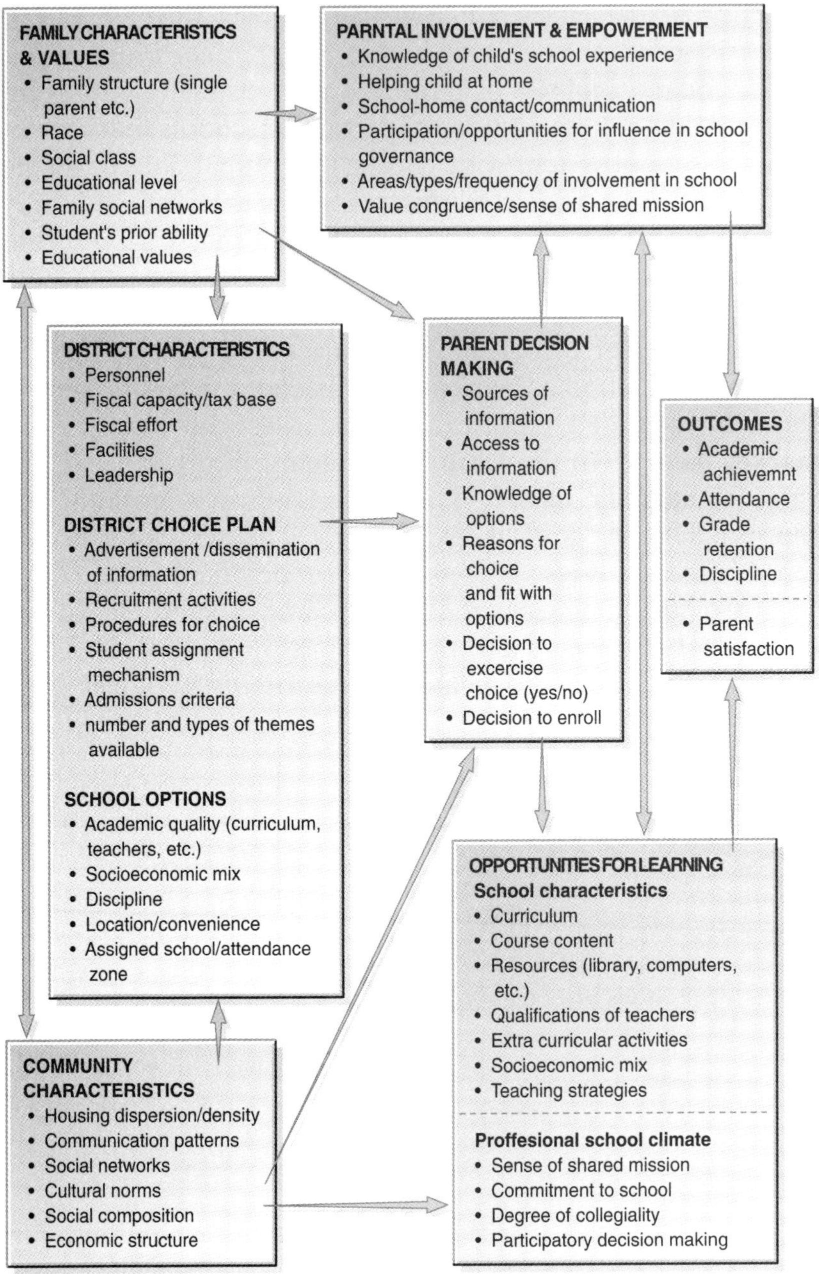

Fig. 14.1. Dynamics of choice.

evidence of the effects of increasing choice through public policy is scant. Virtually all of this research is of four types: (1) secondary analyses of data sets with critically important data missing, (2) case studies of particular schools that cannot speak to school system-wide effects, (3) comparisons of public and private schools, or (4) official reports that deal with only some of the philosophical and design issues that are important to the development of policy and practice across school systems.

The case for increasing the choices that parents have in selecting the schools their children attend appears to rest on three inter-related theories. We refer to these as (a) the **market theory**, (b) the **family sovereignty theory**, and (c) the **value community theory**. The underlying assumptions and the predictions derived from these various theories are represented in Figs 14.2–4. The model depicted in Fig. 14.1 provides a schematic overview and a synthesis of the theoretical arguments and their inter-related concepts. These models are not meant to include all the possible variables in the

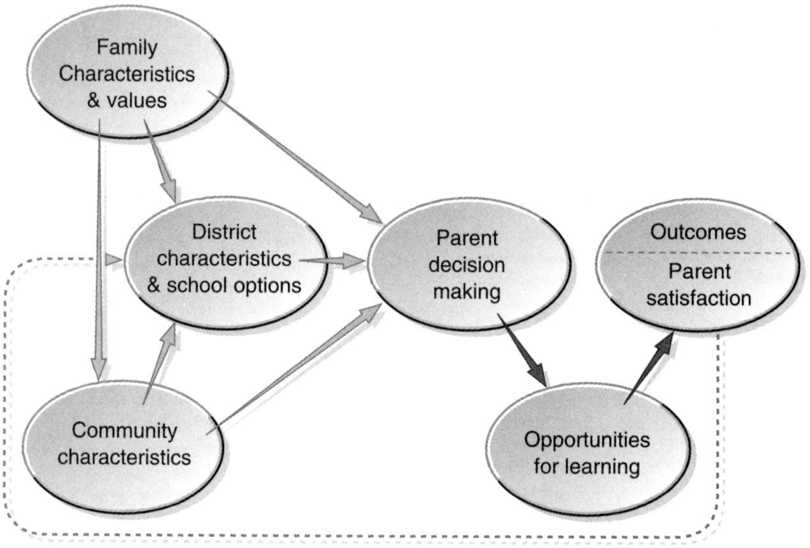

Fig. 14.2. The market theory.

study of school choice, nor do they present all the possible relationships among the concepts. Moreover, the relationships between the different sets of influences and outcomes represented in these theories are recursive and interactive. Thus, these theories should be regarded as overlapping and mutually reinforcing.

The market theory

Contemporary interest in choice may be traced to Milton Friedman's (1955) essay expressing concern over the "indiscriminate extension of government responsibility" in the field of education and his proposal to use a voucher system as a means of "denationalizing" American education. While interest in choice has waxed and waned over the past 35 years, it constitutes the centerpiece of the current education reform debate. A major Brookings study (Chubb and Moe 1990) and other recent research (Coleman, Hoffer, and Kildare 1982; Coleman and Hoffer 1987) have given credibility and legitimacy to choice by asserting that private sector schooling results in higher achievement outcomes for students.

In his recent analysis of the economics of choice, Henry Leaven employs a dichotomous conception of choice to sort out the multitude of proposals currently under discussion (Leaven 1991). Public choice proposals are those in which subsidized choices are restricted to publicly owned and controlled schools. Market choice proposals, on the other hand, are those in which subsidized schooling choices are relatively unrestricted and include private schools. Both categories of choice, however, are based on a set of fundamental economic assumptions about human and organizational behavior in the context of a marketplace for educational goods and services — namely the array of available schools.

On the "demand side" of the market, parents (and students) are presumed to be rational and ambitious consumers. According to choice proponents, parents will make choices among schools in a manner that they expect will be most advantageous to themselves and their children. Thus parents and students, given the resources, will shop around for the best school to meet their needs. It should be noted that nothing inherent in economic theory defines rational parent choices in terms of higher student academic achievement. Any preference — religious values, safety, racial antipathy, and so forth — defines the market in which any given parent wants to consume education.

Nonetheless, most proponents argue that choice will enhance the access of lower income children to educational opportunities that contribute more to their academic learning than those that are currently available to them. Incentives for parents (i.e., consumers) can also be built into the system and/or offered by individual suppliers of educational services (e.g., schools), so as to affect parental perceptions of self-interest and encourage choices consistent with desired social goals (e.g., desegregation) (Rossell 1990).

Theoretically, a number of additional benefits would emanate from an expanded market system for allocating schooling. First, it is argued that a market system would offer an institutional environment that encourages the growth of a number of diverse and innovative schools, because educators would be intent on maximizing their own gains (income, influence, self-esteem, etc.) by responding to the diverse preferences/demands of parents (or guardians) and students who consume educational services. "Suppliers" of education would, moreover, seek not only to respond to demand, but to anticipate and create it. Such a system is able to match more closely the diverse educational needs and preferences that are assumed to exist. The efficiency and effectiveness of a variety of schools depend on the assumption that there is no one best program that is capable of responding to these diverse preferences in a democratic, pluralistic society (Raywid 1989). A centralized bureaucratic system that incorporates a common set of educational goals and programs is seen as less able to match parent preferences as closely as could a more unrestricted system of choice that promotes diversity of educational options. Thus, a choice system would likely result in greater system-level efficiency (Witte 1990a).

In addition to greater diversity among schools, market theory holds that a consumer-oriented system would encourage innovation. Since schools would not have a guaranteed clientele, educators would be forced to innovate and change in response to competitive pressures for students and funding, rather than rely on bureaucratic routines, established standard operating procedures, or professional norms and judgments. Moreover, the competitive dynamic would operate to force greater cost efficiency through minimizing needless administrative procedures and waste at the school level (Witte 1990a).

Tied closely to diversity, innovation, and efficiency is the notion that expanded parental choice would result in greater school flexibility and responsiveness. In light of the increasing complexity of society and of the global economy, proponents argue that schools must be able to respond more quickly to changing needs and demands. A choice system has "a built-in organizational response mechanism" that forces a school to meet new demands and trends or face the loss of parents (i.e., customers) to other schools (Witte 1990a; cf. Hirschman 1970). In theory, this response mechanism is linked to other internal incentives driven by the competitive dynamic which would lead to improvements in educational quality. Schools would seek to hire, promote, and pay more effective staff, who, in turn, would have an incentive to maintain or increase enrollment by achieving outcomes that meet parent and student demands. Staff would be motivated by the satisfaction of collaboration in a school that enjoys clarity of mission and goals, as well as relative autonomy from state and local bureaucratic control.

In sum, the market theory justification for choice holds that the competitive dynamic will ratchet up the overall quality and efficiency of the educational system. The more opportunities parents have to choose their children's school, the more responsive schools will be to parent and student preferences, and the more closely schools, on the whole, will satisfy most parents and students. Choice gives parents the option of potential exit to an alternative school — to drive responsiveness to their preferences and demands (Witte 1990a).

Empirical Research: Market Theory

Empirical tests of assumptions of the market theory rely on considerable inference. This is not surprising given that we do not have a market system to study. The issue is, how reliable are the inferences? Moreover, much of the evidence that is drawn upon relates to efforts which compare public and private schools rather than examining how choice works within the public sector. Thus this discussion of research related to the validity of market theory justifications for choice deals with the broad range of possible choice plans.

The studies that almost certainly have had the greatest influence among supporters of choice are those by Coleman and Hoffer (1987) and by Chubb and Moe (1990). Both studies rely on the High School and Beyond (HSB) national longitudinal data in drawing their conclusions, even though the private school sample of the HSB study is unrepresentative of private schools (Crain and Hawley 1983). Coleman and Hoffer (1987) focus attention on the findings that minority children achieve at higher rates in Catholic than in non-Catholic schools.

Chubb and Moe's 1990 book, **Politics, Markets, and America's Schools**, makes the boldest of the widely cited claims for the positive consequences of choice on educational reform. They conclude that private schools are more effectively organized because of their relative autonomy from external bureaucratic constraints. More effective school organization in the private sector translates into higher student performance, resulting in as much as one additional year of achievement over a four-year high school experience. Their empirical work has been widely criticized. First, as noted above, Chubb and Moe rely on the data from the HSB survey. Second, in identifying effective schools, they identify school characteristics, but many of the schools that they find having these characteristics are public schools. (Whether these schools are schools of choice is not stated.) Third, some of the characteristics they point to may well be the product, not of the superiority of curricula or teaching differences between schools, but of the ability of private schools to exclude or rid themselves of children who misbehave or whose parents do not cooperate with the school.

Economist Henry Levin (1991), examining national data, argues that the statistical difference in student performance in private as compared to public schools is about 0.1 standard deviation and that even so, it is likely that background factors that account for differences cannot be fully controlled and that the resulting bias in the data overstates the effectiveness of private schools (see also Crain and Hawley 1983). Similarly, Witte (1993) concludes that the results of the first two years of the Milwaukee choice experiment, involving private, nonsectarian schools, fail to provide any evidence of success in terms of standardized achievement scores. Jean Merl (1992), after reviewing the existing research and interviewing experts, parents, and school administrators concluded:

> A small proportion — probably no more than 5% — of the nation's 28,000 private schools clearly are better than the average public school They are highly selective and sharply focused, and they spend up to three times as much per pupil, often to keep classes small. But the majority of private schools do not stand out academically.

Fundamental to the claims of choice proponents that the market will lead to higher quality schools are assumptions that parents will choose quality options — defined in terms of academic performance — if offered to them. Maddaus (1990), however, reports in a review of research on parental choice that parents employ a variety of criteria in making choices, some of which may be largely unrelated to school quality. Among the dominant factors cited by parents when asked about their choice of school are academic quality (Nault and Uchitelle 1982; Witte 1993); distinctive curricular options or instructional approaches (Blank *et al.* 1983; Rossell 1985); racial composition (Rossell 1990); and location (Rossell 1990; Bridge and Blackman 1978).

Williams, Hancher, and Hutner (1983) and Darling-Hammond and Kirby (1985) distinguish between **active** and **passive** choosers. They find that academic quality was the most frequently mentioned factor influencing the school choice of parents who **actively** considered other schools at the time of their child's enrollment. In both studies, however, the largest group of parents surveyed were **passive** choosers, or those who did **not** consider other schools when enrolling their child. For these parents, the most important influences on school choice were assignment to a school and transportation/convenience.

In order for parents to make informed choices, they must have access to adequate information about schools and district choice procedures. Questions remain as to whether most parents in fact have the information they need to make choices related to academic quality when choice is available to them. Of course, with limited choices to make, and limited resources to make them, the incentive for parents to acquire (and for schools to provide) information may be too low. In fact, Witte (1993) notes in his study of the first two years of the Milwaukee choice experiment that the most prevalent source of information on the choice program, as reported by parents, is word-of-mouth information. He also notes, however, that parents judged as very satisfactory the assistance they received from the private schools involved in the program and from the Department of Public Instruction. Moreover, judgments of the quality of assistance improved considerably over the first year of the program, a finding that is consistent with a major conclusion from earlier research on the Alum Rock voucher experiment that parent information increased over time (Cohen and Farrar 1977; Capell 1981).

One test of the market theory's prediction that choice would encourage innovation would be to compare the adoption of innovative instructional practices and curricula in schools of choice and in other schools. Sosniak and Ethington (1992) have attempted to apply this test to a national sample of schools. They conclude, in effect, that market theory fails the innovation test; the two types of schools (choice and nonchoice) use similar curricula and modes of instruction. One might consider findings from the Alum Rock voucher experiment, where instructional practices differed little among the minischools from which parents chose (Martin and Burke 1990), as evidence that innovation is not encouraged by choice, but this experiment is peculiar in many ways and generalizations from its outcomes are precarious.

Walker (1971) compared the rates of educational innovation in elite public and private schools in New England and found that public schools were more innovative than private schools. It might be argued, however, that all of these schools were schools of choice. Metz (1986) examined instructional strategies in three presumably different magnet schools and found little variation.

Finally, much of the parental choice debate surrounds the question of whether unleashing the forces of the marketplace will enhance or undermine equality of educational opportunity. Coleman, Schiller, and Schneider (1991) present evidence from their analysis of National Educational Longitudinal Study: 1988 data that indicates a strong response to opportunities for choice among minorities and immigrants across educational and income levels that is not demonstrated among whites of similar backgrounds. Manski's (1992) econometric simulation of a choice system fails, however, to support the argument that poor youth will be better off with a voucher system. He concludes that even under the most favorable assumptions, choice would not equalize educational opportunity across income groups.

In summary, the available evidence is sparse and leaves many unanswered questions for the market theory-based assertions that increasing parental choice will increase student achievement or promote basic educational change. There are at least two major constraints on the capacity of choice to promote practices that enhance students' academic performance, at least in the United States, that need further study[4]. First, relatively few parents may rank academic achievement very high on their list of criteria in selecting a school for their children. Second, among public schools, the effects of choice may be limited by state or district policies related to curricula and process (Metz 1986), and by specification of what is to be measured on tests of student achievement or teacher performance (Smith and O'Day 1990). Specific research questions to be addressed should include:

(1) What is the impact of district choice programs on the outcomes of schooling for all children, and particularly low income and minority children?
- Does student performance differ in magnet and nonmagnet schools?
- Do student attendance and grade retention differ in magnet and nonmagnet schools?

(2) What is the impact of district choice programs on the access to opportunities for learning for all children, and particularly low income and minority children?
- Are there differences in the curricular content of magnet and nonmagnet schools and between magnet schools?

(3) Does the enrollment of children in alternative choice schools typically sort students along socioeconomic lines and/or by race and ethnicity?

(4) What impact do differing systems for admission to choice programs (e.g., the presence or absence of screening criteria, the use of lotteries) and the use of affirmative outreach programs have on the proportions of low income and minority children enrolled?

(5) Does the presence of competition encourage the level of innovation and the diffusion of innovation?

The family sovereignty theory

In most countries, parents are responsible for their children's education through a mandatory education law. It is the parents' responsibility to see that this law is observed.

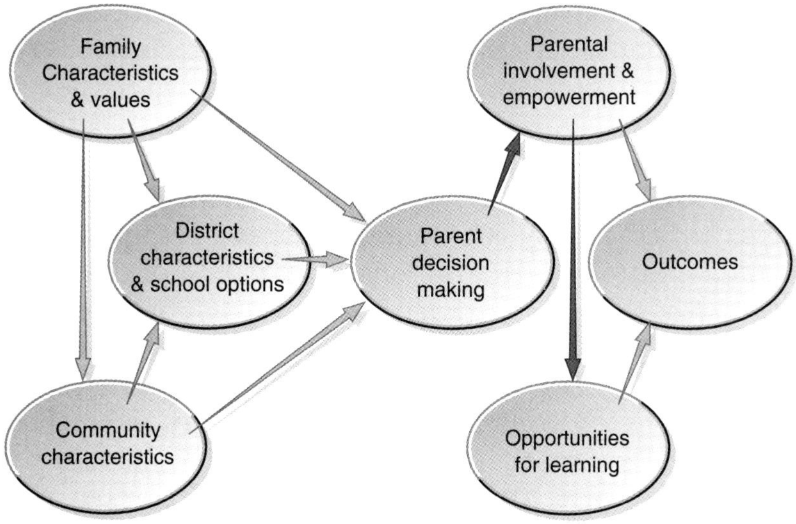

Fig. 14.3. The family sovereignty theory.

Choice advocates suggest that parents should be given authority commensurate with the responsibility. This authority will allow parents to choose their child's school according to their philosophical or pedagogical preferences. Numerous polls have indicated that a majority of parents want this authority to choose the schools their children attend.

Regardless of the reasons and criteria which different types of parents use when choosing a school, there is an assumption among advocates of public choice that parents who choose their schools will be more satisfied with those schools than those parents who do not. The theoretical arguments underlying the relationship between choice and parent satisfaction often center around issues of family sovereignty (see Archbald 1988; Levin 1991).

The family sovereignty position embodies three arguments for choice as an alternative to mandatory assignment. First, legal-philosophical supposition holds that, in principle, parents should have the right and individual freedom to choose a school for their children. Second, because they have exercised choice, parents will be satisfied with and supportive of their school of choice (independent of the character of the choice). Third, it is assumed that when parents choose schools, their choices reflect value-based preferences (e.g., religious or pedagogical) and this, in turn, leads to greater satisfaction and parental support of schools.

The logic embedded in the family sovereignty position, which links choice to parent satisfaction and support for their children's learning, views the individual as a rational decision-maker who acts out of self-interest and who adopts maximizing strategies. This assumption is, of course, the same as the rationality assumption that drives the economic market case for choice. Therefore, public choice theory and the logic of family sovereignty suggest that individual families who exercise their right to make choices weigh the various alternatives rationally in conjunction with their own values and preferences in an attempt to maximize their satisfaction.

Another assumption related to parental sovereignty which links parental choice to satisfaction and support for student learning is related to the investment associated with making a choice. This investment can be defined in terms of time, energy, and/or money. It is generally assumed that parents who invest in their child's education by actively making a choice will view their choice favorably. Even if there are no visible reasons for the choice to lead to satisfaction, many parents may justify their choice and investment by indicating satisfaction with the school and viewing it through "rose colored glasses" (Erickson 1982).

Family sovereignty theory predicts that choice increases parent satisfaction and support for schools. Parents' satisfaction with their children's schools and involvement can, of course, have an impact upon other aspects of schooling, such as the relationship between the school and its community, teacher morale, student recruitment, and student achievement (Henderson 1987).

The act of choice contributes to a sense of ownership and commitment to the child's school (Driscoll 1991). This sense of ownership is often accompanied by a view that, having made a commitment to a particular choice, parents have a greater right and need to be involved (Archbald 1988). Parental involvement in the child's school can be viewed as an important goal of choice as a mechanism to increase parental support of schools, and to increase parental involvement in their own child's education.

Researchers have concluded that the lack of parental participation in many schools is often caused by the unwillingness of schools to invite or include parents, especially those parents who are working class (Lightfoot 1978; Connel *et al.* 1982). Proponents of choice predict that choice would make educators more open to parental involvement and would build parental support for schools (Nathan 1985; Watt 1983). This support can then be translated into a higher sense of satisfaction. Under choice plans the relationship between parents, teachers, and administrators may change to offer parents an opportunity to influence management and school processes (Raywid 1987). The consumer satisfaction requirement may help schools open their boundaries to learn about the needs and desires of the parent clientele and ensure them a role in controlling and influencing school processes (Chubb and Moe 1990).

Critics of choice are skeptical of these claims for a variety of reasons. Cooper (1991), for example, cautions that there may be a tension in expecting parents both to choose schools and to be active partners in education: "Parents-as-consumers may be in conflict with parents-as-participants, since heightened choice may send children off to far away schools, distant from the neighborhood, community and families with whom children grew up" (p. 239). Moreover, some parents may believe that once they have made the critical choice, they no longer need to participate in other ways. The question remains whether these uninvolved parents are less satisfied with their school of choice than involved parents.

In summary, parental choice of schools may offer the opportunity for a change in the nature of school-parent relationships. However, it is not enough to assume that choice, in and of itself, will increase parents' satisfaction with their children's schooling. Choice may act as an impetus for pursuing increased satisfaction through meaningful participation in the life of the school. Whether satisfaction is the outcome of this participation will depend on the responsiveness of the school to parents' interests. According to the family sovereignty theory, the existence of this dynamic creates the possibility of continuous improvement.

Empirical research: Family sovereignty

This argument for choice predicts that the benefits of sovereignty will affect parental investments in their children's education over time because of the psychological effects of choice on parents and because of the logic stating that granting parents the right to greater influence over their children's education should lead schools more actively to seek parental involvement and empowerment in schooling.

As we noted earlier, it is clear that parents want the opportunity to exercise choice, whether or not they are dissatisfied with their child's education (Elam, Rose, and Gallup 1991)[5]. Moreover, Witte (1993) reports significant increases in satisfaction among parents participating in the Milwaukee choice program. We simply do not know much, however, about whether the provision of choice motivates educators to be more open to and encouraging of parent involvement either in schools or their children's education at home[6].

Most of the research on parents and choice has focused largely on parents' reasons for school choice. This literature has established that such factors as social class differences and parents' information-seeking methods have an impact on reasons for choice (Bridge and Blackman 1978; Maddaus 1990).

Other research has started to examine inter-relationships between parental choice and involvement. Research in Israel on the relationship between parents' satisfaction with schools of choice indicated that the more the schools' programs are compatible with the parents' choice, and the more the parents feel that they are empowered, that is, that they are partners in the decision-making process of the school, the more satisfied they are with their school (Goldring and Shapira 1991).

Bauch (1991) found a relationship between parents' reasons for choosing a particular school and participation. Parents who choose for discipline or safety reasons seem to be involved in their children's education, while parents who do not have clear reasons for choosing a particular school seem to be rather "... laissez-faire about their involvement and therefore the potential benefits for parent involvement may be lost to these students" (p. 319). Shapira and Goldring (1990) found a relationship between types of parental involvement and the specialty of schools of choice in Israel. For example, in a magnet school of the arts, the parents were mainly involved in activities directly related to the school's specialty — the arts — while they were not involved in overall school decision making. In contrast, in an ideologically based specialty school the parents were involved in overall policy making and connecting the special ideology to the general school program. Goldring and Bauch's (1993) study of parent involvement in urban school systems found that parent involvement and school responsiveness were more prevalent in Catholic and single-focused magnet schools than in multiple-focus magnet schools ("schools within a school").

One empirical study that has examined empowerment in public schools of choice involves implementation of the Alum Rock voucher experiment that was designed to allow parents to choose among public schools. This voucher system required all parents to make a choice because all parents, even those who chose their neighborhood school, had to choose a minischool within the neighborhood schools. Approximately 18 percent of the parents chose schools outside of their neighborhoods. In this experiment, parent satisfaction increased, but their sense of power and influence did not. The parents were satisfied because they had more educational alternatives, but after exercising their choice, school–parent relationships continued in their familiar manner (Cohen and Farrar 1977).

There is reason to believe that parents who choose private schools over public schools are more likely to be involved in their children's schools (Coleman and Hoffer 1987; Witte 1993). Witte (1993) also found that parental involvement **increased** among parents who participated in the Milwaukee choice experiment, which involves private, nonsectarian schools. It cannot be determined from existing research, however, whether making the choice to send one's children to a **public** school they would not otherwise attend alters the relationship between families and schools. Despite growing legitimacy for consumer choice in education and the documented effects of parent involvement, relatively little is known about their inter-relationships.

The following kinds of questions need to be addressed by future research:

(1) Does the interaction between parents and schools differ across social class, race, and ethnicity in magnet and non-magnet schools, and is this interaction influenced by the social class, race, and ethnicity of parents and schools?

(2) Do magnet schools increase the opportunities schools provide to parents to learn about their children's educational needs and to become involved in schools and their children's education?

(3) Do parents of children in magnet schools differ from parents in non-magnet schools in terms of:

- their knowledge of their children's school experiences and available educational alternatives;
- their involvement in their children's education in school and at home;
- their communication patterns with school;
- their level of satisfaction with their children s school;
- their confidence that they can influence school policy and practice?

The value community theory

Although family sovereignty explains choice phenomena by examining factors associated with individual behavior,

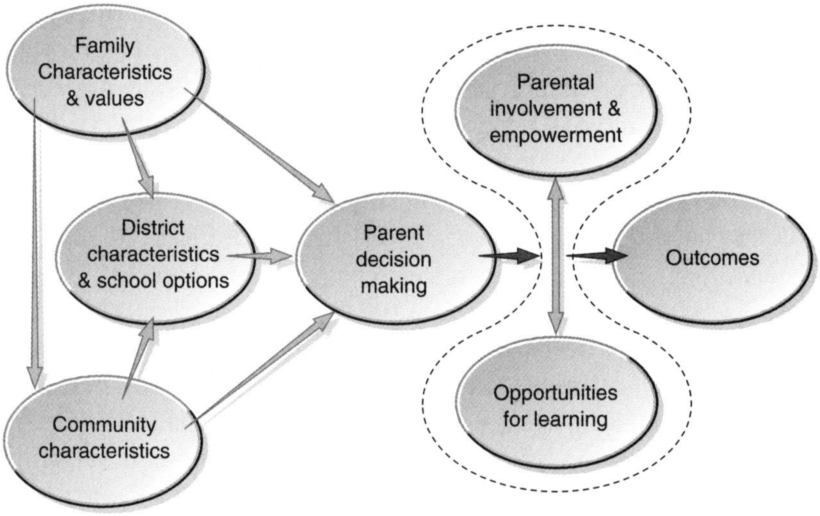

Fig. 14.4. The value community theory.

there is a point where individual self-interests converge with a common good. If all members of a given group, such as a certain group of parents, seek private self-interests, collectively a common good emerges with respect to that group, and that group forms a community (Olson 1965).

While the concept of community has been explored for centuries, and the idea that schools should function as communities is common throughout the research on effective schools, the linkage between community and school choice reflects a new and growing interest among educational reformers. To the degree that school choice promotes the conditions and processes which lead to basic elements of community — commitment, communication, and collegiality — school choice could serve a critical function in enhancing the educational opportunities and experiences of students and their families.

Contemporary notions of community (e.g., Bronfenbrenner, Moen, and Garbarino 1984; Coleman and Hoffer 1987; Newmann and Oliver 1968; Scherer 1972) assert that the community of residence in society today does not reflect the community of psychological meaning for most families (Steinberg 1989). These observers distinguish between a concept of community associated with physical or geographical boundaries and one grounded in social structures and social relations. For example, Newmann and Oliver (1968) include the following criteria in their definition, each of which is viewed as a continuum and indicative of greater or lesser degrees of community: (1) membership is valued as an end in itself, not merely as a means to other ends; (2) members share commitment to a common purpose; and (3) members have enduring and extensive personal contact with each other. The sense of solidarity, membership, and mutual support that results from community is thought to have an impact upon the individual in terms of personal development and integration, and to the larger society in terms of social cohesion and stability (Raywid 1988).

A distinct component of the concept of community is a social network, or a social system of formal and informal organizations and opportunities for information exchange and face-to-face interactions among individuals (Cochran and Brassard 1979; Steinberg 1989). Members who are considered a part of a functional social network are characterized by well-known roles and contexts (neighborhood, relatives, work- or school-mates, people in agencies or organizations, etc.) and are distinguished from a more peripheral social circle (Cochran and Henderson 1986).

The notion of a school as a community embraces the communal associations and personal relationships which are sustained by a school's symbolic and personal dimensions, and the more associative relationships which are driven by the need to accomplish structured activities and maintain particular work relations (Bryk and Driscoll 1988). The vision of the school as a community portrays adults and students linked to one another by a common mission and by a network of supportive personal relations that strengthen their commitment to the organization (Bryk and Driscoll 1988). Three core components comprise this communal school organization (Bryk and Driscoll 1988): (1) a system of shared values among members of the organization which include the "norms of schooling" (Bird and Little 1986), namely, norms for instruction which affect the way teachers' work is conducted and student learning takes place, and norms for civility which affect the relations among individuals in the institution; (2) a common agenda of activities which include those formal and informal events which enable school participants to engage in face-to-face interactions, promote social ties, and encourage communal association; and (3) a pattern of social relations which embody collaboration and extensive involvement and are highlighted by the collegiality shared among the adults in the school and the expanded role of the teacher.

In a comparative analysis of private and public schools, Coleman and Hoffer (1987) examine the impact of community on the degree of social integration between families and schools. The researchers assert that the type and strength of community in schools differentially affect the critical social connections which bond families and schools in the joint enterprise of education. This concept of community refers to two types: functional and value. Functional communities are characterized by structural consistency between generations in which social norms and sanctions arise out of the social structure itself, and both reinforce and perpetuate that structure. Functional communities exhibit a high degree of uniformity and cohesion within geographical, social, economic, and ideological boundaries (Coleman and Hoffer 1987).

In the educational domain, value communities describe a collection of people who share similar values about schooling and child rearing but who are not a functional community; they are strangers from various neighborhoods, backgrounds, and occupations united around an educational organization — their children's school. For example, the families of magnet school students, and the students in schools of choice generally, may constitute a value community if they exhibit a high degree of value consistency — for example, commitment to a particular educational philosophy — but are mainly strangers drawn from a wide collection of city neighborhoods; they have little contact with one another outside school corridors. In contrast, many public, neighborhood elementary schools reflect geographical communities but lack the value consistency or interinstitutional linkages of either the value or functional communities.

Empirical research: Value communities

The theoretical argument for this explanation of the potential role of choice in influencing student performance is particularly strong, though direct empirical work on the theory is limited. James Coleman is probably the most influential proponent of this theory (e.g., Coleman 1988; Coleman and Hoffer 1987). He argues that data from the HSB survey support this theory. We noted earlier that these data are derived from a sample of private schools that is unrepresentative (though Coleman does focus on Catholic schools in some discussions). Moreover, the HSB study does not include data from parents.

The impact of communal school organization has also been examined by Bryk and Driscoll (1988). Their analyses of High School and Beyond data indicate that the level of student social engagement (attendance, classroom order, enrollment) and academic performance are substantially higher in communally organized schools. Likewise, teacher efficacy and satisfaction, staff morale, and teacher absenteeism are influenced positively by a communal school organization. The researchers identify school size and composition as key facilitating conditions for schools which reflect shared values, a common agenda, and collegial, expanded teaching roles.

In an ethnographic case study of a magnet school in

northern California, Smrekar (1991) studied parent–school interactions and concluded that: (1) organizational processes and structures embedded in magnet schools tend to promote social cohesion, commitment, and communication (elements of value communities) among geographically dispersed and socioeconomically differentiated parents; and (2) these elements of community mediate the effects of differences in education and social network ties (elements of cultural capital) which influence family–school interactions with schools.

The research on organizational features which promote school community points to an array of structural and compositional features, including: school size, sector, diversity of student body, and student selectivity (Bryk and Driscoll 1988). The empirical research in this area, however, has focused primarily on the characteristics of "effective" schools in the Catholic and private school sectors (e.g., Bryk, Lee, and Carriedo 1984; Cookson and Persell 1985; Peshkin 1986). The explicit mission or ethos of both sectarian and nonsectarian private schools, combined with a comparatively smaller size and more homogeneous student body, have been considered key elements associated with creating the foundational elements of community (e.g., shared goals, personal ties) within schools. Little research has focused on organizational elements in public sector or public-sector choice schools which may promote or constrain the development of community in these schools.

The research on the dynamic intersection of community and choice is primarily speculative and conceptual. Advocacy of choice on the grounds of potential contributions to value communities raises three questions that warrant empirical study:

(1) What is the impact of school choice programs on the development, character, and maintenance

of value communities among parents and educators?

(2) Do value communities promote educational equity by mediating the influence of social class on family interactions with schools?

(3) Do value communities promote academic achievement by creating a learning environment characterized by high expectations and support?

Conclusion

This chapter presents three theoretical cases and corresponding empirical research regarding the debate surrounding the impact of increasing opportunities for parental choice. It is clear that there are a number of critical questions relating to the validity of these theories that cannot be answered from existing empirical research. The objective of this chapter is not to critique the theories but to specify research questions that emerge from each theoretical perspective that can inform the policy debate regarding the various dynamics and outcomes of efforts to increase parental choice among public schools.

It is also clear that the perspectives presented in this chapter are highly inter-related. Family sovereignty, or individual freedom, is crucial if parents are to have the right to activate markets and make the best possible choices. In addition, "the sum of a great many choices is . . . equal to the best interests of the community as a whole" (Finn 1990: p. 17). Thus, family sovereignty and value communities are also intertwined. We suggest, however, that clearly delineating the conceptual bases for advocating increased parental choice and specifying additional empirical questions for examining the merit of choice programs is essential for a comprehensive view of choice options.

Notes

1 Authors are listed in alphabetical order. The authors acknowledge the comments and insights from our colleagues at the Citizens' Commission on Civil Rights.

2 For a more extensive overview of the arguments for choice, see Paulu (1989); Rayed (1989); Witte (1990a); Witte (1990b).

3 For an extensive overview of the criticisms of choice, see Leaven (1991); Rayed (1990/91); and Witte (1990b).

4 Obviously, many factors influence the capacity and willingness of schools and teachers to change (see, e.g., Cohen 1990; Elmore and McLaughlin 1988; Hawley 1976). We focus here on those that affect the potential dynamics choice might introduce.

5 Advocates of choice often point to this and earlier findings of parent support for choice without also noting that public school parents and non-parent adults overwhelmingly oppose the use of public money to subsidize private school enrollment (Elam, Rose and Gallup 1991).

6 Studies of recent efforts to provide for a greater role for parents through "site-based" decision-making indicate that changes in formal structures do not lead to increased parental influence on educational practice or policy in schools (Clune and White 1988; Malen, Ogawa, and Kranz 1990). These studies, however, do not differentiate between school systems where parents have considerable choice and those where parents' public school options are few or nonexistent.

References

Archbald, D. A. (1988). "Magnet schools, voluntary desegregation, and public choice theory: Limits and possibilities in a big city school systems". Unpublished doctoral dissertation, University of Wisconsin, Madison.

Bauch, P. A. (1991). "Linking reasons for parent choice and involvement for minority families in Catholic high schools". *International Journal of Educational Research*, **15**(3/4), 311–322.

Bird, T. and Little, J. W. (1986). "How schools organize the teaching occupation". *Elementary School Journal*, **86** (March), 493–511.

Blank, R. K. (1989). *Educational Effects of Magnet High Schools.* Paper presented at the Conference on Choice and Control in American Education, University of Wisconsin, Madison.

Blank, R. K., Dentler, R. A., Baltzell, D. C., and Chabotar, K. (1983, September). *Survey of Magnet Schools Analyzing a Model for Quality Integrated Education: Executive Summary* (Report No. 300-81-04320). Washington DC: U.S. Department of Education. (ERIC Document Reproduction Service No. ED 236304).

Bridge, R. G. and Blackman J. (1978, April). *A Study of Alternatives in American Education, Vol. 4: Family Choice in Schooling (Report No. R-2170/4-NIE). Santa Monica, California: The Rand Corporation.*

Brofenbvenner, Moen and Garbarino (1984). Child, Family and Community. In R. Parke (Ed.), *Review of Child Development Research.* Vol. 7. Chicago: University of Chicago Press.

Bryk, A. S. and Driscoll, M. E. (1988). *The High School as Community: Contextual Influences and Consequences for Students and Teachers.* Madison, Wisconsin: National Center on Effective Secondary Schools.

Bryk, A. S., Lee, V. and Carriedo, N. (1984). *Effective Catholic Schools: An Exploration.* Washington, DC: National Center for Research in Total Catholic Education.

Capell, F. (1981). *A Study of Alternatives in American Education, Vol. VI: Student Outcomes at Alum Rock, 1974–76.* Santa Monica, California: The Rand Corporation.

Chubb, J. E. and Moe, T. M. (1990). *Politics, Markets and America's Schools.* Washington, DC: The Brookings Institution.

Clune, W. H. and White, P. A. (1988). *School-Based Management: Institutional Variation, Implementation, and Issues for Further Research (Report No. RR-008).* New Brunswick, New Jersey: Rutgers University, Center for Policy Research on Education.

Cochran, M. and Brassard, J. (1979). "Child development and personal social networks". *Child Development*, **50**, 601–616.

Cochran, M. and Henderson, C. R., Jr. (1986). *Family Matters: Evaluation of the Parental Empowerment Program.* A Summary of a Final Report to the National Institute of Education. Ithaca, New York: Cornell University.

Cohen, D. K. (1990). "Governance and instruction: The promise of decentralization and choice". In W.H. Clune and J. F. Witte (Eds), *Choice and Control in American Education, Volume 1*, pp. 337–386. New York: The Falmer Press.

Cohen, D. K. and Farrar, E. (1977). "Power to parents? The study of educational vouchers". *The Public Interest*, **48**, 72–79.

Coleman, J. S. (1988). Social capital in the creation of human capital. *American Journal of Sociology*, **92**, S95–S120.

Coleman, J. S. and Hoffer, T. (1987). *Public and Private High Schools: The Impact of Communities.* New York: Basic Books.

Coleman, J. S., Hoffer, T., and Kilgore, S. (1982). *High School Achievement: Public, Catholic, and Private Schools Compared.* New York: Basic Books.

Coleman, J. S., Schiller, K., and Schneider, B. (1991, August). "Parent Involvement and School Choice". In *Resources and Actions: Parents, Their Children and Schools. A Report to: National Science Foundation and National Center for Education Statistics.* Chicago: Education Studies Group, NORC/University of Chicago.

Connel, R. W., *et al.* (1982). *Making the Difference: Schools, Families and Social Divisions.* Sydney, Australia: Geo. Allen and Unwin.

Cookson, P. W. and Persell, C. H. (1985). *Preparing for Power: America's Elite Boarding Schools.* New York: Basic Books.

Cooper, B. (1991). "Parent choice and school involvement: Perspectives and dilemmas in the United States and Great Britain". *International Journal of Educational Research*, **15**, 235–250.

Crain, R. and Hawley, W. D. (1982). "Standards of research", [A Critique of James Coleman's "Public and Private Schools"]. *Society*, **19** (January/February 1982), 14–22.

Darling-Hammond, L. and Kirby, S. N. (1985, December). *Tuition*

Tax Deductions and Parent School Choice: A Case Study of Minnesota (Report No. R-3294-NIE). Santa Monica, CA: Rand. (ERIC Document Reproduction Service No. ED 273047.)

Driscoll, M. (1991) *Parents and Public Choice Schools.* Paper presented at the annual meeting of the University Council for Educational Administration, Baltimore.

Elam, S. M., Rose, L. C., and Gallup, A. M. (1991, September). "The 23rd annual Gallup poll of the public's attitudes toward the public schools". *Phi Delta Kappan*, **73**, 41–56.

Elmore, R. F. and McLaughlin, M. W. (1988). *Steady Work: Policy, Practice and the Reform of American Education.* Santa Monica, California: The Rand Corporation.

Erickson, D. A. (1982). "Disturbing evidence about the one best system". In R. B. Everhart (Ed.), *The Public School Monopoly.* Cambridge, Massachusetts: Ballinger Publishing.

Finn, C. E., Jr. (1990). "Why we need choice". In W. L. Boyd and H. J. Walberg (Eds), *Choice in Education*, pp. 3–19. Berkeley: McCutchan.

Friedman, M. (1955). "The role of government in education". In Solo, R. (Ed.), *Economics and the Public Interest.* New Brunswick, New Jersey: Rutgers University Press.

Goldring, E. and Bauch, P. A. (1993, April). *Parent Involvement and School Responsiveness: Facilitating the Home-School Connection in Schools of Choice.* Paper presented at the annual meeting of the American Educational Research Association, Atlanta, Georgia.

Goldring, E. and Shapira, R. (1991). *Empowerment, Choice or Involvement: What Satisfies Parents?* Paper presented at the annual meeting of the American Educational Research Association, Chicago, Illinois.

Hawley, W. D. (1976). "Horses before carts: Developing adaptive schools and the limits of innovation". In S. Gove and F. Wirt (Eds), *Political Science and School Politics*, pp. 1–22. Lexington, Kentucky: D. C. Heath, 1976.

Henderson, A. (1987). *The Evidence Continues to Grow: Parent Involvement Improves Student Achievement.* Columbia, Maryland: National Committee for Citizens in Education.

Hirschman, A. O. (1970). *Exit, Voice and Loyalty*. Cambridge, Massachusetts: Harvard University Press.

Levin, H. M. (1991). "The economics of educational choice". *Economics of Education Review*, **10**, 137–158.

Lightfoot, S. L. (1978). *Worlds Apart: Relationships between Families and Schools*. New York: Basic Books.

Maddaus, J. (1990). "Parental choice of school: What parents think and do". In C. B. Cazden (Ed.), *Review of Research in Education* 16, Washington, DC: AERA.

Malen, B., Ogawa, R. T., and Kranz, J. (1990). "What do we know about school-based management? A case study of the literature. A call for research". In W. H. Clune and J. F. Witte (Eds), *Choice and Control in American Education, Volume 2*, pp. 289–342. New York. The Falmer Press.

Manski, C. F. (1992, May). *Educational Choice (Vouchers) and Social Mobility*. Paper presented at the conference "Poverty and Public Policy: What Do We Know? What Should We Do?" sponsored by the Institute for Research on Poverty, University of Wisconsin-Madison and the Office of the Assistant Secretary for Planning and Evaluation, U.S. Department of Health and Human Services.

Martin, M. and Burke, D. (1990). "What's best for children in the schools-of-choice movement?" *Educational Policy*, **4**, 73–91.

Merl, J. (1992). "Is private education really better"? *The Los Angeles Times*, March 29, p. 20

Metz, M. H. (1986). *Different by Design: The Context and Character of Three Magnet Schools*. New York: Routledge and Kegan Paul.

Nathan, J. (1985). "The rhetoric and the reality of expanding educational choice". *Phi Delta Kappan*, **66**, 476–481.

Nault, R. L. and Uchitelle, S. (1982). "School choice in the public sector: A case study of parental decision-making". In Manley-Casmir, M. E. (Ed.), *Family Choice in Schooling: Issues and Dilemmas*, pp. 85–98. Lexington, Massachusetts: D. C. Heath.

Newmann, F. and Oliver, D. (1968). "Education and community". *Harvard Educational Review*, **37** (Winter), 61–106.

Olson, M. (1965). *The Logic of Collective Action.* Cambridge, Massachusetts: Harvard University Press.

Paulu, N. (1989). *Improving Schools and Empowering Parents: Choice in American Education.* Report based on the White House Workshop on Choice in Education. Washington, DC: U.S. Government Printing Office.

Peshkin, A. (1986). *God's Choice.* Chicago: University of Chicago Press.

Raywid, M. A. (1987). "Public choice, yes; vouchers, no!" *Phi Delta Kappan*, **68**, 762–769.

Raywid, M. A. (1988). "Community and schools: A prolegomenon. *Teachers College Record*, **90**(2), 197–209.

Raywid, M. A. (1989). *The Case for Public Schools of Choice.* Bloomington, Minnesota: Phi Delta Kappa Educational Foundation.

Raywid, M. A. (1990 December/1991 January). "Is there a case for choice?" *Educational Leadership*, **48**, 4–12.

Rossell, C. H. (1985). "What is attractive about magnet schools?" *Urban Education*, **20**(1), **7**, 22.

Rossell, C. H. (1990). *The Carrot or the Stick for School Desegregation Policy: Magnet Schools or Forced Busing*. Philadelphia: Temple University Press.

Scherer, J. (1972). *Contemporay Community: Sociological Illusion or Reality.* London: Tavistock.

Shapira, R. and Goldring, E. (1990). *Principal–parent Relationships: Domain Consensus and Response Strategies. Research Report 1–89*, The Unit for the Sociology of Education and the Community, Tel Aviv University, School of Education.

Smith, M. S. and O'Day, J. (1991). "Systematic school reform". In S. H. Fuhrman and B. Malen (Eds), *The Politics of Curriculum and Testing*, pp. 233–267. New York: The Falmer Press.

Smrekar, C. (1991). "*Building community: The influence of school organization on family–school interactions*". Unpublished doctoral dissertation. Stanford University, School of Education.

Sosniak, L. A. and Ethington, C. A. (1992). "When public school choice is not academic: Findings from the National

Education Longitudinal Study of 1988". *Educational Evaluation and Policy Analysis*, **14**, 35–52.

Steinberg, L. (1989). "Communities of families and education". In Weston, W. (Ed.), *Education and the American Family*. New York: New York University Press.

United States Department of Education (1992). "States, communities giving families more options". *America 2000*, Newsletter #22, Week of March 16.

Walker, J. (1971). Innovation in Public and Private Schools. Unpublished Master's Degree thesis, Department of Political Science, Yale University.

Watt, D. G. (1983). "Some thoughts on educational reform and renewal". *American Education*, **19**, 2–5.

Williams, M. F., Hancher, K. S. and Hutner, A. (1983, December). *Parents and School Choice: A Household Survey*. Washington, D.C.: U.S. Department of Education. (ERIC Document Reproduction Service No. ED 240739.)

Witte, J. F. (1990a). *Choice in American Education*. Charleston, WV: Appalachia Educational Laboratory.

Witte, J. F. (1990b). "Introduction". In W. H. Clune and J. F. Witte (Eds), *Choice and Control in American Education*, Vol. 1. New York: The Falmer Press.

Witte, J. F. (1993, March). The Milwaukee Parental Choice Program: The First Thirty Months. Paper delivered at the American Education Research Association Annual Meeting, Atlanta, Georgia, April 15–18, 1993.

Chapter Fifteen

POSSIBILITIES, PROBLEMS, AND PROGRESS: EARLY LESSONS FROM THE CHARTER MOVEMENT

JOE NATHAN
Center for School Change at the University of Minnesota's Hubert H. Humphrey Institute of Public Affairs

During the past five years the introduction of public charter schools has spread from Minnesota across the country. Mr. Nathan reflects on the evolution of the movement and considers the lessons that can be drawn — both at the school level and at the level of state policy — from the widely varying approaches that have been adopted in the 23 states that have passed charter legislation.

The idea of charter schools is for people who believe that public schools can have a significant, positive impact on youngsters. The charter schools movement brings together three powerful concepts: freedom and choice for families, entrepreneurial opportunities for educators, and explicit accountability for schools. For many advocates, the charter public school movement is an expansion of opportunity, similar to that proposed by people like Susan B. Anthony, Dr. Martin Luther King, Jr., and Cesar Chávez.

This movement is about hope and possibility. It is for people like Janet Fender, a second-grade public school teacher near New Castle, Delaware, who called the state's charter legislation "a chance of a lifetime"[1]. It's about Richard Farias, director of Houston's Tejano Center, who is working with the Houston Federation of Teachers to start a new charter school; it's about Kim Borwege and Nancy Miller, a public school in three formerly empty storefronts on the main street of leSueur, Minnesota; and it's about Bob Stein CEO of an award-winning San Diego public school that decided to convert to charter status.

But educators have heard many reform proposals. In this article I try to answer four key questions about the charter school movement; What is the charter idea? Where did it come from? What can be learned from the charter movement's five years? And finally, what are the key unanswered questions?

Charter proponents readily agree that there are significant problems in society that produce troubled families and challenging young people. It is vital to work toward a more just society that prizes all young people and helps strengthen every family. But acknowledging enormous problems **outside** schools does not mean that educators can not do a much better job helping youngsters, **inside** schools. That's what effective, talented educators have long believed. And the charter school movement gives them the opportunity to carry out their dreams.

A Brief History

In five years, the charter movement has spread from one state and one school to 25 states and hundreds of schools. But the idea is not just to create hundreds or even thousands of new schools. The goal is to encourage widespread improvements in public education.

The term "charter" comes from the contracts that were given to early European explorers. Elsewhere in this book, Ray Budde recalls that more than 20 years ago he suggested that school districts give a "charter" to teachers who were willing to try new approaches. In 1988 Budde wrote a new book on the subject[2] and sent it to many people, including American Federation of Teachers president Albert Shanker, Shanker endorsed the idea and promoting it[3].

Minnesota State Senitor Ember Reichgott Junge, a Democrat, and several local activists heard Shanker's call and refined the idea to fit Minnesota. Public school choice already had a strong base in Minnesota. Minneapolis and St. Paul had offered alternatives and magnet schools since the 1970s. Most of these programs served a cross section of students. Many Minnesota educators had already concluded that giving families options works to reduce the number of unsuccessful students.

In 1985 Democratic Governor Rudy Perpich proposed expanding public school choice for all Minnesota students. Perpich wanted to build on the success of innovative public schools of choice in Minneapolis and St. Paul. He talked with people from award-winning public schools of choice, such as St. Paul's Open School and Baron Open School in Minneapolis.

Governor Perpich knew that, without government action, only affluent families would have choices among schools. He also knew that choice programs would have to be carefully designed to achieve one of his key goals: expanding opportunity and achievement for low- and moderate-income students. Perpich recognized that, in some southern state, school choice had been used to perpetuate segregation. That's why his proposal stipulated that students could transfer across district lines only if doing so did not harm racial integration efforts.

Governor Perpich suggested a "postsecondary option", which would allow juniors and seniors in public high schools to attend colleges and universities. The state per-pupil spending would follow the students and be used to pay for their tuition, books, and materials. Perpich also recommended allowing K-12 students to attend public schools outside the resident districts, if the district had room and if the transfer did not harm desegregation efforts. Both laws were adopted over the following three years.

In 1985 only 33% of Minnesotans favored cross-district public school choice, while 60% opposed it. By 1992, polls by major public education groups found that 76% of the state's residents supported the idea, while only 21% opposed it. Support grew because (1) thousands of students who had dropped out used these laws to return to school; (2) the number of advanced courses in high schools more than doubled, as schools responded to the competition provided by the postsecondary options law; (3) public school choice brought families back into public education; and (4) choice allowed educators to create new, distinctive schools[4].

Minnesotans learned that thoughtful competition could help stimulate improvements. A 1995 study of the post-secondary options law found that more than half of the secondary demonstrators who responded to a survey said that one result of the law was increased cooperation between high schools, colleges, and universities[5].

Unfortunately, teachers in all areas of the state who tried to start innovative programs often encountered intense opposition from fellow educators. Sometimes the opposition came from local school boards as well. AFT president Shanker has described the problems all too often faced by public school teachers who try to create innovative "schools-within-schools":

> Many schools within schools were or are treated like traitors or outlaws for daring to move out of the

lock-step and do something different. Their initiators had to move heaven and earth to get school officials to authorize them and if they managed that, often they could look forward to insecurity, obscurity or outright hostility[6].

By 1989 most Minnesota families had some form of school choice available to them — but without many distinctive choices. So Minnesotans were ready for a plan to increase the number of high-quality public school options.

The character school notion also developed as many progressive educators and political leaders were confronting renewed interest in vouchers. Strong voucher opponents, including President Bill Clinton, Secretary of Education Richard Riley, Colorado Governor Roy Romer, Senitor Paul Wellstone (I-Minn.), and California State Senitor Gary Hart, have endorsed character schools. (Indeed, an article co-authored by Hart appears in this special section.) And State Senitor Reichsot Junge wrote an editorial for Minnesota's largest newspaper, arguing passionately that the charter approach is much better than vouchers[7].

In his 1995 State of the Union speech, President Clinton urged states to adopt charter legislation. "I challenge every state to give all parents the right to choose which public school their children will attend and to let teachers form new schools with a charter they can keep only if they do a good job"[8]. Along with former Senitor David Durenberger (R-Minnesota), the President convinced Congress to allocate millions of dollars to help start charter schools.

Some political leaders who support charters also support vouchers. But, as I point out below, there are important differences between the two concepts. The charter concept, as developed in Minnesota, appeals to people who believe in public education, but who want it to be more dynamic and more accountable for results. In the past five years, 25 states have adopted charter legislation, although the details vary widely. In evaluating such legislation, it is vital to understand the central principles that underlie the charter idea.

Elements of Charter Schools

(1) **The state will authorize more than one organization to start and operate a charter public school in the community.** The state would withdraw what Ted Koldede calls the "exclusive franchise" it has historically given to public school districts[9]. Those districts could continue to offer public schools. But other publicly accountable groups, such as the state board of education, public universities, a new state agency, or a city council or county commission could also sponsor public schools.

The organizers — usually teachers, parents, or other community members — usually teachers, parents, or other community members — could approach either a local board or some other public body to be their sponsor. People would be free to go to any of several places to obtain supervision and sponsorship. This feature distinguishes "strong" from "weak" charter laws. As I explain below, in those states that allow local school boards and other groups to sponsor schools, many more charter schools are created, and there is more positive response from public school districts.

As Kolderie notes, eliminating the "exclusive franchise" given to local districts "removes from the district its ability to take its students for granted"[10]. Many educators try to make improvements in their schools, but they often face strong opposition and active resistance from fellow educators. Professional jealousy and institutional lethargy are real and strong.

(2) **The newly organized (or converted) schools would be public schools.** The would be non-sectarian, and they would not charge tuition. They would be open to all kinds of students, and they would not be allowed to have any kind of admissions test. They would, however, be required to follow all health and safety regulations.

(3) **The schools would be responsible for improved student achievement.** Each school would negotiate a three- to five-year contract (or "charger") with the sponsoring agency, specifying areas in which students would learn more and how that learning would be measured. Schools that failed to achieve their contracted improvements would be closed by the sponsoring organization.

(4) **In return for this accountability for improved results, the state would grant an up-front waiver of virtually all rules and regulations governing public schools.** Aside from the health and safety and other regulations mentioned above, schools would be exempt from state regulations about how to operate.

(5) **The charter school would be a school of choice.** It would be actively selected by the educators and students' families. No one would be assigned to work in or to attend the school who had not chosen to do so.

(6) **The school would be a discrete entity.** Founders could choose any organization available under state laws. The school would be a legal entity, with its own elected board. Teachers could organize and bargain collectively. However,

this bargaining unit would be separate from, and not bound by, the contracts negotiated by any district bargaining unit.

(7) **The full per-pupil allocation would move with the student.** This amount would be roughly equal to the average state per pupil allocation or the average allocation for the district from which the student comes. If the state provides extra funds for students with disabilities or for students from low-income families, these funds would also follow the student.

(8) **Participating teachers would be protected and given new opportunities.** The state would permit teachers to take a leave from their public school systems and retain their seniority. Teachers could stay in local or state retirement systems. Teachers could choose to be employees, to organize a cooperative, or to choose another method of organization available to nonsectarian groups.

Key Early Lessons

Here are the major lessons we have learned about creating charter schools and about effective state legislation. First, there are some school-level lessons.

(1) Charter schools can have a positive impact on student achievement, attendance, and attitude. As Milo Cutter notes elsewhere in this issue, the school board in St. Paul, Minnesota, recently voted 7–0 to renew City Academy's charter because many youngsters who had left larger, traditional schools were flourishing in City Academy's smaller, intensive program. City Academy builds on youngsters' idealism and

undertakes a variety of community service projects, including rehabilitating abandoned buildings. Other charter schools — such as New Visions in Minneapolis Vaughn Next Century Charter in Los Angeles, Bowling Green Charter in Sacramento, and Academy Charter in Castle Rock, Colorado — have also produced achievement gains.

(2) **Many charter proposals focus on low- and moderate-income youngsters.** Opponents feared that most charter proposals would serve affluent students who were already successful. But this has not been the case. A 1995 survey of 110 charter schools in seven states found that most charters were designed at least in part to serve "at-risk" students[11]. Louann Bierlein of Louisiana State University reviewed the six states with the most charter schools and found that "minority youngsters comprise 40% of charter school enrollments, although the same minorities make up just 31% of pupils in the regular public schools in those states"[12].

Some of the strongest support for charter legislation is coming from groups that advocate for low-income families and youngsters, particularly those representing "communities of color", such as the Urban League, the Urban Coalition ACORN (Association of Community Organizations for Reform Now) and so on. It is heartening to talk, as I have, with African-American and Hispanic parents who have found that charter school educators are more open to their ideas and suggestions, in part because the schools need their help to improve achievement and stay open. As one inner-city parent told me, "I went to lots of meetings for years. They ran me all around the mulberry bush. But at the Skills for Tomorrow Charter School, the teachers make me feel like they really want me involved — and not just in fund raising"[13].

(3) **Many talented educators will accept responsibility for results, in exchange for considerable autonomy.** Hundreds of charter school proposals have been made in the past five years, and more than 260 charter public schools have opened their doors. This figure includes both public schools that decided to convert to charter status and brand-new charter schools.

(4) **Good ideas aren't enough.** Charter schools need people with "start-up" skills and operational skills. Charters are small businesses as well as schools, and successful charter schools obtain accounting, legal, and other business services, often through subcontracts. Charter schools must work out the business details or face major problems.

(5) **Charter schools can be located in places other than a traditional school building.** One charter school occupies a former VFW building in downtown Denver. Others share space with a low-income housing project, with social services agencies, and in a city recreation center that was underused during regular school hours. One charter school is located in a former grocery store. Some use formerly empty storefronts.

(6) **One of the biggest challenges facing charter schools is where to obtain start-up funds.** Most districts can go to local voters and sometimes to the state legislature for money to erect a new building or to alter an existing one. Most states

do not give this opportunity to charter schools. A 1995 survey of 110 charter schools identified this as one of the biggest problems facing charter schools[14]. A 1996 study agrees: "Without doubt, the absence of capital funding, access to conventional school facilities, and start-up money to cover initial equipment, planning, etc. is the heaviest cross charter schools bear today"[15].

We have also learned some important policy lessons.

(1) **The charter idea encourages school districts to improve.** In Rochester, Minnesota, home of the Mayo Clinic, a private nonsectarian Montessori school asked the local school board for a charter. The board decided instead to create its own Montessori elementary school. The Forest Lake District had rejected parents' request for a Montessori school. A LeSueur group of teachers and parents were frustrated by the indifference of some local educators to recent advances in technology. They established the "Minnesota New Country School" in three formerly empty storefronts on Main Street. Now teachers in the district go to this charter for lessons about using technology in their classrooms. As Linda Nathan and Larry Myaft report elsewhere in this issue, the Massachusetts charter law helped convince the Boston Public Schools and Boston teacher union to establish a "within-district" program to help teachers create new schools. This is exactly what charter proponents hope will happen as legislation expands to allow more charter schools.

(2) **The strongest state charter laws produce much more activity.** Strong laws, such as those in Michigan, Massachusetts, New Jersey, Delaware, North Carolina, and Arizona, allow charter school developers to ask for sponsorship from some group other than a local school board. A 1996 study found that, as of December 1995, there were 222 charter schools in the six states that allowed more than one kind of organization to sponsor a school, while only 14 charters existed in the six states with the weakest legislation[16]. The advice charter school developers most frequently offered to state policy makers in a 1995 survey was that the state law permit them to go to groups other than the local school board for permission to establish the charter school[17].

Dozens of well-developed charter proposals have been rejected by local school boards in California, Colorado, and Minnesota. In one notorious Minnesota case, a group of teachers at an alternative school applied to their local school board for permission to convert their school to charter status. The school board turned them down in a 4–3 vote and then voted 7–0 to tell them what fine teachers they were. Minnesota's legislation requires either local board approval or two votes from local board members (out of 7), allowing an appeal to the state board. This is extremely frustrating.

(3) **Many charter opponents have shifted legislative strategies.** Originally, they opposed any charter law. Now they support laws that limit charter sponsorship to local school boards. As noted above, however, the experience in many states with such provisions is that very little happens.

(4) **Charter school legislation offers important opportunities for unions.** Some unions are discovering that they can help establish and run charter schools. Unions can also subcontract with charters and sell, for example, their legal, accounting, and curriculum expertise. Charters offer a new form of teacher empowerment, employee ownership, and school governance. Charters established as cooperatives offer a new professional option to educators frustrated by district officials or school board members.

Many charter school teachers maintain their union membership. Both the American Federation of Teachers and the National Education Association are helping members start charter schools. Byron King of San Diego's O'Farell Charter and Milo Cutter of City Academy in St. Paul are two of the teachers involved in this effort.

(5) **The charter school movement can learn from the sometimes frustrating experience of successful inner-city alternative public schools.** Even though many successful inner-city alternative schools have operated for more than 20 years, they are often ignored or marginalized. A widely circulated 1995 Harvard report is only the latest to overlook their record. The report's summary asserts that "actual choice experiments remain young and modest ... Very little is known about school choice programs"[18]. This kind of comment reminds alternative school educators of Ralph Ellison's observation in **Invisible Man**, "I am invisible, understand, because people refuse to see me"[19]. Critics often ignore such researchers as Mary Anne Raywid of Hofstra University, who has

> for more than a decade reported the success that alternative schools have with students[20]. Raywid has pointed out that public school choice is not just about competition; it is also an acknowledgment that there is no one best kind of school for educating all students.

The goal of many alternative school educators — to serve all kinds of students — was distorted and marginalized by many school districts over the last two decades. Many alternative schools were pushed into being as "last chance" programs, serving only those students with serious behavior problems rather than being allowed to serve a cross section of students.

This was a great mistake. Many alternative schools modeled the kind of small, family-like program now recommended by various reports. Roseanne Wood, principal of an alternative high school in Tallahassee that was judged by the Florida Department of Education to be one of the outstanding high schools in the state, wrote that the more choices and alternative programs districts and states offer, "the fewer programs we'll need for problem students because fewer students will be exhibiting problems. We'll have more students with 'schools that fit' "[21].

In states where only school districts can authorize charter schools, many school boards seem to be most open to proposals that would serve primarily troubled youngsters. Certainly, there is a need for such programs, and they ought to be made available. But the idea of charter school is to provide an array of options, not to be limited to one particular focus.

Some critics insist that many low-income families and families whose home language is not English do not know how to make an informed choice of schools. This is not the first time people have argued that certain groups do not know enough to make good decisions. The same argument was used against women's suffrage, and the opponents of the civil rights movement insisted that "those people" did not know enough to vote.

Unanswered Questions

The charter school movement is young. And it faces a number of important, unanswered questions.

- What will be the impact of charter schools on their students? Will charter school educators be able to work with evaluation specialists to develop new, reliable, and valid methods of assessing student achievement that go beyond what is now available?
- How much impact will charter school programs have on existing public schools? Will other local school boards establish policies like the pilot school program in Boston that Larry Myatt and Linda Nathan describe in this issue?
- Will policy makers be willing to adopt "strong" laws that permit groups other than local school boards to sponsor charter schools? What other groups will emerge as sponsors of charter legislation? How will the challenge of finding startup funding be resolved?
- How will colleges and universities react to charter schools? Will many colleges of education ignore charter schools, as they often did with alternative schools over the last 20 years? Will colleges involve strong charter schools in the preparation of teachers and administrators?
- How much will charter advocates learn from previous school reform efforts? Will they study mistakes as well as successes of such previous efforts as progressive education, multiage classrooms, open schools, etc.? Will

charter advocates learn from somewhat similar efforts in other countries?

- Will powerful education groups continue to push for weak charter laws? Or will they seize the opportunities that the charter idea offers both the organizations and their members?

It's too early to know all the answers. but some things are clear. Because admissions tests are prohibited and accountability for results is demanded, the charter idea is a controlled, not an unlimited, choice plan. And every day more educators and parents learn about the idea and see it as a chance to create the schools they think make sense The charter idea is clearly spreading.

The rapid growth of the charter schools movement shows that the idea touches something very deep in Americans. As people have come to our nation from all over the world, America has grown, and many Americans have prospered, in part because we give visionaries a change to carry out their ideas. That is the spirit behind the charter school movement — a deep respect for innovative educators and a strong belief in hope and possibility. Charter schools are about real empowerment for educators.

Notes

1 Fender, J. (1995). Quoted in Sandy Dennison, "Charters Charm Frustrated Teacher", *Wilmington Sunday New Journal*, 19 November, p. 1-B

2 Budde, R. (1988). *Education by Charter*. Andover, Mass: Regional Laboratory for Educational Improvement of the Northeast and Islands.

3 See, for example, Shanker, A. (1988). "A Charter for Change", *New York Times*, 10 July 1988, sec. 4, p. 7 and idem, "Charter Schools: Option for the Other 80 percent", *School Administrator*, November, p. 72.

4 See Nathan, J. and Yaseldyke, J. (1994). "What Minnesota has Learned about School Choice", *Phi Delta Kappan*, May, pp. 682–688.

5 Legislative Auditor (1996). Postsecondary Enrollment Options Program. St. Paul: Office of the Legislative Auditor.

6 Shanker, op. cit.

7 Junge, E. R. (1996). "Charter Schools Will Work Better Than Private School Vouchers", *Minneapolis Star/Tribune*, 23 January, p. A-11.

8 Clinton, B. (1995/1996). "State of the Union", 23 January 1996; see also "Remarks by the President Aboard Air Force One", 22 September 1995, mimeographed.

9 Kalderie, T. "The States Will Have to Withdraw the Exclusive", Center for Policy Studies, St. Paul.

10 Ibid, p. 14.

11 Medler, A. and Nathan, J. (1995). *Charter Schools: What Art They Up To?* Denver: Education Commission of the States and Minneapolis: Center for School Change.

12 Bierlein, L. (1996). In Chester E. Finn, Jr., Louann Bierlein, and Bruno V. Manno (Eds). *Charter Schools in Action: A First Look*, p. 2. Washington, D.C.: Hudson Institute.

13 Nathan, J. (1995). "At Charter Schools, Parents Feel Needed", *St. Paul Pioneer Press*, 27 March, p. 1-C.

14 Medler and Nathan, op. cit.

15 Finn, Bierlein, and Manno, p. 5.

16 Bierlein, L. (1996). *Charter Schools: Initial Findings*. Denver: Education Commission of the States.

17 Medler and Nathan, op. cit.

18 Fuller, B. (1995). *Who Gains, Who Loses from School Choice: A Research Summary*, pp. 1–2. Denver: National Conference of State Legislators.

19 Ellison, R. (1952). *Invisible Man*, p. 3. New York: Random House.

20 Raywid, M. A. (1989). *The Case of Public Schools of Choice*. Bloomington, Ind.: Phi Delta Kappan Educational Foundation, Fastback, No. 283.

21 Wood, R. (1989). "SAIL, A Pioneer for Schools of Choice in Florida". In Joe Nathan (Ed.), *Public Schools by Choice*, pp. 181–202. Minneapolis: Free Spirit Press.

Chapter Sixteen

SCHOOL AUTONOMY AND PARENTAL CHOICE: CIRCUMSCRIBED REALITIES OF REFORM

VIVIAN WILLIAMS
Department of Educational Studies, University of Oxford

This review of the bounded realities of the interplay between contemporary concepts of school autonomy and parental choice is a reflective perspective of the development of political ideologies and their somewhat pragmatic implementation by various versions of the Conservative government in England and Wales subsequent to its election in June 1979.

It is virtually impossible to meet anyone within the education community in the UK able to offer a dispassionate, objective, and balanced view of changes in the education system of England and Wales mandated assiduously during the years of continuous Conservative administration since 1979. Clearly, it will be some time before such assessment will be possible. However, the temptation for many education commentators is irresistible, and instant dismissive or supportive judgments are commonplace — as seismic organizational and curricular changes occur almost weekly. To provide a measure of understanding of the systemic background to recent and current changes it might be useful to provide an outline of relevant government strategies in an attempt to convey some of the ostensible reasons for those policies involving "parental choice" and "school autonomy".

The Context For Legislative Change

The foundations of the postwar educational system in England and Wales were laid in the Education Act of 1944. As all education systems respond to significant socioeconomic change, many subsequent developments occurred — perhaps most notably the introduction of comprehensive, nonselective secondary (high) schools during the 1960s and 1970s. These changes were accomplished within the existing national legislative framework, whereas the reforms considered in this brief review have required fundamental systemic and curricular restructuring. Retrospectively, the first politically significant thrust towards the current reforms is conventionally, but not universally, attributed to an initiative taken in 1976 by James Callaghan, then Prime Minister of a Labor government (Chitty 1993).

In a major policy speech at Oxford, Prime Minister Callaghan initiated a "great debate" on the existing educational system to explore public unease over perceived failures to deliver the considerable expectations stemming from earlier extensions of educational opportunity, predicated on assumptions about education as investment and as a highway to a more egalitarian and economically successful society. There had been a general retreat from that confident educational expansion as deteriorating national economic conditions in the mid-1970s, with demographic decline accompanied by rapidly rising levels of public sector expenditure, occurred concurrently with mounting public concern and much dissatisfaction with the declared aims and outcomes of the education system, frequently (and simplistically) perceived as a decline in educational standards and pupil achievement. Those concerns were steadily fueled through media output providing adverse publicity about falling achievement standards and rising behavioral problems in primary (elementary) and secondary (high) schools. The confluence of many public anxieties led to political demands for the introduction of some form of educational accountability.

Although Callaghan avoided used of the term "accountability" his intentions were explicit and the text of his speech was deliberately leaked in advance to the press to ensure full media attention. His determination to reflect, and align with, public perceptions of the shortcomings of the education system was uncompromising:

> It is thought some people would wish that the subject matter and purpose of education should not have public attention focused on it; or at any rate that profane hands should not be allowed to touch it ... I take it that no one claims exclusive rights in this field. Public interest is strong and will be satisfied. It is legitimate ... so there will be discussion. If everything is reduced to such phrases as "educational freedom versus state control" we shall get nowhere. I repeat that parents, teachers, learned and professional bodies, representatives of higher education and both sides of industry, together with the Government, all have an important part to play in formulating and expressing the purposes of education and the standards that we need ... To the teachers I would say that you must satisfy parents and industry that what you are doing meets their requirements and the needs of their children. If the public is not convinced then the profession will be laying up trouble for itself in the future ... It will be an advantage to the teaching profession to have a wide

public understanding and support for what they are doing[1].

The call for public discussion was crystallized in a government Green Paper (**Education: A Consultative Document** 1977), and the subsequent "great debate" focused on four main themes. Now the term "accountability" was used and the Paper included trenchant criticism of existing educational practice in schools. It was transparent that the government intended to introduce evaluative policies to monitor the activities of schools. The four themes identified for critical public discussion were: curricula; standards and assessment; schools in relation to occupational life and the community; and initial and postexperience training of teachers.

During the following two years, considerable attention was devoted to research into concepts and models of educational accountability. Product and, especially, process evaluation procedures in schools were adopted by many local education authorities (LEAs) and a general acceptance emerged within the "educational establishment" of the inevitability of the introduction of educational accountability to new constituencies and stakeholders — largely client groups, notably parents and employers. However, changes occurred only in small incremental steps and at glacial pace — for example, whole-school evaluation undertaken through school self-reporting audits augmented by LEA advisory service mechanisms. Before any significant systemic reform could be mandated the Callaghan government lost office in 1979 to a Conservative government, with Margaret Thatcher as Prime Minister.

Almost immediately, the Thatcher-led government mandated educational reforms that, **inter alia**, reflected the salient concerns of the Callaghan government between 1976 and 1979. The thrust of government reformist rhetoric was, thus, unbroken although the nostrums for educational success now came from radically different political ideologies. In addition, curtailment of public sector expenditure developed as a government priority throughout the 1980s. With the decline in school population totals and a generally perceived failure of existing educational provision as a mainspring of technological change and success, the public sector of education was susceptible to retrenchment expenditure policies and systemic reform, particularly as these disquieting trends coincided with the ascendancy of politically influential "New Right" paradigms (Bondi 1991). Successive government White Papers, Education Acts (e.g., 1980, 1981, 1986, 1988, and 1992) and new Statutory Regulations mandated policy objectives that sought to raise standards in schools for all levels of ability and to introduce curricula and programs perceived as relevant to social and, especially, economic development. Efficient and effective returns on public sector resources invested in the education system were demanded and governmental measures to secure demonstrable and public accountability from teachers and schools were vigorously pursued; for example, legislation for the appraisal of the performance of teachers in the 1986 Education Act (MacLure 1989).

To consolidate the many piecemeal legislative changes implemented during the 1980s, and create a radically new systemic framework, the Thatcher government's 1988 Education Act was "... the most important and far-reaching piece of educational law-making for England and Wales since the Education Act of 1944" (MacLure 1989, p. 23). The government's intention was foreshadowed in 1987 in a parliamentary speech by the then Secretary of State for Education (Kenneth Baker): "... we need to inject a new vitality into the system. It has become producer-dominated ...", criticisms which echoed those made by Prime Minister Callaghan a decade earlier (Sayer and Williams 1989).

Influenced by politically fashionable trends in the 1980s for "market" ideologies and individual "consumerism", and seeking to reduce traditional distinctions between public and private sector organizations, the government developed policies to increase choice and diversity in education. Parents were regarded as both consumers and decision-makers. For example, in the Education Act of 1988 the government launched a determined policy to extend both choice and diversity with its main targets being to remove exclusive control over provision for the education service from Local Education Authorities (LEAs), and to wrest curriculum content and development together with stewardship of attainment standards from educators — Baker's "producers". The pivotal strategy was to remove most of the existing legal powers of LEAs for educational provision within their geographically defined areas, as LEAs were perceived by central government as independent subjective decision-making bureaucracies, inefficient resource allocators, and unaccountable to central government authority. Devolution of financial and personnel control to individual school governing bodies — representative of parental and local community interests — were the most significant of the reforms and other former LEA educational duties and powers have been similarly transferred. In addition, various servicing and maintenance school functions have been contracted out to the private sector.

The political nostrums of "consumer choice" and "school autonomy" have been extended to other systemic educational functions. In addition to a small cluster of City Technology Colleges (Secondary schools specializing in science and technology and financed by central government and sponsors from industry and commerce), a new sector of grant-maintained schools has been created, school inspections have been "privatized" and there are now (June, 1993) indications that through the application of local school "autonomy" and parental "choice", social equity and open access to educational provision are at risk and may be replaced by concepts of selection and elitism in many schools.

The planned removal of LEAs as major mediators of

educational policy and development has also produced other problems. The longitudinally measured policy perspectives of LEAs are being replaced by shorter terms consumerism and subjective priorities of many articulate parents. Education is manifestly not a free-market zone as the vast majority of schools, providing education for some 90 percent of pupils between the ages of 5 and 18 years, are funded from government public sector allocations and are under central control. Thus, unless parents are able to afford significantly high fees for their offspring to attend independent schools, choice is largely confined to maintained state schools required to observe centrally prescribed curricular and financial objectives. Nevertheless, within a state controlled and funded system there are issues of autonomy and choice which have been sharpened both in terms of focus and effect following the introduction of local management of schools (LMS) in 1991.

Through its policies and subsequent legislation for sweeping reform, the government assembled a politically attractive package, ostensibly to provide greater opportunities for parental responsibility and diversity. Direct control over educational provision and finance was to be achieved via the exercise of centrally "constrained choice" through devolved powers prescribed under mechanisms for the local management of schools. Consideration of these mandated government restructuring provisions is considered in the following sections of this chapter in relation to the realities of autonomy and choice through strategies for, and experience in, their implementation.

Parental Choice: Issues of "Exit" and "Voice"

Prior to the 1988 Education Act, parents were not considered powerful stakeholders in educational provision in the public sector. Although the 1944 Education Act conferred a degree of parental choice it was heavily circumscribed through unilateral interpretation by LEAs of their duties and powers[2]. Formal parental complaints about academic, social, and resource provision of education in schools could be made to the school governing body, the LEA, and, as a last resort, to the central government department, which constantly supported decisions taken by the LEA. Further, requests for transfer of pupils to alternative schools were granted only when resource implications were unproblematic and permission to move out of defined catchment areas secured from the LEA. Thus, unless parents were articulate, knowledgeable, persuasive, and persistent their "voice" was generally subdued and ineffective. Overwhelmingly, "exit" choice was available only to those who could afford to pay fees at independent schools (Hirschman 1981).

The present government's determination to "... carry the concept of parental choice to the heart of our educational system ... to move the balance of discretionary power away from providers of education to its clients ..." is explicit in the legislation over reforms for LMS and grant-maintained schools (GMS) in the 1988 Act (Sayer and Williams 1989, p. 149). In the view of Kenneth Baker, then Secretary of State for Education:

> If the product is not all that it should be, parents should not be put in a position of having to like it or lump it. Grant-maintained schools will be a threat to the complacent and to the second best. They will also challenge all within the service to do better ... Their [GMS school's] success will depend upon their management and upon parental commitment and that recipe, through financial delegation and more open enrollment, is available to all schools ... I want to give people a chance to press for excellence; I want to give them the means to demand excellence; and I want to create a spur which will oblige all LEAs to deliver excellence.
>
> (Sayer and Williams 1989: 149)

Since 1988, central government reformist policies for local governance and management of schools have provided parents with new opportunities for, and dilemmas over, choice. Following the introduction of the Parents' Charter in 1991 and new statutory requirements of the 1992 Education Act, information is being made available to parents on relative attainment levels of schools through the publication of "educational performance tables" based on unrefined public examination and standardized test outcomes, pupil attendance/truancy ratios, destinations of school-leavers, and via summaries of external school inspection reports on planned four-year cycles. Parental dilemmas may be summarized as a choice now between "exit" to another state-funded school (under new open enrollment policies) or to an independent school, or to adopt, perhaps as a parent governor, an active partnership "voice" role through seeking to find remedies for perceived dissatisfaction with the neighborhood school. With unprecedented access to information about local schools, it is argued that parents are now able to make informed choices about their options: choosing among schools within their area or actively engaging in "improvement" activities at a particular school. However, information available to parents presents other dilemmas — through statistical overload, challenges over validity, accuracy, significance, and ambiguities over presentation and interpretation of published data.

These developments have led to related general anxieties that with the new freedoms of choice and increasing diversity of educational provision some schools will be "losers" in a competitive market for pupil enrollments. Parental choice has led to particular schools becoming oversubscribed while other schools are disadvantaged through decline in pupil enrollments and face consequential difficulties over providing an entitlement curriculum and in sustaining achievement standards considered important by parents. Inevitably, perceptions that schools are underachieving have led to problems over their viability as pupil enrollments fall through parental exercising of "exit" choice to other publicly-funded schools. It is difficult to sustain argument has mounted sustained pressures one LEAs over required reduction in the number of surplus school places during the past decade. Currently, it is estimated that approximately one million surplus places exist throughout England and Wales and reduction in their number through the exercise of "exit" choice, together with a loss of demonstrable educational viability in schools perceived by parents to be failing, was politically attractive to the government.

The concept of choice is also susceptible to other instabilities. Government strategies were predicated to a substantial degree on assumptions that its policies for parental choice and local management of schools would be attractive to parents and, consequently, gain widespread support for the entire package of its mandated reforms. Characteristically, the government failed to test these assumptions in the early years of reform and, as change has proceeded in hyperdrive since 1988, clear signs of innovation fatigue and resistance to unbridled political change are now emerging. For example, a major problem has arisen over the proposed testing regime at Key Stage 3 (comprising a group of assessment tasks within the mandated National Curriculum for 14-year-old pupils). Although opposition and concern had been expressed by teachers and some parent groups in 1990 when Key Stage 1 tests were introduced for seven-year-old pupils, it was not sustained. Similarly, when "raw" 16+ examination results were published in late 1992, protests were voiced, but not orchestrated, over the failure to include "added value" variables. However, at that time, incipient resistance began to emerge over the conceptual framework and content of the proposed tests in English for 14-year-old pupils — scheduled to be undertaken during early summer, 1993. Somewhat maladroit governmental handling of the issues involved led to mounting resistance to these tests and rejection of their use by teachers in schools and by national teachers' associations; during March and April 1993 protests escalated sharply and the entire field of testing and assessment has been subjected to ridicule by all the major teacher's associations; by many school governing bodies throughout the country; by national associations of parents' organizations; and, perhaps most damaging to government strategy, through the resignation of three major policy leaders and proponents of its assessment policies.

Concurrently, and disturbing from a government perspective, parents made unanticipated and politically unwelcome choices. National surveys undertaken in May 1993, by major polling organizations such as Gallup, indicated that two-thirds of parents were opposed to testing at the age of seven years[3]. And this emerged following three years of testing at age seven and in direct contradiction to assumptions and claims made by the government. Further, a narrow majority of parents (51%) supported the decision taken by all teachers' associations to boycott the national tests for pupils of 14 years of age, and more than 60 percent believed these tests should be abandoned for 1993. It is ironic that key government policies for unprecedented extension of parental involvement in educational decision making, for access to and supply of information about educational processes for their offspring, and the ways in which legislation implemented, was created widespread antipathetic reaction to government strategies. At that stage, it was difficult to envisage a successful retrieval of the government's position without considerable political sensitivity, imagination, and, above all, consultation with those most closely involved in the policies of change — the teachers and the parents — leading to substantial policy modification.

Nevertheless, political rhetoric apart, it is clear that the government tilted decisively the balance in the governance of schools away from its earlier LEA political and professional control towards client interests. The 1988 Act, together with the foreshadowed Education Act of 1993, provide for rapid gains in the momentum for unprecedented structural school reform. For many educators the grant-maintained (GM) category, although numerally small at the present time, represents a potential threat to the traditional maintained public sector of education. For others it represents an exciting innovation to promote excellence through self-managing schools that are responsive to parental, community, academic, and vocational demand.

"Opting Out": The Grant-Maintained Sector of Schools

Another thrust in the government's determination to end the earlier dominant role of LEAs and initiate corporate parental choice within the maintained system of public education was legislated in the 1988 Act. A substantial section of the Act deals with school opting-out procedures to create a new category of autonomous grant-maintained schools, formerly within LEA control, and directly financed by central government[4]. From their inception, GM schools have received larger recurrent and capital financial allocations than schools remaining within the LEA sector. Although delegation of control to maintained schools under local management of schools (LMS) attracted widespread approval, the creation of an autonomous GM category of maintained schools has aroused considerable antagonism. The reason both for support and opposition may be glimpsed from a government publication outlining its proposals for GM schools. They:

> ... will add a new and powerful dimension to the ability of parents to exercise choice within the publicly provided sector of education ... parents and communities have new opportunities to secure the development of their schools in ways appropriate to the needs of their and in

> accordance with their wishes, within the legal framework of the national curriculum.
>
> (DES 1987)

> The message was clear that the government, intent on increasing its centralizing power in partnership with parents and school communities, proposed the removal of any direct or significant intermediary role for the LEA in school governance. Undeniably, an early motive for seeking GM status was, at least partially, fear of LEA intentions. A headteacher of a GMS claimed "Many of these 'opted out' schools appear to have taken the decision to escape from the clutches of their LEAs to avoid closure, amalgamation, unwanted reorganization, or some other dire Fate".
>
> (Stevens 1991: 222)

The process for opting for GM status is relatively simple. If, in a ballot, a majority of the parents entitled to vote apply for the GM category, the governing body of the school publishes proposals to acquire GM status and submits these to the Secretary of State for consideration and approval. If approved, school buildings, property and staff contracts are transferred from LEA control to the governors of the GM school. Under Conservative government policy achieving GM status represented an irreversible decision. Parents of pupils attending a school when the decision to seek GM status is taken operated a one-way valve; there were no possibilities of successor parents exercising their choice to revert to a school within an LEA framework. At the time of this review (June 1993), some 600 schools have secured or are under consideration for, GM status — a substantial underachievement on the initial government expectation of more that 2,000 GM schools.

The operation of the GM "market" over choice suggests it is developing in ways other than those envisaged by Chubb and Moe (1990). In a recent study of GM schools, it was revealed that only 4 percent of parents were advocates of GM status (Coleman, Bosh, and Gloves 1993). From this study and other reported experience it appears that, unless holding representative office, neither parents nor teachers have been closely involved in decisional processes leading to GM status. Although limited in its scope to an initial group of a hundred GM schools, this study indicates that the most influential advocates of GM status were the headteacher and the school's governing body. Further, GM school admission policies have been questioned. In some areas, GM schools have introduced local admission policies which include pupil-selection variables such as interviews, consideration of reports from primary schools (on both academic achievement and social acceptability criteria) as well as more conventional admission criteria such as siblings already attending the school and residence within the neighborhood. Thus concern exists that the concept of parental choice may be turned on its head in that oversubscribed schools and others through the application of restrictive school-specific subjective criteria will supersede intended freedoms of parental choice, for example, attainment, geographical proximity, acceptable pupil behavior.

Although this new category of GM schools represents a relatively small proportion of the approximately 25,000 schools in England and Wales, the creation of GM schools has aroused disproportionate hostility. LEAs perceived GM schools as defectors from the collectively maintained system and initially feared serial opting out: a situation that has now occurred in some areas and is likely to increase sharply under new provisions in the 1993 Education Bill currently in transit in parliament[5]. Teachers' organizations, unequivocally supportive of the secondary schools comprehensive system, are suspicious of incipient fragmentation of the existing framework that may lead to privatization of education (consistent with similar government policies in other public sector organizations) and the reintroduction of selective education with its inherent danger of creating differentiated social class institutions. Certainly, existing GM schools enjoy levels of direct funding from central government at significantly higher levels (approximately 15% for recurrent allocations), than available to the overwhelming majority of secondary schools within the LEA sector of the maintained system. Unsurprisingly, for GM schools a significant degree of institutional satisfaction exists, especially over corporate independence and autonomous school governance and with existing levels of recurrent and capital resourcing.

It was possible, indeed considered inevitable by some commentators, that if the Conservative government had been returned to office at the general election 1997, a rapid expansion in the GM sector would have occurred and have become an irreversibly established sector in the educational landscape because of the presumed degree of community and parental support and public antipathy likely to exist over any proposal for the reintroduction of an LEA framework (Chubb and Moe 1992).

Local Management of Schools: The Reform Strategy

Superficially at least, the concepts of school autonomy and organizational choice underpinning the strategy of LMS appear to be the most logical and desirable innovation of the 1988 Education Act. The devolution of wide powers and decision-making authority to local community and school-site levels was regarded as an appropriate and justified development by a government apparently determined to reduce the stifling effects of local bureaucracies on initiative, choice, and efficiency. Earlier widely publicized pilot projects by a few LEAs in the delegation of budgetary control (termed local financial management) to school governing bodies and headteachers indicated widespread approval by both lay and professional interest groups. In addition, the development of site-based program for self-managing schools in Australia and

Canada during the early 1980s suggested that an LMS strategy offered solutions to many of the problems inhibiting educationally effective and financially efficient provision in schools (Caldwell and Spinks 1988).

However, cynically inclined observers detected adventitious political stratagems. From its election in 1979, the Conservative government had been in conflict with several left-wing Labor-controlled, predominantly urban, education authorities over its new education policies and associated expenditure controls. With the introduction of LMS and mandated delegation of control for fiscal and management decisions to schools in 1991, the political power of LEAs was immediately severely curtailed.

In turn, the government's controlling role over educational policy innovation was enhanced through more direct expectations of compliance by individual school governing bodies (with increased parental representation) in connection with their new explicit responsibilities for the implementation of the government-prescribed National Curriculum and its associated assessment regime. Opportunities for direct central control over the new formula-driven site-specific budgets and presumed cost-benefit considerations at individual school level were also anticipated. The responsibilities of LEAs with broadly focused binocular overviews of continuity and adequacy of educational provision in their geographical areas were currently being replaced by more closely focused monocular perceptions of educational priorities at individual schools. Further, through such local autonomous arrangements it was intended that the influence of the wider "education establishment" will become inconsequential.

Thus, the intention was that LMS would enable individual schools to plan more effectively the use of allocated resources, including teaching and support staff: to prioritize expenditure through the flexibility offered by financial virement in determining ways of attaining their selected goals and objectives. Through this fundamental new structural reform the government's intention was that schools became more responsive to local client groups — parents, community, and employers. But schools might also become narrowly parochial in outlook through limited ambition and bounded opportunities for their pupils. This possibility may perhaps be an unsought, unintended outcome of planned increased central government control but for many parents and teachers it is a real and an unwelcome one. Some commentators believe that the notion of "power to the people", in this instance parents and local communities, is largely illusory and masked the Conservative government's determination to tighten its grasp on the education system through holding the steering reins of educational development — the curriculum, assessment, and public sector expenditure.

To facilitate that tightened control, the powers of LEAs were extensively reduced and their formed direct responsibilities for local schools effectively withdrawn. Although weakening of LEA responsibility occurred, LEAs retained certain residual tasks, including the disbursing of central government grant allocations to schools in accordance with schemes of financial delegation submitted to, and approved by, the Department for Education. However, independent and discretionary financial decision making by LEAs was largely removed other than for a limited number of collectively provided services for which LEAs may, or must continue to, retain responsibility for maintained schools in its geographical area, for example psychological assessment of pupils, provision of school meals, transportation to and from school, capital expenditure program and hypothecated grant funding. An important retained duty of the LEAs is to monitor and evaluate the devolved LMS arrangements within a framework of approved performance indicators. Eventually, it was anticipated that at least 93 percent of government allocations for schools' funding would be delegated to schools and the LEAs would become, if they continued to exist, merely agencies of central government.

Some understanding of the implementation of these policies can be conveyed by thorough consideration of the practical constraints on LEAs under the framework of change in school governance. Following legislation in 1988, each LEA was required to submit to the government department a scheme for delegation of powers to its schools for the introduction of LMS. With the planned introduction of LMS, in the financial year 1991/92, LEAs applied a government formula to calculate the distribution of hypothecated funding to individual schools. The formula allocations were based on various elements, the most important being a capitation component termed the age-weighted pupil unit (AWPU). The weighting is applied to the age range of enrolled pupils attending each school. The total of centrally provided resources allocated through the AWPU formula amounted to a required minimum of 75 percent of individual school allocations. Other factors in the formula of the schools' delegated budges included recognition by LEAs of historic funding elements such as the age and condition of school buildings, the number of pupils with designated special educational needs, assessment of variables such as poverty, linguistic and ethnic criteria. Other discretionary elements include provision of services such as school meals, peripatetic and advisory teacher supply, child guidance, and welfare support. As the range of LEA discretionary nonmandatory services might reduce the totals of direct allocation of funds to schools, the government has placed an upper limit on LEA central office retention with the requirement that such retention would taper to a maximum of 7 percent by 1995.

Within these policy constraints, LEAs have applied the prescribed general formulas required to be "simple, clear and predictable" to all primary, special, and secondary schools predicated on an inexact assumption that a common standard of provision would be achieved (DES 1988). Clearly, this is an unrealistic objective as schools very considerably in age, size, condition, equipment, and teaching staff profiles. For example, smaller schools require larger, enhanced capitation allowances than those for larger schools — especially noticeable in relation to the curriculum entitlement model adopted for delivery of the prescribed National Curriculum.

Central to the National Curriculum educational model in schools, under LMS a major problem exists over the determination of the teaching establishment of schools. Within broad tolerances over effective and efficient provision of education every school is at liberty to establish and employ its own complement of teachers — both in terms of total size and local emphases over curricular and pastoral opportunities. However, under the formula-driven funding calculations, the central government department has assumed an "average teacher" salary formula rather than "actual" staffing costs. As some 75 percent of delegated school budgets are required for teachers' salaries, serious problems of choice are posed for schools. For example, many governing bodies and headteachers have been required to make unpalatable choices when replacing older, experienced and, thus, more expensive teachers with successors who are younger, inexperienced, but relatively inexpensive. Because of financial constraints, selection criteria may become unprecedentedly skewed towards noneducational considerations — and some difficult decisions have to be made when appointing staff in schools — a matter of considerable concern for school governing bodies and headteachers. Examples of such myopic considerations are not uncommon in weekly correspondence published in weekly educational journals[6].

LMS: Adjusting to the Realities of Autonomy

An illustrative example is provided below to indicate the ways in which under LMS requirements many LEAs are adjusting to competitive market situations where the roles of provider and client have been largely reversed. One LEA (anonymously named here as "Brokenshire"), experiencing discomfort over loss of former roles and powers, is confronted with major financial problems over retention of central office staff and its own viability. Stemming directly from the statutory delegation of management autonomy under LMS its 400 schools now control some 90 percent of the total funding available for the provision of education. Conscious that most schools acknowledge the absence of direct experience, expertise, and adequate training in their new responsibilities on which to base informed managerial judgments, the LEA has offered a range of its services on an annual contract basis. Based on its own unrivaled expertise and experience prior to the phased introduction of LMS in 1991, the LEA has devised a contractual scheme for annual purchase of services by schools on an individual or neighborhood group basis.

Within the Brokenshire scheme outlined below are several "key" areas for service and support offered by staff from the LEA central office. Illustrative details of costs to schools for these LEA contracted services are provided in the Appendix.

Quality schools partnership (QSP)

Undoubtedly regarded by both the LEA and schools as the most important educational service, QSP is managed by an ad hoc board, representative of the LEA and schools in contractual membership, to provide LEA services and deployment of staff and resources. As a collaborative enterprise between the LEA and schools its main objectives are to continue to develop the curriculum (content, pedagogy, management), the assessment and recording of pupil progress, and to organize provision for the professional development of the teaching and support staff in member schools. Each member school prepares and prioritizes its own planned program varying in focus and pace in response to identified needs. From the LEA, member schools receive consultancy services on perceived needs, and have access to professional networks, and centrally provided in-service training programs, as well as a range of advisory support services for curricular and pastoral development. To provide an equitable baseline, each member school has an entitlement to allocations of inspectorial and advisory teacher time commissioned by individual schools. Examples of annual subscription school allocations are:

Primary and special schools	12 half days
Middle schools	16 half days
Secondary schools	20 half days

Quality assurance is monitored through the QSP Board which formally evaluates LEA provision for consultancy, adequacy of support networks, training programs, and other curricular services.

Management and administrative services (MAS)

Another key area essentially designed to provide uninterrupted continuity is in the sphere of management and administrative support, together with specialist expertise over school governance business (e.g., legal and personnel matters), school internal financial control, and premises management (including health and safety). Under existing legislation, the LEA is required to monitor these concerns and provide finance for associated activities from the retained element in its centrally held budget. Within funds delegated to schools a variety of responsibilities for financial, personnel, and premises responsibilities have passed to the governing bodies of individual schools. However, the joint LEA-member schools provision offers support and expertise for various nonroutine services such as legal advice on education and employment legislation and employee grievance procedures. In addition, important routine school administrative workloads may be contracted to the central office under the joint MAS scheme to secure cost-effective arrangements, for example, employment contracts for both teaching and support staff, salary assessments, processing medical and other claims together with payroll management. For these additional administrative services a subscription fee is payable, as indicated in the Appendix.

Special educational needs (SEN)

Similar arrangements within the joint provision scheme are made to provide specialist teacher expertise and support for pupils who are categorized as having special educational needs. Under a negotiated purchase scheme, member schools have access to the services of an LEA team of support teachers for diagnostic assessment and group teaching of special needs pupils, advice on appropriate specialist resources and equipment, and a variety of SEN liaison and training programs.

Miscellaneous services

Other important areas of management expertise and procedures hitherto beyond the experience of the majority of schools are those of bulk purchasing and protective insurance policies. The advantages of economies of scale are self-evident through bulk purchasing schemes for school books, stationery, and materials. In this illustration, Brokenshire has for several years been in membership of a regional consortium and also negotiates advantageous contracts for heating fuels and for the maintenance of school property and grounds.

Specifically, the legal complexities of school insurance policies are of great significance to all schools. The importance of adequate protection for schools and teachers is a major consideration. In recent years in the UK several widely publicized accidents involving pupils engaged in outdoor activities organized by schools have occurred. In other respects insurance policy protection for schools is also of paramount importance — not least because of a rapid increase in school burglary, arson, and vandalism. Under the joint LEA-schools scheme, insurance cover is provided for employee legal liability, personal accident, maternity leave, official transport, premises and equipment, and engineering work. As some 65 percent of premium costs are premises related, the advantages of a comprehensive group scheme for the insurance of personnel and buildings are likely to be more attractive to schools rather than alternative unilateral arrangements.

Brokenshire appears to have judged the launch of its offer of contractual services to school rather well. During 1993/94 a further tranche of LEA-retained government funding is scheduled to be allocated to schools. Brokenshire has allocated these funds to schools on formulas which closely correspond with the subheadings of the Joint Provision scheme (see Appendix). In effect, the offer of the Joint Provision scheme has been calculated to ensure that school subscription fees for any or all of these services do not exceed those additional funds to be made available under the new 1993/94 increase in delegated budgets. It also represents natural progression of governmental intention that by 1995 all ELAS, including Brokenshire, will have delegated at least 93 percent of government funding allocations to schools under the LMS mandate.

Significantly, LEA services being offered on a contractual basis are currently attractive to the majority of schools whether within LEA or GM sectors. The positive response of the majority of schools to enter contracts for LEA services, such as those offered by Brokenshire, rather than seek open market competitive alternatives is attributable to several factors. These include the efficiency of LEA services provided by experienced officials who are knowledgeable about specific school requirements, especially through close cooperative contact during the transition period leading to the onset of LMS. Others include the development of a new sense of "autonomy fatigue" experienced by some schools through initiation into the tedium of purchasing school supplies, contract tendering, and various personnel technical procedures[7]. Further, financial and management institutions, such as banks, accountants, solicitors, and consultants have generally perceived the recently autonomous schools' sector as an underdeveloped and, as yet, financially unattractive commercial market. Finally, recent experience indicates that the use of "expensive" hours of headteachers and senior teaching staff on noneducational management tasks is demonstrably not cost-effective and many schools have appointed bursars and business managers to undertake complex administrative tasks (Sayer and Williams 1989, p. 21).

Within total budgets delegated to its schools, Brokenshire has apportioned additional funds received from central government provided to facilitate implementation of mandated policies under LMS. For all these central government-funded facilitation initiatives, the LEA allocation to individual schools under delegated LMS funding arrangements are calculated on pupil capitation formulas. Further, as these government grants are provided for specific purposes, virement is not possible among the hypothecated allocations. Nevertheless, within each grant-aided program schools have considerable autonomy to prioritize operational objectives. In Brokenshire, specific grant-aid funding falls into four categories:

(1) Support for the development of the National Curriculum and mandated pupil assessment. This, the largest of the government funded allocations, is intended for expenditure in connection with teacher in-service training in these two areas but also permits individual school prioritization of expenditure on, for example, curriculum texts and other similar publications, the purchase of school equipment, and

provision of optical mark readers for assessment procedures.

(2) School management support funding to facilitate training for headteachers and senior staff in connection with new administrative and managerial responsibilities under LMS. In addition, expenditure on the training of elected members of school governing bodies is permissible; for example, over their new legal and community responsibilities under LMS.

(3) To secure effective administrative and procedural implementation of LMS, central government grant aid is provided for the training of support staff in schools and the provision of hardware and software equipment for the installation of management information systems in every school.

(4) Specific funding is available to support the introduction of mandated systems for the performance appraisal of teachers; for example, costs of classroom observation, appraisal training programs, and monitoring and evaluating the effectiveness of individually developed school systems.

Concluding Observations

In any prognostication about the scale and direction of education reform into the 1990s for England and Wales, it may be safely assumed that further legislation will build on the foundations laid in the blockbuster Act of 1988 and the voluminous 1993 Education Act (based on the lengthiest Education Bill ever presented to Parliament). The dismantling of earlier organizational structures and the emergence of new institutions and patterns of relationships with an emphasis on client choice and community focus under a carapace of central government prescriptions suggest that further legislation will follow in the governance and management of schools.

In April 1992, the Conservative government (with John Major as Prime Minister and John Patten as Education Minister) was re-elected for a further maximum five-year term. Further education legislation during the span of the new administration appears inevitable as the government seeks to strengthen and extend its reformist education policies. Certainly, an undeviating linear ideological philosophy to promote increased parental choice and autonomous self-managing schools will lead to the demise of LEAs. For example, in the 1993 Educational Bill provision already exists for the establishment of a new government funding agency for schools, which will allocate central government funding directly to schools in those LEAs where 75 percent of schools have achieved GM status. In such areas, former LEA education planning functions will be subsumed by the funding agency. It is also possible that a parental "voucher" system might be introduced as it is attractive to many government supporters.

Since 1980, through a process of legislative attrition encapsulated in 12 Educational Acts of continuously mandated structural change, government policies for increased centralized control, wider parental influence, and extensive school autonomy have required the education system to accept new ideological prescriptions and consequential reformist strategies, and to grapple with unprecedented problems associated with radical and frequently pragmatic innovation. A wholesale and rapid dismantling of a system that evolved gradually and thoughtfully during the past century is vigorously under way, much of it without consultation with, or support of, the "educational establishment", the LEAs and the teachers, and to the bewilderment of many parents. Undoubtedly, many influential interest groups in the nation are entirely supportive of government action, but it is believed that until very recently the majority of parents have been largely unaware of the impact of the application of economic market forces and supermarket attitudes on the education system. Indeed, many educators fear that the rapid pace and dramatic scale of current seismic changes will become meaningful for individuals only at the educational "check-out desks" on leaving schools at the ages of 16+ and 18+, or from colleges/universities at 21+, and at other subsequent critical points in the occupational, social, and cultural lives of pupils and young adults.

It is of course much too soon to reach definitive conclusions about the systemic effects of the 1988 Education Act, or the impact of the imminent 1993 Education Act, and continuing implementation of radical change. Nevertheless, it is devoutly to be hoped that the extended period of reformist zeal will lead to enduringly positive gains in the continuity of achievement, humanity, and culture of the nation rather than to outcomes of irretrievable individual and collective loss as a consequence of many earlier and traditional educational values being supplanted by apparently determined philistinism (Chitty and Simon 1992). Encouragingly, recent opposition by all national teachers' organizations to the impracticable burden of National Curriculum prescriptions and to fundamentally flawed government imposed testing regimes has been widely supported by parents' associations and many school governing bodies. With the combined weight of professional, parental, and lay interest groups it appears doubtful if the government can continue to espouse concepts of school autonomy and parental choice in education without accepting that both are circumscribed realities at ideological and executive levels.

Finally, policy decisions requiring successful systemic change and leading to cultural shifts in exercising responsible choice and organizational autonomy are dependent upon sustained discourse and consultation among disparate clients and providers. Regrettably, these processes have been ignored during recent years in the United Kingdom under prescriptions of governmental determinism sedulously pursued through application of the politics of derision to the counsel of teachers and LEA officers. But these are the educators who are unequivocally committed to progressive improvement in the effective provision of educational opportunities for all pupils on behalf of parents and other client groups in society. They are the interpreters of educational policies and providers of services through practitioner responsibilities for the realization of values; not only those of "choice" and "autonomy" but also for others such as "access", "equity", and "justice". It is through the bounded realities of individual lives in schools that educational values are immediately perceived and most sharply experienced. For these compelling reasons a new focus is required in educational reform involving understanding of the complex dynamic inter-relationship of processes in the initiation, acceptance, and implementation of planned change.

Recent history of legislated educational change in England and Wales demonstrates that the exercise of unilateral central government political self-interest is a fundamentally flawed strategy to achieve effective reform. The charting of new pathways towards that educational destination is both overdue and urgent.

Notes

1 From a *Times Educational Supplement* report, October 22, 1976. An overview of the period 1976–85 is provided in Williams (1985).

2 Education Act, 1944: Section 76.

3 Report on parent surveys, in the *Independent*, May 10, 1993.

4 Education Act, 1988: Sections 52–104.

5 The 1993 Education Bill provides for small, mainly rural

primary schools to elect for collective GM status through group opting out of LEA control.

6 For examples, see letters published in the *Times Educational Supplement*, September–October 1992.

7 The 1988 Local Government Act extended competitive tendering to schools, and the majority of individual school contracts are in connection with school cleaning services. However, the 1993 School Teachers Pay and Conditions document, effective from September 1993, poses particularly complex problems for schools in the equitable application of the new statutory salary structure.

Appendix

Brokenshire LEA-Schools Joint Provision Contract Scheme Formula Program Allocations to Schools, 1993/94

	LEA service offered	School allocation[1]	Per pupil allocation
1	Quality schools partnership	P = £1,800	£3.00 (Years 1–6)
		S = £3,000	£1.50 (Years 7–13)
2	Management and administration	P = £730	£2.00
		S = £985	£2.00
3	Payroll management	P = £40	£2.00
		S = £40	£2.00
4	Resources	P = £300	£6.50
		S = £520	£3.00
5	Special needs	P = £1,000	£40.00
		S = £1,000	£40.00
6	Maternity leave	P = None	£3.00
		S = None	£3.00
7	Other insurances	P = None	£5.00
		S = None	£5.00
8	Bulk purchases	P = £100	£1.00
		S = £100	£1.00

9	Information technology	P = £300	£1.50
		S = £300	£7.00
10	Property services	P = £100	£1.50
		S = £100	£1.00

[1]P = primary, S = secondary.

References

Bondi, L. (1991). "Choice and diversity in school education: Comparing developments in the United Kingdom and the USA". *Comparative Education*, **27**(2), 125–134.

Caldwell, B. J. and Spinks, J. L. (1988). *The Self-Managing School.* London: Falmer.

Chitty, C. (1993). *The Education System Transformed.* London: Baseline Books.

Chitty, C. and Simon, B. (Eds) (1992). *Education Answers Back: Critical Responses to Government Policy.* London: Lawrence and Wishart.

Chubb, J. E. and Moe, T. M. (1990). *Politics, Markets, and America's Schools.* Washington, DC: The Brookings Institution.

Chubb, J. E. and Moe, T. M. (1992). *A Lesson in School Reform from Great Britain.* Washington, DC: The Brookings Institution.

Coleman, M., Bush, T., and Glover, D. (1993). "Researching autonomous schools: A survey of the first 100 schools". *Educational Research*, **35**(2), 107–126.

Department of Education and Science (1987). *Grant Maintained Schools: Consultation Paper.* London: DES.

Department of Education and Science (1988). *Circular* 7/88. London: HMSO.

Education: A Consultative Document. (1977). Government Green Paper, Cmnd. 6869. London: HMSO.

Hirschman, A. O. (1981). *Essays in Trespassing.* Cambridge: Cambridge University Press.

MacLure, S. (1989). *Education Re-formed.* London: Hodder and Stoughton.

Sayer, J. and Williams, V. (Eds) (1989). *Schools and External*

Relations: Managing the New Partnerships. London: Cassell.

Stevens, D. B. (Ed.) (1991). *Under New Management*. London: Longmans.

Williams, V. (1985). *Accountability in Education: Public Confidence and Professional Development*. Oxford: University of Oxford Department of Educational Studies.

Chapter‘ Seventeen

RAISING THE EFFECTIVENESS OF SCHOOLING AND LEARNING BY ENLARGING SCHOOL AUTONOMY: THE CASE OF THE NETHERLANDS

HANS G. L. C. LODEWIJKS
Psychology of Development, Learning, and Culture, Tilburg University, The Netherlands

Introduction

During the last five to ten years a lively public debate has been going on in the Netherlands about the question whether and how to increase the autonomy of schools and school organizations. Enlarging autonomy of schools is meant to make schools more responsible for developing education policies of their own and is expected to enable them to improve their schooling and teaching practices systematically.

Discussions about an issue like school autonomy almost necessarily show complicated, even delicate and sometimes remarkable characteristics in a country where Article 23 of the Constitution (see Appendix) explicitly defines the general framework for the division of responsibilities between government and the schools. Article 23 of the Netherlands Constitution is quite specific about what should be the relation between government and public as well as private schools in terms of funding, but also in terms of who is considered responsible for education policy matters.

To explain the Dutch case with respect to school autonomy and freedom of choice more fully, it is necessary to describe in some detail how this constitutional framework works, and why, despite the "holy principle" of freedom of education and the sharply defined division of responsibilities, it is not that easy to put government at a distance in order to increase the autonomy of schools. The main reason is that just because government has to safeguard fully the equal treatment (financially, as well as substantially) of private and public schools, it has to put pressure on schools to meet rules, requirements, and regulations that are meant to guarantee this equality.

Pillarization, a Striking Feature of Dutch Society

One of the most striking features of Dutch society is the relative autonomy that is granted to the different societal and religious groups in the country to found, to supervise, and to direct their own organizations. This phenomenon is named denominational autonomy, or "pillarization". Denominational autonomy is either guaranteed by the Constitution, by law, or simply by tradition. Under the principle of denominational autonomy, relations between (central) government and the different "denominational organizations" are — generally speaking — defined as follows: All organizations of whatever denomination are treated alike by the government in all matters, including the funding and financing of their activities and infrastructural provisions. Government always and unconditionally keeps out of policy matters regarding the organizations, except in some cases of setting general quality criteria and financial regulations.

What this means is that in Dutch society we find groups and organizations which have almost the same social or political aims, but which are based on a different ideology. This pillarization is found in many organizations, such as broadcasting, sports, and education. Schools may be Protestant, Catholic, state or independent. These distinctions begin at nursery school and continue through universities. State and private schools have the same status. Both are 100 percent government-subsidized. The government inspectorate checks the standard of education in all types of school.

Education and Schooling: General Characteristics of the Dutch System

The principle of "denominational autonomy" is most prominently reflected in the Dutch education system. This system is complex and its structure and functioning are not easy to explain to someone looking at it from abroad (Lodewijks 1995).

The unique structure of the Dutch education system goes back to parliamentary discussions in the nineteenth century, during which fundamental issues were raised about who should have the final responsibility concerning the founding and financing of schools, the setting of standards, and the recruitment and qualification of teachers. The different religious groups ("pillars"), namely, the Roman Catholics and the Protestants, played a prominent role in this debate. These groups were striving for freedom to provide education according to their religious belief. Furthermore they required from government that both private (Protestant and Catholic) and public-authority schools would be financed on an equal basis and from public funds. The parliamentary debate about these issues resulted in what was called the "educational pacification", which was translated into a constitutional article laying down "freedom of education". From that moment, any societal group was entitled to set up its own schools, and to decide the manner and the matter of the teaching provided.

Furthermore, it was laid down that central educational policy concerning financing, structuring, and legislation could only be developed and implemented after extensive consultation — and, as it worked out, at least general approval — of the organizations representing the different, initially mostly denominational, national groups of school boards.

At the same time, ceritral government and the municipalities were charged with the responsibility for public education, in the sense that they had to see to the establishment, financing, and administration of public schools. This meant that the central and local authorities had to offer sufficient public schooling facilities to those groups of parents that preferred this kind of schooling for their children.

What makes the system unique is that public schools and **bijzondere scholen** (private schools, i.e., those founded and run by the nongovernmental societal groups) are treated equally by the government in financial matters. Consequently, the government fully finances all types of schooling, public as well as private.

Evidently, any understanding of how the autonomy of schools can be increased in the Netherlands should start from a consideration of the special way in which the Dutch education system is organized, how it is financed, and who bears the ultimate responsibility for the content and the development and improvement of schooling.

Structural Aspects of the System

The Dutch education system is considerably differentiated with respect to types of school. In current government policy there is, however, a strong tendency to change the system by integrating different types of school into more comprehensive types.

In 1985 a new Primary Education Act came into force,

through which kindergarten and the primary school were integrated and became the new **basisschool**. A new curriculum was developed for the **basisschool** (which has eight grades), to cater for pupils from five (when compulsory education begins) to 12 years old. It seems that the next step in this field will be the reintegration of some forms of special education (especially for children with learning and behavioral problems) within regular primary education. Society today no longer seems to consider the segregation of children with these kinds of problems in separate and specialized schools an acceptable way of giving them special attention. Instead, it is felt that regular primary education should try harder to keep these children within the **basisschool**. Of course, it is not only idealism that has promoted the idea of the reintegration of special education. Compared to the regular **basisschool**, schools for special education are far more costly, and special education spending shows a relatively strong increase every year, mainly because of the ever-increasing influx of pupils into special education.

Moreover, in 1993 a new Bill passed Parliament that focused on the integration of the different school types and curricula for the first phase in secondary education, covering four grades and the rest of compulsory education (for pupils up to 16 years of age). The new Bill is about the integration of the curricula of the five different types of school in this phase into one new type of school that is to offer what has been named **basisvorming** (impossible to translate, but "continued primary education for all" comes close). These five types of schools are at present:

- Junior vocational education (LBO, ages 12–16);
- Elementary general secondary education (LEAO, ages 12–16);
- Junior general secondary education (MAVO, ages 12–16);

- Higher general secondary education (HAVO, ages 12–17);
- Preuniversity education (Atheneum, Gymnasium, ages 12–18).

The Constitutional Principle of "Freedom of Education" and the Practice of Centralized Educational Policy

The principle of freedom of education, the outcome of the "school debate" in the nineteenth century, which was laid down in the Constitution of the Netherlands, permeates strongly any thinking about education and any educational policy. Since the "educational pacification" it is the responsibility of the government to stimulate, to coordinate, and even — but only after consultation with the school organizations — to initiate educational improvement and development.

What comes out of all this is a picture of an education system that, in terms of its policy and development, can be characterized as centralized as well as decentralized at the same time. Because all schooling in the Netherlands is publicly financed, the central government defines (at a national level) all kinds of specific regulations with respect to the financing and the general structure of the school system. The same holds for the qualifications required of teachers and for general standards regarding the quality of education: all these are defined at a central level.

On the other hand, the individual schools are free — within the limits of the general structure of the system — to define their own teaching methods and to choose what particular content they are going to teach. However, the central government dictates in general the subject matter areas that must be covered by all types of school within the system. (There is only one exception to this rule: The government has no direct influence on the subject matter areas taught at universities and colleges, though it has considerable influence on the structure, duration and financing of university programs.)

The educational policy process in the Netherlands is quite complicated. On the one hand, central government has a strong influence on this process. The government can help improve and change the educational process and its structure either through legislation, or by defining general rules and regulations. An altogether different way in which the government can stimulate educational development and improvement is by setting up nationwide innovation and stimulation projects. These projects are generally fully state-financed and are aimed at national educational priorities. Examples are: the improvement of primary education with particular emphasis on the implementation of the new **basisschool**; the educational priority policy, which is aimed at immigrant and indigenous children from disadvantaged background; the national assessment of educational achievement; the introduction of computers and new information technology in schools.

On the other hand, for any legislatory act of the Minister of Education to take effect, ample and serious consultation is required with the school organizations that represent the "main streams" of school boards. There are three of these so-called "umbrella organizations", each representing one of the main "pillars" in Dutch society: The Dutch Catholic Schools Council, the Dutch Protestant Schools Council, and The Dutch Nondenominational Private Schools Council. Furthermore, public schools are represented through the Association for the Advancement of Public Schooling, together with the Union of Netherlands Municipalities. All these organizations, whose membership comprises teacher unions, parent organizations, organizations of school administrations, and so forth, play a part in defining the route through which educational development and policy can be accomplished.

Improving the Quality of Schooling by Further Expanding the Autonomy of Schools

The question of school autonomy arose again, when, in 1988, the former Minister of Education and Sciences, Mr. Wim Deetman, presented the memorandum "Taking schools into the 21st century — a view of management for the nineties" (Ministry of Education and Sciences 1988) to the Lower House of the States General (parliament).

This memorandum described the social and cultural developments expected to take place in the near future. It was stressed that these rapid changes in the social and cultural environment of the schools would have a major impact on the way individual schools would function. Rapid changes and increasing diversification make it essential for schools better to pursue their own policies: "... a school requires the scope to balance social developments against its own specific situation and to make certain choices on that basis. In this sense, the primary and secondary schools of the nineties will be autonomous".

Despite this scope, schools will still be required to adhere to certain rules, and to operate within a general framework. As far as the optimal deployment of resources and the achievement of the highest possible educational standards are concerned, however, schools will be free to take most decisions themselves. Each school will be run by a professional management that will determine the number of staff to be taken on and their various functions, according to the restrictions imposed by the budget. Decisions on how to deploy staff and material resources will be taken, as far as possible, in a way that meets the school's specific requirements.

While schools will be given greater autonomy and a measure of responsibility for financial matters among others, the concern for educational standards at schools will not be decreased. Government authorities will continue to monitor the ways in which these standards are maintained. The emphasis will shift, however; rather than seeking to guarantee equality by laying down prescriptive regulations, this will be done by retrospective assessment. Tools to be used for this purpose will include attainment targets, policy on further training, and the development of educational indicators. In those cases where standards are found to be too low, talks will be held between the inspectorate and the school concerned, and sanctions might be considered, affecting the allocations of funds.

In short, autonomy for schools does not mean that governmental authorities will no longer be playing an important role, but that schools will be called to account on the basis of their responsibilities.

The Present System of Management in Primary and Secondary Education

The education system in the Netherlands is regulated by a great variety of centrally imposed regulations which often present serious hindrances to greater autonomy in the way schools are run. The types of regulations involved concern the following issues:

(1) accommodation,
(2) maintenance of material resources,
(3) legal framework, and
(4) teaching qualifications.

As for primary education, regulations governing the organization of teaching do, theoretically, provide sufficient scope for schools to operate freely. Schools are severely hindered in their endeavors actually to profit from this freedom, however, by the detailed systems of rules which exist in other areas: particularly those governing the legal position of staff and certain sections of the regulations governing staff numbers.

Secondary education is characterized by dissimilar target groups and educational institutions with equally divergent structural features. The efforts to achieve increased autonomy and to improve standards must consider these differences.

A number of bills which will contribute towards increasing autonomy within secondary education have already been introduced into parliament. The bills concerned deal with school accommodation and the maintenance of material resources. The present regulations governing staff numbers, combined with those which deal with the legal status of staff and with teaching qualifications, constitute a major obstacle to enlarging school autonomy. The regulations governing the organization of teaching itself provide less scope for secondary schools to operate freely than is the case for primary schools, but they do allow certain margins within which independent decision making is possible.

It was concluded that in many respects the existing funding system proves to be the determining factor in the details of regulations which govern the various areas. Furthermore, the system according to which costs are declared and reimbursed does not provide schools with any incentive to take advantage of the scope available for them to take the initiative.

This means that it is extremely doubtful whether creating more freedom, for instance in the realm of the content and organisation of teaching, would lead to the desired changes in behavior unless the funding system is adjusted as well.

Increased Autonomy by Changing the Funding System: Opportunities to Conduct Individual Staff Policies and to Improve the Quality of Teaching

The Dutch Ministry and parliament decided that the funding system must be changed if schools are to be enabled to function autonomously. The obvious solution is to switch to a system of lump sum funding. A major element of greater autonomy in schools is that they will be enabled to conduct their own staff policies; an essential part of this is that schools should be able to determine the numbers and composition of their staff. School management must be consolidated if an active staff policy is to be pursued which aims to improve the quality of teaching. The costs involved can be met by reallocating the staffing resources within a school.

In secondary education the competent authorities can lay down the number of teaching periods each working week shall comprise. Though the table which lays down the minimum number of hours to be spent on each subject will be retained, various measures will introduce a degree of flexibility into the application of the table in practice, and the number of hours left for the school itself to allocate will be increased.

As for primary schools, the introduction of lump sum funding and the opportunity of conducting their own staff establishment policy will enable them to make better use of the opportunities which the Primary Education Act and the Special Education Interim Act already offer schools to pursue a policy of their own choosing. The freedom for a school to deploy its means as it sees fit can be further increased by allowing reallocation of funds between staff costs and material resources expenditure.

The corollary of schools acquiring greater autonomy is that central government will have a smaller role to play, for instance in the area of staff policy. No great differences should be allowed to grow, however, in the primary legal status of staff working at different establishments, for example in public-authority schools as compared to private schools. Neither should unequal fringe benefits and pension rights be allowed to hinder the mobility of staff among schools. This means that central government must still lay down regulations providing a general framework for the legal status of staff. Within this framework drawn up by the state, schools or clusters of schools will have considerable freedom to determine details of the position of staff in their own way and to conduct their own staff policy.

School Autonomy and the Quality Issue: Some Findings

The main reason for the Dutch government to increase autonomy for schools is to enable them to develop individual school policies. It is expected that being enabled to do so will affect the quality of teaching and learning in schools. There is some empirical support for the assumption that diminishing bureaucratic hindrances for schools affects the quality of schooling. Important studies in this respect are the ones being conducted in the United States by Coleman, Hoffer, and Kilgore (1982) about high school achievement of public, Catholic, and private schools, the study by Chubb and Moe (1990) on institutional contexts and school achievement, and the study by Rowan, Raudenbush, and San Jin Kang (1991), analyzing the relations between organizational structure of high schools and school achievement.

Chubb and Moe claim to have shown that private schools in the United States do better because they are organized more effectively and this is "... a reflection of their greater autonomy from external (bureaucratic) control" (p. 24). In their research they found that Catholic private schools exercise much more control over principal educational matters (e.g., the curriculum, teaching methods, staff, and disciplinary matters) than public schools are able to do. As Catholic schools turn out to be more effective in general, they conclude that this relates to the opportunities for these schools to develop their own education policies: "... Bureaucratic influence is much stronger — and autonomy much weaker — in ineffective schools than in effective ones" (p. 155).

Rowan, Raudenbush, and San Jin Kang (1991) also found a relation between the way Catholic schools are organized and their effectiveness. They show that Catholic schools manifest much more supporting leadership (in educational matters), a stronger participation of the teaching staff concerning decisions about school matters, and more effective cooperation among the staff than public schools. These results are comparable to the Coleman, Hoffer, and Kilgore (1982) study, in which it was shown that Catholic schools, much more than public schools, reflect "organic" management styles. One explanation for this might be that Catholic schools are less subordinated to external regulations and cumulating external goal settings.

Though the Dutch education setting for "denominational schools" is quite different from the American situation, there is some (weaker) comparable evidence available in the Netherlands about this issue. For instance van Wieringen (1992) examined the relations between the institutional context of a school, the degree of autonomy, and the effectiveness of secondary schools. In the Netherlands the institutional context consists of a private and a public sector. Van Wieringen (1992) found some evidence of differences in school performance between private and public schools: private schools, mainly those outside the (sub)urban areas of the country, doing better than public schools. One of his explanations for these differences relates to the autonomy granted to schools in both sectors. Schools with lower degrees of constraints and more autonomy show a higher level of performance. Research on autonomy given by school boards or school governing bodies to their schools shows no difference in areas like curriculum and instruction and in areas of personnel and student policy. Differences are shown with regard to current expenditures. As a whole, striking differences between the public and private context could not be observed in the Netherlands. Some differences, nevertheless, are to be seen and they might be increased due to national policies aiming at enlarging the scale of operations of school boards and at increasing the autonomy of schools.

Sleegers, Bergen, and Giesbers (1992) examined the following questions: (1) Is it possible to distinguish types of schools on the basis of their policy-making capacity? (2) Do schools differ in their use of the scope for policy-making because of differences in policy-making capacity? (3) Do schools in a situation of greater autonomy differ in their use of the scope for policy-making because of differences in policy-making capacity?

Because a fully autonomous school system did not exist at the time of the execution of their research, the intentional, not the actual, behavior of schools has been investigated. In a survey among 57 school managers and 143 middle managers of 23 comprehensives, the above questions were tackled. The research concerning the first question resulted in a threefold typology: schools with a hierarchically oriented policy-making capacity (type 1); schools with a reduced colleague-oriented policy-making capacity (type 2); schools with a colleague oriented policy-making capacity (type 3). Regarding the second question it appeared that type 1 schools exercise the least instructional and administrative policy, type 3 the most, whereas type 2 schools occupy a middle position. The results of this research do not allow answers to the third question unambiguously: relations lacked significance and school managers and middie managers were almost unanimous in their views regarding the intentional behavior of schools. The main outcome then is that it has been shown again that there is a relation between "the organizational/institutional context" of a school and its capacity to make use of it for education policy making.

However, some contradictory evidence may be derived from a study by Vermeulen *et al.* (1992) about school autonomy, personnel policy, and educational innovation. One of the main assumptions underlying the idea of increasing school autonomy is that this will lead to better personnel planning (geared to the demands and necessities of optimizing schooling and teaching) and — henceforth — to higher degrees of effectiveness. Vermeulen *et al.* (1992), however, found no evidence that schools with higher autonomy had developed "coherent personnel policies". They only found some isolated aspects of staff development. For example, some schools had developed regular planned dialogs between school management and individual staff about the functioning of teachers. Most schools, however, had not developed consistent policies in this respect.

Only some situational or context variables related significantly to personnel management, educational innovation, and organizational development of the school. This has to do mainly with variable school size, much more than with allowed autonomy: Larger schools more often engage in activities related to personnel and staff development.

In those schools Vermeuien *et al.* (1992) found many more examples of formal training schemes, policy plans, and terms of references for the various employees. Furthermore, they found a relation between school complexity (schools with different streams) and the presence of pupil counseling, guidance, and monitoring systems.

Though the findings of Vermeulen *et al.* (1992) could not confirm the assumption that increasing autonomy directly stimulates educational improvement, school autonomy might still be an important factor in raising the quality of schooling and learning in Dutch schools. Getting opportunities as a school to develop individual policies might be a necessary (e.g., the studies by Chubb and Moe (1990), by Coleman *et al.* (1982), and by Rowan *et al.* (1991)), but not at all a sufficient, condition for the improvement of schooling. It very much depends on the way schools make use of it, whether autonomy leads to improvement. Schools who have not learned how to make use of newly gained opportunities and freedom will not do any better, unless they decide on their own to develop coherent views and consistent practices to raise their performance.

APPENDIX: ARTICLE 23 OF THE CONSTITUTION OF THE NETHERLANDS[1]

(1) Education shall be the constant concern of the government.

(2) All persons shall be free to provide education, without prejudice to the authorities' right of supervision and, with regard to forms of education designated by law, their right to examine the competence and moral integrity of teachers, to be regulated by Act of Parliament.

(3) Education provided by public authorities shall be regulated by Act of Parliament, paying due respect to everyone's religion or belief.

(4) The authorities shall ensure that primary education is provided in a sufficient number of public-authority schools in every municipality. Deviations from this provision may be permitted under rules to be established by Act of Parliament on condition that there is opportunity to receive the said form of education.

(5) The standard required of schools financed either in part or in full from public funds shall be regulated by Act of Parliament, with due regard, in the case of private schools, to the freedom to provide education according to religious or other belief.

(6) The requirements for primary education shall be such that the standards both of private schools fully financed from public funds and of public-authority schools are fully guaranteed. The relevant provisions shall respect in particular the freedom of private schools to choose their teaching aids and to appoint teachers as they see fit.

(7) Private primary schools that satisfy the conditions laid down by Act of Parliament shall be financed from public funds according to the same standards as public-authority schools. The conditions under which private secondary education and preuniversity education shall receive contributions from public funds shall be laid down by Act of Parliament.

(8) The government shall submit annual reports on the state of education to the States General.

References[2]

Coleman, J., Hoffer, T., and Kilgore, S. (1982). *High School Achievement: Public, Catholic and Private Schools Compared.* New York: Basic Books.

Chubb, J. E. and Moe, T. M. (1990). *Politics, Markets and America's Schools.* Washington, DC: Brookings Institution.

Lodewijks, J. G. L. C. (1995). "Functies en structuur van het Nederlandse onderwijssysteem: een schets van ons scholenstelsel". (Functions and structure of the Dutch education system). In: N. L. Dodde and J. G. M. Leune (Eds), *Het Nederlandse Schoolsystem,* Groningen: Wolters-Kluwer.

Ministry of Education and Sciences (1988). *De School op Weg naar 2000* (Taking schools into the 21st century). Memorandum (in Dutch). The Hague: Staatsuitgeverij.

Rowan, B., Raudenbush, S. W., and San Jin Kang (1991). "Organizational design in high schools: A multilevel analysis". *American Journal of Education,* **99**, 238–266.

Sleegers, P. J. C., Bergen, Th. C. M., and Giesbers, J. H. G. I. (1992). "School en beleidsvoering". (School and policy making). *Pedagogische Studiën,* **69**, 177–199.

Vermeulen, M., Visser, H. J., van Ojen, Q. H. J. M., and Karstanje, P. N. (1992) "School autonomy, personnel policy and educational innovation". In Tj. Plomp, J. Pieters, and A. Feteris (Eds), *European Conference on Educational Research, Book of Summaries, Vol, 1.* Enschede: University of Twente.

Wieringen, A. M. L. van (1992). "Richting, autonomie en kwaliteit in het onderwijs". (Institutional context, autonomy and quality of education). *Pedagogische Studiën*, **69**, 162–176.

Notes

1 Text of Article 23 was taken from a translation by the Dutch Ministry of Education and Sciences.

2 Throughout the article translations of technical, policy, and official terms and expressions are borrowed, or derived from, several documents provided by the Dutch Ministry of Education and Sciences.

Chapter Eighteen

CONSTRAINTS ON PARENTS' INVOLVEMENT IN SCHOOLS

A. J. C. King
Social Program Evaluation Group,
Queens University, Canada

The purpose of this chapter is to provide background information on the issue of community involvement in schools. While the concept of community can be viewed as quite broad, for the most part the focus here is on the involvement of parents in schools and the expectations held by employers and postsecondary educational institutions of schools. This chapter illustrates that while there may be gains for children as a result of the direct involvement of their parents in school life, there may also be some costs with regard to instructional and organizational effectiveness. These must be taken into consideration when designing strategies to increase parent involvement in schools.

Two main themes are examined in some depth. The first is concerned with what parents really want for their children from the school experience, how parents' expectations can shape the form and content of school programming, and what conflicts this can create between teachers and parents. The second theme is associated with the relationship between parents and children, and the effect of that relationship on school achievement. Threaded through the discussion is the reminder that schools not only develop necessary skills, knowledge, and values in children, but also play an important role in determining who will have access to economic opportunity and status. One of the most critical issues in educational programming is the tendency for schools to differentiate among children in terms of educational destinations, and that this differentiation tends to occur disproportionately along social class, racial, and ethnocultural dimensions. This issue is mentioned because of its association with the controversy of whether or not to stream students into courses and programs of different levels of ability. More specifically, the view of the role of the school held by representatives of the wider school community — parents, employers, and postsecondary institutions — is examined. This is done from the perspective of how students are evaluated and how these evaluation decisions are utilized by employers and postsecondary institutions.

Data Sources

Data from two studies are used for this analysis. The first is a current study by King and Peart (1994) designed to identify issues associated with the evaluation of student achievement. Information collected by survey and interview from 8,000 elementary and secondary students, 500 teachers, 520 parents, and 48 employers was used to obtain an interpretation of the role of the school. The second major source of information is a secondary analysis of data from a survey of young people aged 11, 13, and 15 in 11 countries (King & Coles 1992). This survey was originally designed to obtain self-reported information on young people's health and behavior and attitudes, but it also provides student perspectives on the relationship they have with their parents. A series of regression analyses were conducted using these data to examine factors influencing the extent to which young people can talk about their problems with their parents.

The Canadian Educational System

Canada is experiencing considerable strain at the present time with regard to the quality of its elementary and secondary school programs. It might be useful to provide a brief outline of the structure of Canadian education and to summarize contentious issues influencing the educational system in order to provide a framework for the analysis.

The responsibility for elementary and secondary education in Canada rests with the provinces. Most provinces have similar systems, starting with kindergarten for five year olds and going to grade 12. Ontario, the province from which most of the data for this study were collected, has an additional year. In Ontario there are usually two distinct school systems, elementary (kindergarten to Grade 8) and secondary (Grade 9 to the fifth year), but most other provinces have a junior high at some point in the middle years. Quebec is structured quite differently, with elementary school going from kindergarten to Grade 6, secondary school from Grades 7 to 11, and an intermediate stage prior to university called Collèges d'enseignement général et professionel (Cégep). The Cégeps act as two-year preuniversity programs and three-year professional programs designed to prepare young people for work. All the other provinces have some form of community college system which acts as an alternative to university and emphasizes professional programs.

While there are many similarities in curriculum from province to province, there are differences in the degree of curriculum standardization and in the use of standardized testing. Some provinces have standardized examinations at different points through the system, including graduation exams which are part of the entry process to postsecondary education. Others have none. Nevertheless, in all provinces standardized testing to ensure quality and consistency in program delivery is an issue.

There are three major concerns with the educational system that are being raised across the country. First, there is an overall secondary school drop-out rate of 30 percent, which is considered inefficient and unfair to minority groups. Second, there is concern about the level of achievement in basic skills as reflected in the number of plans to increase standardized testing within and across provinces. Both the drop-out rates and concerns about the quality of education are related to the concern that Canada is not competitive in the world economy, and that it is the role of schools to give young people the skills to make Canada more competitive. The third issue is as political as it is educational, and is referred to in the two themes noted in the introduction to the chapter. There is widespread feeling that students are being streamed into dead-end programs and these programs contain disproportionate numbers of students from certain races and ethnic groups, and that perhaps the best way of dealing with the problems of streaming is to keep all students together in the same program. The widely divergent German and Japanese educational systems are held up as models. Not surprisingly, the destreaming issue is also getting much attention in the United States.

Research Questions

Parents want what is best for their children. This means success in school, which later translates to access to post-secondary education and improved economic opportunities. Schools, then, sort young people into categories, each with its own status and economic implications. Involvement of parents in school planning and even program delivery influences this sorting process. The following analysis is designed to identify parent priorities in some depth and considers the likely impact of their views on the sorting process. Under what conditions will increased parent involvement increase the quality of learning by reinforcing the school experience with encouragement from home? Will greater involvement in schools exacerbate the confrontations among the cultural and racial groups related to equity issues because of the inevitability of differences of achievement in school occurring among the various groups?

If children are streamed to make the system more efficient, the differences among its groups will become more apparent. If all students are integrated and receive the same content through elementary and secondary school, then preparation for the next stage in life and the overall quality of learning may be adversely affected. Streaming at some stage is inevitable, and the outcomes with regard to student backgrounds patently clear. To what extent does parent involvement in schools touch on this issue?

The next research questions are concerned with the expectations employers have of the educational system, and the requirements that must be met for entry into post-secondary education. Finally, it is necessary to examine the relationship between young people and their parents and how this relationship either reinforces performance in school and/or discourages it when parental expectations are not met. Obviously, not all young people will be successful in meeting their parents' expectations. What is the cost to these young people who are not successful?

Parent Interpretation of School Marks

To what extent do parents view the purpose of education as the acquisition of the knowledge, skills, and attitudes important to adapt to adult society, and to what extent do they see school as providing their child with the opportunity to go on to further education? These questions provide the focus for our survey of 520 parents of secondary school students. We asked them to interpret two sets of marks as they might relate to their children. One set of marks was moderately high, averaging just over 80 percent, the other moderately low, averaging just over 60 percent. The parents were asked to indicate what they would communicate to their children if they had obtained these sets of marks. We assumed that the parents' comments would fall into three broad categories. The first would be concerned with the relationship between the marks obtained and the marks they expected their children to get. The second broad category of interpretation would relate to the implications of the marks for postsecondary educational opportunities. The third would be concerned with what their children had learned, what they were able to do as a result of the courses they had taken in terms of skills developed and knowledge acquired. However, it was also our assumption that this last concern would probably receive little attention from parents.

The parents' expectations with regard to marks obtained by their children are based on their view of their child's academic potential. So the first set of marks, which is above average, would be quite acceptable to most parents, but less so to others. A substantial number of the comments from parents were concerned with the extent to which expectations they held for their children were being met. Table 18.1 presents sample comments concerned with the two sub-themes: the first comments by parents about children who were meeting or exceeding expectations, and the second about children who were not meeting expectations.

Table 18.1. Parents' comments regarding expectations

Child meeting expectations	Child not meeting expectations
Very proud/pleased	Improvement necessary
Keep up good work	Try harder
Congratulations/doing well	Working below potential
Marks that I would expect	More concern for your future
	Improve study habits

The first group of responses also reflect a summary view of the marks: that is to say, the marks are viewed in aggregate, and the overall view is that the child is performing well. The second group of comments all assume a level of potential achievement that is not being met. This perspective on the relationship between what is being achieved and what is not being achieved tends to be shared by teachers as well, and to a lesser extent, the students. The main solution of the problem of closing the gap between what is being achieved and what should be achieved is that more effort is required.

Certainly the most powerful force at play in the parents' assessment of their children's marks relates to the implications of those marks for future career decisions, especially those related to attending university, but also, to a lesser extent, to attending community college. Table 18.2 again illustrates the two main themes. The first, that the student is on track toward meeting his or her career goals. The second, that there is a risk of not meeting his or her future career goals.

Most parents were concerned with the relationship between marks obtained and admission to university. A subset was concerned about getting access to particular programs, and the high marks necessary for such programs

Fig. 18.2. Parents' interpretation of marks in regard to career implications

Student is on track	Goal at risk
All options open	Need higher marks for chosen future
All goals still available	Remember marks need to get to university
All doors still open	Forget university with these marks
Any postsecondary future is O.K.	These marks will not get you into the university of your choice.

Table 18.3. Parents' responses to low set of marks

Try harder	Help required	Blame someone/something	Change aspirations
Devote more time	Look into tutoring	Could be teachers' fault?	Consider changing the level of difficulty of courses
Do more homework	Go to the Guidance Department	Are drugs involved?	Learn a trade
More studying required	Set up a plan to improve	Part-time work	Consider community college
Pay attention more		Extracurricular activities?	Marks will be good enough for college
Try to do better			
I know you can do better			

as architecture, engineering, and medicine. For this latter group of parents, marks in the low 80s could be seen to put their children at risk.

It is interesting that the lower set of marks was almost universally derided as inappropriate for the children of the responding parents. The low set of marks did not represent the child' potential and, therefore, there was need for some form of major remediation. Table 18.3 shows that the responses tended to fall into four broad categories.

Marks in the low 60s are not acceptable to parents. Therefore, there must be something wrong with their child or the system. Some parents blame other things. Time is not available to give school work full attention because the student is involved in part-time work or extracurricular activities; but for the most part, parents accept the mark, blame the child, and talk about the need for greater effort. Some are quite specific with regard to how they would accomplish this, for example, "I would require an intense discussion with my child to decide on a course of action", or "tightening up of general rules". Some parents attempt to find answers within the school: the Guidance Department, or some remedial support or a tutor. The few who are accepting of the lower set of marks say such things as, "It may be necessary to change the level of difficulty of courses taken", "Reduce aspirations from university to college", or "Consider taking a trade".

While there are different views represented among parents regarding the appropriateness of marks in the low 80s, there appears to be near consensus on the negative aspects of receiving marks in the low 60s. Students who obtain marks in the low 60s must be under considerable stress because of its impact on relationships with their parents, as well as teachers. Also, these marks confound the student's sense of direction with regard to what career he or she may aspire to — closing some doors, but leaving uncertain what doors are still open.

None of the parents saw the marks as indicative of having learned or not having learned particular knowledge and/or skills. They did not see areas that required particular attention or remediation, and rarely did they mention a particular subject requiring much greater effort. They saw the marks as representative of academic potential and, more specifically, as indicators of entry eligibility into specific university programs.

Interpreting the Report Card

Report cards vary in form and substance from the early years of elementary school to secondary school. In the early years of elementary school, the emphasis is placed on effort, attitudes, social skills, and achievement in relation to ability. Secondary school marks essentially reflect performance, and medians are provided to allow parents to compare the performance of their children with those of others. Both the marks and the medians provide guidance for career decisions. Provision is usually made for comments. These comments typically are designed to explain gaps that might exist between students' achievement levels and their ability levels, the difference being explained in terms of amount of effort put forth.

We asked parents a number of questions about the reporting system. Over 90 percent indicated that they would prefer to receive more information regarding the meaning of the marks received in relation to future educational plans. Ninety percent indicated that they would like to see more information on how students compared with their classmates and how they work with other students. There is no question that parents typically view marks as indicators of where their children stand in relation to other students and how successful their children are in competing for access to university. Parents are also concerned with effort, behavior and social skills, but they consistently do not ask for more information on what is being learned. The competition, or sorting, appears to be the most important element of the secondary school experience from the point of view of parents. This is much less true in elementary schools because of the importance given to the acquisition of basic communication and computation skills by young people.

Standardized Testing

Parents were asked a number of questions about national and provincial standardized testing programs currently being proposed. Over 60 percent of the parents agreed "to a great extent" or "to some extent" that the proposed standardized testing programs would improve the quality of education and the effectiveness of teachers. It is clear that, for parents, standardized testing provides more explicit indicators of where students stand in relation to each other and can be used to facilitate career decisions for their children.

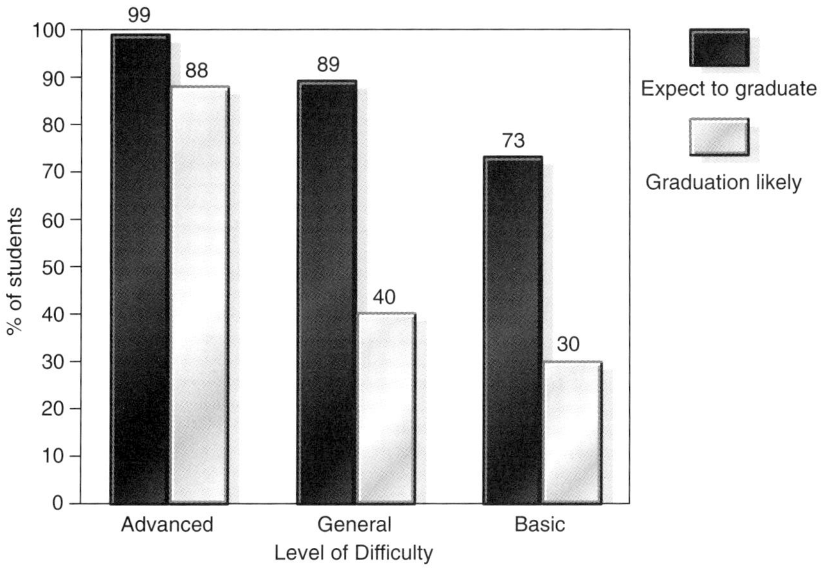

Fig. 18.1. Grade 9 graduation expectations and reality (Source: King and Peart 1990).

Influence of Universities

Do more students want to graduate from secondary school and attend university than circumstances will make possible? Simply stated, yes. Far more students would choose to attend university than will ultimately attend. Figure 18.1 indicates the proportion of Ontario students in Grade 9 by level of difficulty who expect to graduate and who are likely to graduate. It is very clear that a substantial number of students will not achieve this goal of graduation.

A limited number of places are available in postsecondary institutions, and more students would choose to attend than there are places. Marks assigned by teachers are the principal data for making decisions on accessibility to these institutions, and therefore marks take on more importance in this process of "sorting" than they do in providing insight into what students have actually learned. For teachers, the great challenge in evaluating students is to provide the students and their parents with the information needed for the "sorting" decisions, to indicate what has actually been learned, and somehow maintain the motivation of all students throughout this intensively competitive process.

This influencing of students' aspirations through marks is done over a period of time, not only during the years in secondary school, but also in elementary school. Most teachers make an effort to maintain students' self-esteem through this competitive process. Other countries often cited as having high levels of academic achievement sort much more harshly. For example, in Japan, a set of examinations at the end of secondary school is used to determine who will go to what postsecondary institutions, inducing a great deal of stress before and after. This approach to evaluation (and sorting) emphasizes comparison across students rather than what is being learned.

Table 18.4 summarizes the responses of 6,000 students in 10 schools regarding the influence of various factors on course

Table 18.1. Importance of factors in selecting optional courses, by level of difficulty*

	Very important (%)		Important (%)		Somewhat important (%)		Not important (%)	
Factor	A	G	A	G	A	G	A	G
How difficult course is	21	28	38	40	28	25	13	7
Preparation for future job	63	58	26	29	8	9	2	4
Interest/enjoyment	49	49	37	37	12	11	2	3
Friends take same	7	11	16	20	36	34	41	35
Preparation for college	27	42	24	29	15	15	34	14
Preparation for university	70	32	17	17	6	16	8	35
Parent influence	16	17	30	30	30	27	23	26
Sibling recommend	6	8	17	17	26	26	51	49
Teacher recommend	13	16	38	36	33	30	16	18
Guidance counselor	15	17	36	32	29	27	20	24
Who teaches the course	18	26	28	25	28	25	25	25

*Based on preliminary analysis.

selection. It can readily be seen that for students in both advanced-level and general-level courses, preparation for a future job is a very important factor in their deliberations. For students in advanced-level courses preparation for university is even more important, and for students in courses at the general level preparation for community college is quite important. Interest also shapes course selection, as might be expected, because interest and success influence career directions.

The marks students obtain and the courses they select are used by universities for most admission decisions. Since attendance at university represents a goal of the majority of students, it is not surprising that marks become the primary motivation (as opposed to what knowledge and skills might be learned). An overall average of 60 percent in six final year courses is normally the minimum required by a university applicant for consideration. These minimum requirements, however, do not guarantee admission since some universities, for example, will only consider applicants with much higher averages. Specific programs, especially those that are oversubscribed, may have even more stringent admission requirements. The admission average in effect at the time of application is very much dependent on supply and demand, that is, the number of available spaces and the number of applicants. The average required by a specific university or a specific program may therefore change from year to year.

The requirements of high marks for admission to university exerts extraordinary pressure on students and, in turn, on parents and teachers. Teachers, especially of senior courses, are acutely aware of the implications of evaluation for university-bound students and of the anxiety of most parents. One English teacher explained:

> . . . the parents want the students to get into university and the students need such high marks to get in . . . and parents question me because the student is not achieving too well.

> You know, the level the student feels he should be at or the level the parent feels he should be at. And, I simply say . . . this is what I expect and all of the students have to meet this standard. This is what I am doing, trying to prepare them for university.

Admission requirements were described by one mathematics teacher as creating a yardstick effect:

> Students know from a university standpoint they need 90 percent to be a doctor and 90 percent to be an engineer. The universities really define what the marks mean more than anybody else. It's who marks are for. If we didn't have a "mark measuring stick", we would use something else, words or something similar, but numbers are easier to use than words. We get a hundred graduations with marks that we don't get with words.

Employers' Expectations

In our interviews with 48 employers from a wide range of businesses and organizations, we asked two major questions. The first was concerned with how they interpreted the marks and diplomas of job applicants, and the second with what were the qualities that they thought were the most important in potential employees. We found that employers tend to use diploma completion rather than marks or completion of specific courses for employment decisions. However, they often require the demonstration of specific skills such as word processing. Since marks are rarely used for employment decisions, students who plan to go directly to work from secondary school are difficult to motivate to achieve high marks.

Employers assume that a diploma represents proficiency in school, particularly in English and mathematics, but they almost never ask for a transcript, except perhaps to check for attendance. Personal traits such as manner, attitude, initiative, and interpersonal skills are very important in the hiring process. Employers are usually willing to train secondary school graduates for specific jobs.

Diplomas tend to be used as an employment screen, to reduce the number of eligible applicants. The learning represented by the content is rarely used in the employment decision. To the question regarding skills desired, employers consistently mentioned ability to work with others, communication skills, and responsibility as evidenced by showing up on time and dressing appropriately. These skills are regularly reinforced by national business organizations.

Parents, Children, and School

Our Canadian survey of 11, 13, and 15 year olds (King & Coles 1992) involved approximately 1,700 young people at each age level. We conducted a number of factor and regression analyses that clearly restated the strong relationships among problems in school and at home, adjustment problems at home, health-risk behaviors, and self-esteem. For this study, we did a series of regression equations related to the extent to which young people can talk to their father and mother about their problems, and to both parents about things they are interested in. We included 20 variables in the analysis, including the following: self-esteem; attitude towards school; alcohol use; smoking behavior; health status; happiness; physical problems such as headaches, backaches, depression, and trouble sleeping; involvement in sports; physical fitness; bullying behavior; loneliness and feeling like an outsider; ease of making friends, and time spent with friends, as well as the three dependent variables. The findings are summarized in Tables 18.5, 18.6 and 18.7. Variables are only presented if they are significant at the 0.005 level. The direction of the variable is suggested by the wording.

Table 18.5. Regression on "I can talk to my parents about things I am interested in"

	Boys	Girls
Grade 6	Happy Involved in sports Good health Enjoys school	Happy Enjoys school Good health Involved in sports
Grade 8	Happy Involved in sports Rarely depressed	Happy Positive self-esteem Physically fit
Grade 10	Happy Makes friends easily Involved in sports	Happy Involved in sports Positive self-esteem

Table 18.6. Regression on "I can talk to my father about my problems"

	Boys	Girls
Grade 6	Positive self-esteem Happy	Happy Positive self-esteem Rarely lonely
Grade 8	Happy Positive self-esteem Enjoys school	Happy Positive self-esteem Enjoys school
Grade 10	Happy Makes friends easily Positive self-esteem	Positive self-esteem Physically fit Makes friends easily Rarely depressed Never been drunk

Table 18.7. Regression on "I can talk to my mother about my problems"

	Boys	Girls
Grade 6	Positive self-esteem Happy No sleeping problems	Happy Positive self-esteem Rarely lonely
Grade 8	Enjoys school Makes friends easily	Happy No sleeping problems Physically fit
Grade 10	Happy Makes friends easily Positive self-esteem	Positive self-esteem Makes friends easily Never been drunk Rarely depressed

Children who feel they can talk to their parents are much more likely to be well-adjusted in all respects than their peers who have difficulty talking to their parents. Success in school is a vital component of a positive relationship between parents and children. The relationship among the factors in the regression equation is quite remarkable. The importance of the relationship between "being able to talk to parents" to other critical life measures such as general happiness, self-esteem, enjoyment of school, and avoidance of health-risk behaviors cannot be underestimated. The almost universal occurrence of increased happiness with ease of talking to parents indicates how basic the relationship between parents and their children is, and how important it is, not only in early adolescence, but also for 15 and 16 year olds. In fact, those young people who experience problems talking to parents throughout the critical teenage years are far more likely to have health problems and be involved in health-risk behaviors such as drinking, smoking, and early sex.

Neither should the importance of school be underestimated in this relationship. Young people tend to like school when they achieve successfully in school, and meet their parents' expectations. When they achieve at a lower level, and do not meet parents' or schools' expectations, there is considerable stress. This stress contributes to alienation and the reforming of friendship groups that are often characterized by antisocial behaviors. In the follow-up to this study employing a more comprehensive survey, similar relationship were found in the 24 participating countries (King, Wold, Smith, and Harel 1996).

Summary

This chapter was designed to show the powerful expectations shared by parents, employers, and postsecondary institutions of the "sorting" role of the school. School is viewed as a highly competitive environment, and the more parents are directly involved in the process, the greater the opportunity they have to improve the life choices of their offspring. The competitive process is driven by the specific evidence of achievement (marks) required by universities and the credentials required for employment decisions.

Certainly there are aggregate differences in school populations by social class or ethnocultural background. These differences spawn dillerent parent groups which act as advocates for particular groups of children. Conflicts among these groups is added to an already competitive process. Elementary teachers, in particular, attempt to mediate the impact of competition on students by emphasizing individual development and broadening the role of education. Parents, employers, and postsecondary institutions emphasize basic knowledge and skill development, as evidenced by their promotion of standardized tests which reinforce this competitive stance.

Reduce drop-outs, improve the quality of education,

make students more competitive, and reduce the alienation experienced by students. Are these goals compatible? Probably not. Overt competition produces winners, but also losers. Losers create their own world — a world of disaffected and alienated youth. Can parent involvement in the day-today life of schools improve the quality of education? The findings clearly show that success in school is related to positive parent–child relationships, but the converse is also true. Lack of success in school is related to poor parent–child relationships. Involvement in school-related discussions can increase the sense of power and responsibility of parents and effectively link the home and school for the child. But this must be done with sensitivity reflecting the requirement that every child must benefit from his or her school experiences.

References

King, A. J. C. and Coles, B. (1992). *The health of Canada's youth.* Ottawa: Health and Welfare, Canada.

King, A. J. C. and Peart, M. J. (1990). *The good school.* Toronto: Ontario Secondary School Teachers' Federation.

King, A. J. C. and Peart, M. J. (1994). *The Numbers Game.* Toronto: Ontario Secondary School Teachers' Federation.

King, A. J. C., Wold, B., Tudor-Smith, C., and Harel, Y. (1996). *The health of youth: A cross-national survey.* World Health Organization.

AUTHOR INDEX

Adler, C. 183
Alexander, K. 227
Anthony, Susan B. 390
Archbald, D. A. 370
Aylon, H. 189
Aylon, H. *et al.* 183

Baker, Kenneth 411, 412, 414
Ball, S. J. 178, 227
Bauch, P. A. 373, 374
Bensman, D. 302, 303, 313
Bergen, Th. C. M. 451
Bierlein, Louann 397
Bird, T. 378
Blackman, J. 366, 373
Blank, R. K. 355, 358, 366
Bondi, l. 411
Booth, Charles 329
Brassard, J. 377
Bridge, R. G. 366, 373
Bronfenbrenner, G. 376
Bronfenbrenner, M. 376
Bryk, A. 247, 377, 378, 379, 380
Buddle, Ray 391
Bulgarian Constitution 328
Burke, D. 367
Bush, President 318
Bush, T. 420

Caldwell 422
Callaghan, Prime Minister James 408, 409, 410, 411
Camarena, M. 247
Capell, F. 367
Carl, J. 178
Carmon, N. 183
Carnegie Foundation 177, 178, 180
Carriedo, N. 380
Carron, G. 179, 195
Chau, T. N. 179, 195
Chávez, Cesar 390
Chen, M. 189
Chitty, C. 408, 432
Chubb, J. E. 177, 204, 218, 271, 31819, 327, 361, 3645, 372, 419, 421, 450, 453
Clark, Kenneth 330, 333
Clinton, President Bill 393
Clune, W. H. 382
Cochran, M. 377
Cohen, D. K. 367, 374, 382
Coles, B. 458, 472
Connel, R. W. *et al.* 371
Cook, M. 227
Cookson, P. W. 177, 178, 271, 272, 380
Coons, J. E. 348
Cooper, B. S. 218, 372
Corbett, H. D. 173

Crain, R. 271, 364, 365
Cutter, Milo 396, 401

Darling-Hammond, L. 366
Davis-Fishman, Katherine 305
Deetman, Wim 446
DeMare, R. D. 227, 247
Department of Education and Science (DES), UK 419, 424
Dionne, E. J. Jr. 332
Dougherty, K. 293
Doyle, D. 218
Driscoll, Mary 271, 2779, 283, 371, 37780
Durenberger, David 393

Echols, F. H. 271
Edelman, Marian Wright 329
Efrat, E. 179, 183, 198
Elam, S. M. 355, 373, 382
Ellison, Ralph 401
Elmore, R. F. 382
Erickson, D. A. 371
Ethington, C. A. 367
Etzioni, Amitai 342

Faires-Conklin, N. 168, 172
Farrar, E. 367, 374
Finn, C. E. Jr. 381
Fliegel, Seymour 336
Foster, G. E. 327

Gaebler, T. 334, 343
Gallup, A. M. 355, 373, 382
Gamaron, A. 227, 247
Gans, Herbert 338
Gendler, T. 327
Gibbs, Nancy 273
Giesbers, J. H. G. I. 451
Glasman, N. S. 173
Glenn, C. L. 325, 327, 345, 348, 349
Glover, D. 420
Golan, S. 173
Goldblat, E. 166, 169
Goldring, E. 373, 374
Guba, E. G. 167
Gutmann, A. 209

Hancher, K. S. 366
Harel, Y. 475
Hart, Gary 393
Hawley, W. D. 364, 365, 382
Henderson, A. 371, 377
Hentoff, Nat 349
Herman, J. L. 173
Heyns, B. 227
Hill, P. T. 327
Himmelfarb, G. 329
Hirschman, A. O. 363
Hirschmann, A. O. 411, 414
Hodgkinson, H. 273
Hoffer, T. 231, 337, 344, 361, 364, 374, 376, 3789, 4501, 453
Hutner, A. 366

Inbar, D. 163
International Labor Organization 197
Israeli Ministry of Education and Culture 179, 188, 197

Jenkner, Siegfried 349
Johnson, Vivian 349
Junge, Ember Reichgott 391, 393

Kilgore, S. 361, 450, 451, 453
King, A. J. C. 458, 472, 475
King, B. 401
King Jr, Dr Martin Luther 390
Kirby, S. N. 366
Koldrie, Ted 394
Kozol, J. 273

Leaven, Henry 361, 365, 382
Lee, V. 247, 271, 380
Levin, H. M. 370
Lewy, A. *et al.* 199
Lightfoot, S. L. 371
Lincoln, Y. S. 167
Lipshitz, G. 183
Little, J. W. 378
Lodewijks, J. G. L. C. 441

Mackie, 270
MacLure, S. 411
Madaus, G. 173
Maddaus, J. 366, 373
Major, Prime Minister John 432
Malen, B. 382
Manski, C. F. 367
Martin, M. 367
McDill, E. 227
McLaughlin, K. 345, 382
Mead, L. 329, 349
Meier, Debbie 305
Menahem, G. 271
Merl, Jean 365
Metz, 367, 368
Meyer, John 278
Ministry of Education 163
Moe, T. M. 177, 204, 218, 271, 319, 327, 361, 364, 365, 372, 419, 421, 450, 453
Mortimore, P. 332
Myaft, L. 399
Myatt, L. 403

Nathan, J. 371
Nathan, L. 399, 403
Nault, R. L. 366
Neill, A. S. 258
Netherlands, Constitution 4545
Nevo, D. 166, 167, 168, 169, 173
Newmann, F. 376, 377

Oakes, J. 227, 247
O'Day, J. 368
Ogawa, R. T. 382
Oliver, D. 376, 377
Olson, M. 376
Osborne, D. 334, 343

Patten, John 432
Paulu, N. 382
Peart, M. J. 458
Peled, E. 163
Perpich, Rudy 391, 392
Persell, C. H. 227, 247, 380
Peshkin, A. 380
Plank, S. *et al.* 271
Powell, Arthur 257
Power, S. *et al.* 177
Pressman, J. 350

Ratteray, Joan 350
Raudenbush, S. W. 450
Raywid, M. A. 306, 361, 372, 377, 382, 401, 402
Reshef, S. 163, 179
Richardson, J. 333
Riley, Richard 393
Romer, Roy 393
Rose, L. C. 355, 373, 382
Rossell, C. H. 361, 366
Rowan, B. 450, 453
Rutter, M. 332
Ryan, K. 332

Salganik, L. 345
San Jin Kang 450
Sayer, J. 411, 414, 430
Scherer, J. 376
Schiller, K. 271, 367

Schneider, B. 271, 367
Schorr, D. 335
Schorr, L. B. 335
Scriven, M. 171
Sever, R. 183
Shachar, A. 183
Shanker, Albert 391, 392
Shapira, R. 179, 373, 374
Simon, B. 432
Sizer, Theodore 257, 306
Slater, D. 179
Slavin, R. E. 209
Sleegers, P. J. C. 451
Smith, M. S. 368
Smrekar, C. 380
Sosniak, L. A. 367
Sostre, L. 293
Spinks, J. L. 422
Stake, R. E. 167, 168
Steinberg, L. 377
Stevens, D. B. 419
Stiggins, R. J. 168, 172
Stufflebeam, D. L. *et al.* 168
Sugarman, S. D. 348
Swirski, S. 179

Thatcher, Prime Minister Margaret 410, 411
Times Educational Supplement 434, 435
Tocqueville, A. de 339, 340
Tudor-Smith, C. 475

U. S Department of Education 355
Uchitelle, S. 366

Vermeulen, M. *et al.* 452, 453

Walker, J. 367
Walzer, Michael 343
Watt, D. G. 371
Weber, Max 249, 250
Wells, A. S. 178, 271
Wellstone, Paul 393
White, P. A. 382
Wieringen, A. M. L. van 451
Wildavsky, A. B. 350
Williams, M. F. 366
Williams, V. 411, 414, 430
Willms, J. D. 271
Wilson, B. L. 173
Wilson, W. J. 324, 325, 331, 344, 346
Wison, W. J. 339
Witte, J. F. 271, 346, 3636, 373, 374, 382
Wold, B 475
Wolfe, Tom 350
Wood, Roseanne 402
Wynne, E. A. 332

Yogev, A. 179, 183, 189

SUBJECT INDEX

ability groups
  Canada 461
  USA, student segregation 211–12, 215–26, 227–33
accountability, UK 410, 411
ACT tests, USA 266
A. D. Gordon School, Israel 84
affirmative action, USA
  ability grouping 221, 222
  tracking 228
Alternative Schools Program, East Harlem 306–15
Alum Rock, voucher experiment 374
  Arab education 50, 136–7, 142, 143, 152–4
  church schools 137, 143, 144
  demographic changes 137–41
  school enrollment 149–52
  socioeconomic divisions 154–5

basisschool, Netherlands 443
Better Education Through Alternatives (BETA) School, East Harlem 301–2

Canada
  ability groups 461
  education system 459–60
  pupil testing 461–6, 467
  report cards 466–7
  social problems 472–5
  universities 468–71
central control
  Israel 79, 80, 82, 161–2, 183
  New York 318–20
Central Park East School, East Harlem 302–3
Central Park East Secondary School, East Harlem 305–6
charter schools, USA
  elements of 394–6
  movement 390–4
  operation of 397–9
  policy implications 399–403
  student attitude 396
Cheled, Israel 32
church organizations, inner cities 346–7
church schools *see* religious education
City Technical Colleges, UK 412
civilizational heritage, Israeli schools 41, 44, 45, 46

communities
action by, inner cities 338–9
concepts of 376–7
schools 377–8
community schools, Israel 52–3, 84, 125–7, 145, 181
compensatory education
Israel 195–6
USA, public schools 218–19
Compulsory Education Act (1949), Israel 47–8, 81, 110
Conservative School, Israel 111, 115–16, 119–23, 124–5, 127–9, 131–2
controlled choice
Fall River, USA 280–2
peripheral communities, Israel 194–8
Shoshany program 93–6
Cultural Basket program, Israel 188, 194
cultural capital, Israel 136

discrimination, USA
equality standards 207–8
racial 221, 222
diversity, USA
public/private schools contrast 218, 219–20
students 213–14
Druze school, Israel 50

East Harlem
Alternative Schools Program 301–4, 306–13
socioeconomic conditions 299–301
East Harlem Performing Arts School, East Harlem 302, 303
Eastern Europe, community involvement 345
Education Act (1944), UK 408
Education Act (1988), UK 411
Education Act (1992), UK 415
education vouchers, USA 374, 393
education welfare, Israel 183
educational efficiency, school choice 45
educational equality
Israel 54–5
USA 207–11, 233–45
within schools 211–33
educational shares
see also educational trust funds distribution 290–2
educational systems
hierarchical authority systems 259–60
Israel 83, 89–90
central control 79, 80, 82
sectarianism 46
market vs public institutional control 243–5, 364–5
Netherlands, teaching quality 450–3
organization of 334
organizational change, UK 431–4
output-driven 260–1, 265–8
USA
central control, New York 318–20
institutional reform 204–5
public/private school contrast 223–4, 226, 233, 242–5

educational trust funds 284–6, 286–90, 292–3
El-Ha'Maayan schools, Israel 50
European Convention for the Protection of Human Rights and Fundamental Freedoms 328
evaluation
  output-driven schools 203, 266
  schools
    Israel 163, 164–74
    USA 210, 267–8
expectations
  employers 471–2
  motivation 267–8
  pupil ability 256–7, 330
  standards 260

Fall River, controlled choice program 280–2
family-school links, USA 333

gifted children
  see also magnet schools
  expert teachers 33–4
  Israel
    Cheled 31–2, 36–8
    parental choice 39–40
    special schools 39–40
GM schools *see* grant maintained schools
Gonen School, Israel 51, 66–7
grade point average (GPA), USA 261–2
grant maintained schools (GM), UK 412, 418–21

Hebrew education
  Arab enrollment 156
  choice patterns 47–8
  Israeli school sectors 58–60
homework, USA 259

inequality
  see also educational equality
  Israel
    higher education 136
    secondary school education 16–18, 22–5
  school choice 45
information
  information centres, Israel 97
  parent information centres 279–83
  Parents' Charter, UK 415
  school choice 366–7
inner cities
  church organizations 346–7
  public schools, community involvement 338–9, 342–4, 344–6
institutions
  see also educational systems
  educational control, USA 204–5, 212, 223–4, 226
integrated schools, Israel 48, 52, 54–8, 88
interdisciplinarity, Labor Movement School, Israel 117–18
International Covenant on Economic, Social and Cultural Rights 327–8
Israel
  A. D. Gordon School 84
  administrative districts 182
  Arab education 50, 136–7, 142, 143, 144, 152–4
  autonomous schools 82
  central control 79, 80, 82, 161–2, 183
  community schools 52–3

Compulsory Education Act (1949) 47–8, 81, 110
cross-district schools 401
curriculum development 13, 92, 117–19
curriculum differentiation, secondary school education 17–18
Druze schools 50
education welfare program 183
El-Ha'Maayan schools 50
elementary schools 83–4
ethnicity 18, 22–5, 43
gifted children 31–40
Jaffa 87–90, 137–44, 152–4
Jerusalem 31–40
Kedma schools 51, 57
kibbutz 69
Knesset Record 36 (1963) 81
Labour Movement School 125–7
Labour party 66–7
Law of State Education (1953) 47, 79, 81
local authorities 77–8, 80–2
magnet schools 50–1, 110, 111
Megamot 17–18
Ministry of Education and Culture 19–22, 31, 32, 33, 34, 40, 43
Muslim schools 143
National Religious Party 64, 65
Noam schools 50
peripheral schools 186–8
politico-cultural divisions 63–4
religious schools 60–2, 137, 143, 144, 147, 149, 155, 156
school autonomy 161–4
school choice 144–9
school evaluation 163–74
Siyasat Tajhil 145
Tali schools 50
teacher professionalism 163
Tel Aviv-Jaffa 83–103
Yeshivot schools 61

Israeli Supreme Court, school choice 43–4

Jose Feliciano School *see* East Harlem Performing Arts School

Kedma schools, Israel 51, 57

Labor Movement School, Israel 111, 113–15, 117–19, 123–7, 129–31
language medium, Israel 149–52
LEA *see* Local Education Authorities
leadership, Alternative Schools Program, East Harlem 315–16
learning environments
Israel 155
USA
academic/nonacademic track contrasts 229
public/private school contrast 224–5
LMS *see* local management of schools
Local Education Authorities (LEAs), UK 412–13
GM schools 418, 419, 420
LMS 425–31
reduced powers 423–4
local management of schools (LMS), UK 413, 418

LEA adjustments 425–31
reform strategy 421–5

Madares Ahliya, Israel 145
magnet schools
gifted children 51
Israel 48, 49, 50–1, 84–5, 110, 111, 112–13, 128
USA 274–5, 325, 357–8
management
Netherlands 447–8
UK 427–8
Management and Administrative Services (MAS), UK 427–8
marketplace 44
educational innovation 367
school choice 68, 204, 361–9
MAS *see* Management and Administrative Services
matching, student/subject 13–15
mathematics grouping, USA 221–3
Minnesota
charter school movement 390–4
school choice programs 274–5
mixed ability
Israel 87, 90
USA 215

National Curriculum, UK, testing regimes 416–17
National Educational Longitudinal Study (NELS), USA 206
National Religious Party, Israel 64, 65
Neighborhood Rehabilitation program, Israel 183
NELS *see* National Educational Longitudinal Study
Netherlands
basisschool 443
centralized educational policy 444–7
denominational schools' autonomy 441–2
education system, denominational autonomy 441–2
educational systems 450–3
management 447–8
pillarization 440–1
Primary Education Act (1985) 442–3
school funding 448–50
school integration 442–4
Noam schools, Israel 50

organizations, general characteristics 249–54, 262–4

parent information centres 279–83, 292
parental attitudes 96–100
parental choice, UK 413–14, 420
parental involvement
East Harlem 316–17
Israel 70–1, 129–32
school choice 371–2
UK 415
Parents' Charter, UK 415
performance gains
output-driven schools 267
student achievement 263
peripheral schools, Israel 178–80, 186–94, 195–6
pillarization, Netherlands, societal organization 440–1

pluralism, Israel 86
poverty action, USA, 1960s–70s 340–2
Primary Education Act (1985), Netherlands 442–3
private schools
  Israel 143–4
  socioeconomic status, performance effects 239–40
professionalization, teachers 163
pupil testing
  Canada 461–6, 467
  UK 417

QSP *see* Quality School Partnership
Quality School Partnership (QSP), UK 426–7
quality thresholds, USA 209–10

religious education
  Israel 60–2, 121–3, 137, 143, 144, 147, 149, 155, 156
  USA 335–7
report cards, Canada 466–7
rights allocation, organizational 263

SAT tests, USA 266
school autonomy
  decision making capacity 37–8
  Israel 38, 181
    assumptions 161–4
    peripheral communities 194–8
  Netherlands, centralized educational policy 444–7
  Shoshany program, Tel Aviv-Jaffa 91, 93
  UK, LEA services 425–31
  USA
    factors affecting 234–5
    political vs market system 204–6, 233–6
    public/private school contrast 233–4
school choice
  commitment 278–9
  community-based strategy 178
  cultural values 44
  East Harlem 298, 313–17
  empirical research on 360
  European Convention for the Protection of Human Rights and Fundamental Freedoms 328
  family sovereignty theory 369–75
  vs integration debate 54–8
  International Covenant on Economic, Social and Cultural Rights 327–8
  Israel 42–7, 67–72, 144–9, 181–6
    controlled choice 51–3, 181
  market place 177–8, 204, 361–9
  motivation for, Israel 113–17
  opposition to quality 45
  organization of 327
  ownership 271–2, 312
  peripheral regions, community-based strategy 178–80
  process of, Israel 59–60
  rights to 327–8
  sector competition, Israel 60–1
  Shoshany program, Tel Aviv-Jaffa 91–2

theoretical case for 358
Universal Declaration of Human Rights 327
USA 68, 282–3, 325
controlled choice 178
criticisms 205–6, 356–7
residential segregation 324
states support 355–6
value community theory 375–81
school cultures 277
school design, organizational innovation 254
school diversity, performance equalization, public/private school comparison 238
school drop-outs, Israeli Arabs 152–4
school funding
Netherlands 448–50
school choice programs 285–6
student maintenance, Israel 53, 186–8
UK 424–5
school governing bodies, UK 412
school integration
Israel 54–8, 181
opposition to 56–7
Netherlands 442–4
school performance
see also evaluation
educational performance tables 415
USA, socioeconomic status 237–40
school size, Alternative Schools Program, East Harlem 314–15
school viability, UK 416
sectarianism, Israel 41–72
SEN *see* Special Educational Needs
Shoshany program, Israel 90–6
Siyasat Tajhil, Israel 145
social capital 264
social conditions, children, USA 273–4
social institutions
Israel 64
USA, inner cities 342–4
social problems, Canada, home/school relationships 472–5
social values, USA, school expectations 331–2
socioeconomic status
poor schools
public/private school comparison 239–40
public/private student comparison 240–2
Special Educational Needs (SEN), UK, LEA services 428
standards
see also evaluation; expectations
external 263
extra-school demands 258, 260
internal 263
Japanese/American comparison 258–9
output-driven schools 265–6
teacher responsibility 261
variations in, USA 261–2
State Education Act (1953), Israel 47, 49, 79, 81
stratification
social 14

subjects 13–14, 15–16, 25
streaming *see* ability groups
student diversity, intra-school, USA 213–14
student integration, Shoshany program, Israel 92
student performance, USA 255–6
subject status
academic/utilitarian 14, 15–16, 19–21
cultural capital 16, 18

Tali schools, Israel 50, 67, 111
teacher commitment
Israel, professionalization 163
public/private school comparison 244
teaching methods
Israel
heterogeneous classes 90
peripheral/central schools contrast 187
teaching quality, Netherlands 450–3
Tel Aviv-Jaffa *see* Israel
tracking *see* ability groups

UK
accountability 410, 411
City Technical Colleges 412
Education Act (1944) 408
Education Act (1988) 411
Education Act (1992) 415
educational reforms, debate 408–13
GMS 412, 418–21
LEA services 425–31
LEAs 412–13, 418–20, 423–4, 425–31
LMS 413, 418, 421–31
MAS 427–8
National Curriculum 416–17
parental choice 413–14, 420
Parents' Charter 415
pupil testing 417
school funding 424–5
school governing bodies 412
school viability 416
SEN 428
Universal Declaration of Human Rights, school choice 327
universities
Canada, influence on school pupils 468–71
Israel 136
Urban League Street Academy, Minnesota 275–6
USA
ability grouping 221–3
affirmative action 221, 222, 228
Alternative Schools Program 301–4, 306–13
charter schools 390–403
compensatory education 218–19
discrimination 207–8, 221, 222
education vouchers 374, 393
educational diversity
public/private school contrast 218, 219–20
students 213–14
educational equality 207–11, 207–45
educational systems 204–5, 223, 226, 233, 242–5

family-school links 259, 333
learning environments 224–5, 229
NELS 206
poverty action 340–2
SAT tests 266
school choice 45, 68, 282–3
controlled choice 178
school evaluation 210, 267–8
school integration 58
student performance 255–6

volunteers, gifted children 34–5
vouchers, Arab schools 155–6

White Plains, school choice program 282–3

Yeshivot schools, Israel 61